SELLING THE AIR

A Critique of the Policy of Commercial Broadcasting in the United States

THOMAS STREETER

THE UNIVERSITY OF CHICAGO PRESS
CHICAGO & LONDON

Thomas Streeter is associate professor of sociology at the University of Vermont.

The University of Chicago Press, Chicago 60637
The University of Chicago Press, Ltd., London

Printed in the United States of America

05 04 03 02 01 00 99 98 97 96 1 2 3 4 5

ISBN 0-226-77721-9 (cloth)
ISBN 0-226-77722-7 (paper)

Library of Congress Cataloging-in-Publication Data

Streeter, Thomas.
Selling the air : a critique of the policy of commercial broadcasting in the United States / Thomas Streeter.
p. cm.
Includes bibliographical references and index.
1. Broadcasting policy—United States. 2. Broadcasting—Law and legislation—United States. 3. Broadcasting—United States—History. I. Title.
HE8689.8.S77 1996
384.54′0973—dc20 95-35028
CIP

♾ The paper used in this publication meets the minimum requirements of the American National Standard for Information Sciences—Permanence of Paper for Printed Library Materials, ANSI Z39.48-1984.

To Robyn and Seth

Contents

Acknowledgmen

Many people have helped me through the writing of this book. Over the years, David Sholle has provided me abundant supplies of that rarest and most valuable of commodities in academia: detailed, astute, and helpful criticism of my work whenever asked. Michael Curtin and David Kennedy provided invaluable readings of my work and, on many levels, gave me the encouragement and support that made it possible to keep going through some hard times. Duncan Brown, Andrew Calabrese, Richard Maxwell, Robert McChesney, Beth Mintz, John Peters, Robert Taylor, and Wendy Wahl read much or all of the book manuscript at various stages and provided important and frequently crucial reactions and suggestions for improvement. At a very early stage in this project, David Trubek and Hendrik Hartog helped me see the importance of the category of property, and pointed me in the direction of some of the more useful ways of thinking about it. Robert Gordon helped me see the importance of understanding history to understanding liberalism, and vice versa. Roberta Astroff, Patricia Aufderheide, Julie D'Acci, John Fiske, Theodore Glasser, Hanno Hardt, Peter Jaszi, Leah Jacobs, Lisa McLaughlin, Willard D. Rowland, Jr., James Schwoch, Joan Smith, Lynn Spigel, and many others provided comments, criticisms, and intellectual attention that have proven invaluable. And thanks to Douglas Mitchell of the University of Chicago Press for all his support, especially for his enthusiasm for the tone and mode of inquiry of this book, and to Joann Hoy for her first-rate copyediting.

I am grateful to my colleagues in the Sociology Department of the University of Vermont for understanding the importance of a humane, supportive environment to good intellectual work, and for acting on that understanding. Also contributing to that supportive environment were Lynn Carew, Dawn Wales, and Kathy Morris, who provided friendly, generous, and resourceful administrative support, and Jenny Hyslop, who provided excellent and eager research assistance.

My spouse, Robyn Warhol, changed my life while I was working on this book. She made it possible for me to complete this project, not only

by providing abundant intellectual and emotional encouragement and support, but also by helping me learn how to care about what really matters in life.

I would like to thank the University of Vermont Committee on Research and Scholarship for research support in 1990, 1992, and 1994, and the National Endowment for the Humanities for making it possible for me to attend the faculty seminar "Liberal Ideals in American Law, 1870–1940" at Stanford Law School in the summer of 1990. And thanks to the *Cardozo Arts and Entertainment Law Journal* for permission to use my article "Broadcast Copyright and the Bureaucratization of Property" (vol. 10, no. 2 [1992]: 567–90) in chapter 7.

Introduction

This book looks at the ways that ideas, habits of thought, have figured in the creation and maintenance of American commercial broadcasting: ideas about markets, property, individuals, social process, politics, work, and the home. In exploring the sometimes contradictory roles those ideas have played in the life of the institutions of radio and television, the book also becomes an inquiry into the ideas themselves.

Although the book does question the necessity and value of commercial broadcasting, the word "critique" in the title does not mean that the book is critical in the simple negative sense. This book is a critique more in the Kantian sense of the word, in the sense of getting at the constitutive conditions of something. My central question is not, Is it good or bad to organize electronic media on a commercial basis? but rather, What does it mean to organize broadcasting that way? What are the conditions that make it possible to take the practice of broadcasting—the reproduction of disembodied sounds and pictures for dissemination to vast unseen audiences—and constitute it as something that can be bought, owned, and sold?

This question, I believe, is of more than purely intellectual interest. Broadcasting was one of the first industries to deal largely in electronic intangibles, and certainly the first such to become part of the everyday lives of almost all Americans. The complexities involved in the social construction of broadcasting may have bearing on contemporary concerns about intellectual property, culture, and politics in the development of new communications technologies. This book suggests, moreover, that the legal and political problems facing computer communications today are much less unique and much more like those associated with broadcasting than the oft-heard euphoric rhetoric of "the information revolution" would imply. If our future social and economic systems will be increasingly characterized by the technologically enabled commodification and exchange of intangibles, commercial broadcasting provides us with an example of one way that such commodification can be accomplished, and of the problems that are likely to be encountered if that path is taken.

In contrast with most of the literature on these issues, however, I emphasize the imaginative character of the institution. Commercial broadcasting is the product of communities of people acting in accordance with collectively developed visions, ideas, and hopes; it is, in the first instance, a kind of social philosophy in practice. To be sure, it is not simply a blueprint come alive; as I will show, the history of U.S. broadcast policy is littered with mistakes and misconceptions. But one can see, even (and perhaps especially) in the misconceptions, a consistent pattern of broad social vision, of ideas about the way human life is and ought to be. I call the social vision underlying those patterns "corporate liberalism," and put considerable effort in this book into demonstrating that commercial broadcasting is not best explained as simply the product of impersonal forces such as technology or economics or some behind-the-scenes conspiracy; it is explainable only by taking into account corporate liberal habits of thought. Ideas do matter.

But ideas often matter in peculiar, unexpected ways. If there is a central criticism advanced in this book, it is not that unregulated free markets in electronic media are morally a bad idea, but that the effort to enact them in commercial broadcasting has been deeply contradictory, has come into conflict with its own principles. Unlike many critics of commercial broadcasting, I do not see the problem simply as the public interest being overrun by private greed, or openness suppressed by monopolies and centralization. One of the principal arguments of this book is that these ways of framing the issue rest on a dubious assumption: the natural or perfunctory character of "private" economic forces. Both critics and supporters often speak of commercial broadcasting as if it were simply the product of an absence of political or social control, as if it were the result of some elemental state where you simply take the lid off and let greed—or enlightened self-interest—run its course.

But commercial broadcasting, I believe, is more a product of deliberate political activity than a lack of it. It is political, not just in the sense that it requires spectrum regulation and similar regulatory activities, but because its organization as commercial, as a set of marketplace activities, is itself dependent on extensive and ongoing collective activities, activities that typically involve favoring some people and values at the expense of others. Commercial broadcasting exists, in other words, because our politicians, bureaucrats, judges, and business managers, with varying degrees of explicitness and in a particular social and historical context, have used and continue to use the powers of government and law to make it exist. The effort to create a free open marketplace has produced an institution that is dependent on government privileges and other forms of collective constraints. Although constructed in the name of the classical ideals of private property and the free marketplace, American commercial broadcasting, under close inspection, calls the coherence of those ideals into question.

Though this book is about ideas, therefore, it does not focus just on ideas in the abstract; in this sense it is not Kantian at all. It looks at the interaction of ideas with social practices and structures. It is a study of ideas in the trenches, so to speak, more a work in the sociology of knowledge than of philosophy. If I were more interested in theoretical precision than in accessibility, I might have replaced the word "ideas" in my opening sentence with Foucault's "discursive practices," by which he means collective habits of talk, action, and interpretation embedded in historical contexts that establish and enact relations of power and resistance.[1]

The kinds of analysis used here are quite diverse. At various points, I engage forms of philosophy, sociology, political science, discourse analysis, and legal analysis, and to varying degrees draw on concepts and concerns from Marxism, feminism, poststructuralism, cultural studies, and the law and society and critical legal studies movements. And throughout the book, I frequently discuss history, both because of the importance of historical context and because ideas about history so often play a role in justifying collective actions and institutions. (Although I do on several occasions make use of some original historical documents, much of my history comes from secondary sources and my concern is generally with variations in interpretations of what happened, not with historical facts in isolation; some might therefore prefer to say that I engage in historical interpretation rather than history.) This wide-ranging interdisciplinarity is called for by the character of the topic. Doing as much justice as possible to such diverse bodies of knowledge and methods, however, has proven to be the most difficult part of this project.

The book is organized into three parts. Part 1 makes the case that American commercial broadcasting, from its beginnings to the present day, embodies a set of principles and social structures usefully called cor-

1. One of the better definitions of discursive practices is "interpretations of conduct that produce and affirm actions and their concomitant subjects and objects that are institutionalized because the interpretation is oft repeated and accepted" (Michael J. Shapiro, *Language and Political Understanding: The Politics of Discursive Practices* [New Haven: Yale University Press, 1981], 130).

porate liberalism. Part 2 discusses problems with conventional ways of thinking about broadcast regulation and suggests an alternative framework. Part 3 pursues the line of inquiry suggested by that framework by rethinking classic issues such as markets and free speech and introducing new issues. In particular, this section argues that commercial broadcasting is constituted in the ongoing creation of three types of ephemeral property—stations, copyright, and viewership—which in turn define and allocate the most basic power relations in the industry and thus ought to be understood as the core issues of broadcast policy.

The first chapter lays out the epistemological approach of the book, situates it within contemporary currents of thought, and suggests an approach to the electronic media that is at once interpretive and law- and policy-oriented. Much of what is puzzling about television and radio, I suggest, stems less from what we don't know, from a lack of facts *about* television, and more from confusion about what we do know, from confusion about the fact *of* television. In concert with interpretive theories in general, then, I argue that useful inquiry, instead of obsessing over gaps between belief and fact, is better off searching for the ways that human beings interrelate beliefs and facts as they go about life, the ways that interpretations create, shape, and transform the facts of human institutions and vice versa. Although this approach links me to cultural studies, within the movement I take the side of those who, like Tony Bennett, argue that we need to "put policy into cultural studies," that is, to shift the focus of cultural studies toward the study of and intervention in the politics of the production of culture, and to overcome the common bifurcation between political economy and textual analysis.[22] Toward this end, I suggest that radio and television should be understood, not as just technologies or cultural forms, but as legal inscriptions on technologies.

Chapter 2 sketches the outlines of the dominant mode of thought called liberalism and its twentieth-century corporate variant, wherein classical liberal terms (rights, property, legal formalism, etc.) are qualified by contingent administrative procedures and the values of expertise. One of the characteristic habits of talk and action within corporate liberalism, I suggest, is a kind of "popular functionalism," a tendency to imagine human life tautologically as a functionally interrelated system. The fact that commercial broadcasting appears to be functional for, say, the corporate consumer economy or for monopoly capitalism, then,

2. Tony Bennett, "Putting Policy into Cultural Studies," in *Cultural Studies,* ed. Lawrence Grossberg, Paula Treichler, and Cary Nelson (New York: Routledge, 1992), 23–37.

may be less the result of underlying sociological or economic laws than of the fact that broadcasting's organizers imagine their world and broadcasting's place in it in functional terms, and act accordingly. Functional, systemic visions of broadcasting have played an important role in its construction.

Chapter 3 presents a revisionist history of the origins of commercial broadcasting, a reinterpretation of existing historical accounts of broadcasting's beginnings. The traditional ways of telling the tale emphasize technological and economic forces and tend to trace broadcasting's beginnings to the early 1920s. My account, in contrast, emphasizes the importance of social and political visions in shaping what broadcasting became, and focuses more on the pre-1920 period, both to illustrate the deeper corporate liberal roots of commercial broadcasting and, by pointing to the paths not taken, to emphasize the contingent, political character of the path that was.

Part 2 provides a critique and analysis of the terms and practices of corporate liberal broadcast policy. Chapter 4 looks at the closest thing to ongoing public debate about broadcast structure in American society: the peculiar world of broadcast law and policy, particularly the activities surrounding the Federal Communications Commission (FCC). It approaches those discussions as the product of an interpretive community: it looks at how broadcast policy experts talk, how their careers are organized, the explicit and implicit beliefs embodied in their behavior and discourse, and what can and what can't be done and said in the "broadcast policy arena." It argues that what is called broadcast policy in the United States is not really about the decision to make broadcasting commercial, about the policy of commercial broadcasting (hence this book's awkward subtitle: "the policy of" instead of "broadcast policy"). Broadcast policy takes that decision completely for granted, and avoids discussions of it by dismissing them, in an act of self-fulfilling prophecy, as "impractical." Rather, broadcast policy is an institutional enactment of a central corporate liberal hope and operating assumption: that expertise can solve the dilemmas of liberalism in a corporate consumer economy, that it can square the principles of individualism, private property, and a neutral rule of law with the fact of collective, bureaucratic institutions and the need for shifting, contingent forms of decision making.

Chapter 5 critiques both the traditional policy discourse of competition and monopoly in industry regulation and conventional progressive emphases on free speech and the public interest. The chapter suggests that part of what keeps those discourses afloat is a kind of ideological slippage akin to what postmodernists call simulation. As an alternative to the traditional discourses, the chapter proposes one grounded in economic sociology, which calls attention to the contingent political activities upon which markets, competitive or otherwise, depend. One of the directions that economic sociology points, the chapter argues, is toward a politicized understanding of property.

Part 3 applies the insights of the previous chapters to the forms of property constitutive of commercial broadcasting. Chapter 6 explores the creation of the industry's fundamental building blocks, marketable broadcast stations, with an emphasis on the ongoing government intervention necessary to that creation and the problems of legitimacy that intervention entails. Chapter 7 explores certain facets of the creation of property relations in broadcast content, primarily through copyright, with an emphasis on the tensions between the principle of authorial uniqueness upon which copyright law rests and the collective, "unauthored" character of broadcast production. Taken together, the ways in which the broadcast licenses and copyright are created and enforced not only determine who can communicate over and profit from the broadcast airwaves; they also shape the character of cultural production by defining what it means to be a communicator.

Chapter 8 investigates the third, and perhaps most peculiar, element necessary to market exchange in U.S. broadcasting: the creation of a form of property in the broadcast audience, which then can be sold to advertisers. The "commodity audience," this chapter argues, is best understood neither as a democratic polity faithfully served by the industry, nor as a mass of unwitting factory workers on the assembly line of the culture industries. Borrowing from feminist theory, this chapter suggests that the audience bears a relation to commercial broadcasting similar to that of unwaged domestic "women's work" to the political economy: a precondition to the economic system, but invisible and unthinkable from within its constitutive ideology (except as a kind of natural process happening outside of human will). In a legal and economic sense, then, the broadcast audience is less like a public or a factory worker than it is like a housewife.

The concluding chapter offers some suggestions for a new politics of commercial broadcasting. It shows the continuing relevance of corporate liberalism to current debates, and charts some of the interrelated theoretical and strategic issues suggested by the book's analysis. It concludes with a call for refiguring traditional policy terms within a framework that acknowledges the political character of law, markets, and property, which would enable a more direct and democratic debate over the structure of the electronic media.

PART ONE

Liberal Television

The Fact of Television: A Theoretical Prologue

The Fact of Television

The philosopher Stanley Cavell once described something he called "the fact of television." The "fact of" television, Cavell suggested, was not the same as "facts about" television: facts about its economic structure, its technology, the size of its audience, and so forth. Rather, he was talking about "something like the sheer fact that television exists," which he took to be on the one hand obvious but on the other among "the most mysterious facts of contemporary life."[1] Most of this book is devoted to discussions of the laws and policies that shape and constrain commercial television and radio in the United States. But the goal is to inquire into the "fact of" television and radio, not simply to provide facts about television and radio or facts about media law.

This book is about the fact of television because, like Cavell, I find there to be something mysterious about the sheer fact of the existence of television, about its presence in our lives. Unlike Cavell, however, I do not approach television primarily as a collection of texts or programs, as something that is simply watched. I am interested in television as a *practice.* Television is something people do. It is not just a thing or a collection of symbolic works. This book approaches the fact of television, then, from the perspective of television as a set of social activities. And it focuses on the large degree to which these social activities involve law and politics. A central thesis of the book is that television as a practice is usefully understood, not just as a technology, not just as a cultural form, but as a kind of legal inscription on technology.

This chapter explores the conceptual implications of looking at television and radio this way—in terms of the "fact of" instead of "facts about," and as a practice, not a thing. In doing so, it lays out a theoretical foundation for the chapters that follow, and situates this book within contemporary theoretical discussions. After discussing the implications of approaching broadcasting as a practice, the chapter makes the case

1. Stanley Cavell, "The Fact of Television," *Daedalus* 111 (fall 1982): 75.

for the centrality of law and legal liberalism to the practice of broadcasting, and then elaborates the relevance of this kind of analysis to contemporary critical and cultural theory.

The American system of broadcasting is almost seventy years old. As we will see, the basic structures developed in the 1920s at commercial broadcasting's birth—advertising, the network system, government licensing in the public interest—remain in place today. Those structures have survived the Great Depression, a world war, and at least one complete technological metamorphosis (the shift from radio to television). Most of the same corporations that dominated its creation continue to shape its activities today: General Electric, RCA, NBC, CBS, Zenith, and Westinghouse are still prominent names both inside and outside the industry. The institution of U.S. commercial broadcasting has outlasted the average twentieth-century nation-state.

Yet in our culture, talk about so stable an institution is peculiarly unstable. It's often asserted in almost the same breath, for example, that television is "simply" a commercial product, no different from any other item available in stores, yet also a special public institution akin to a school or a New England town meeting. A chair of the FCC once said for example that "television is just another appliance. It's a toaster with pictures." Yet he also felt it necessary at times to claim for television a special role in embodying hallowed constitutional principles of free speech and democracy—hardly the kind of claim one makes for toasters.[2] Academics often only add to the confusion. Some academics have discussed television as if it were a kind of literature, even if only to demonstrate that it is literature of an inferior sort. And it has become fashionable of late to counter such negative comparisons by drawing alternative analogies: television is a new art form that supersedes literature, or television is countercultural much like medieval carnivals. Carnival, artwork, town meeting, commodity: taken together, these characterizations don't add up.

There are two ways to respond to the incoherence of our commonsense ways of describing broadcasting. One is to assume that the problem lies in a lack of facts *about* television. "We don't know what television is, therefore we need to conduct research that will give us a better

2. For the "toaster" quote, see Richard Stengel, Peter Ainslie, and Jay Branegan, "Evangelist of the Marketplace: The FCC's Mark Fowler Wants to Strip Away TV Regulations," *Time,* November 21, 1983, 58; for the reference to hallowed constitutional principles, see Fowler's statement of August 7, 1985, printed in FCC, "In the Matter of Inquiry into Section 73.1910 of the Commission's Rules and Regulations concerning the General Fairness Doctrine Obligations of Broadcast Licensees," Docket 84-282, 102 F.C.C. 2d 145, 58 Radio Regulation 2d 1137, Release Number FCC 85-459 (released August 23, 1985, adopted August 7, 1985).

idea." Most of the academic literature on television and radio adopts this admirably modest, careful stance. It is in this literature that many of the facts about television can be found.

But there is another, equally reasonable, response to our confusion: to explore the possibility that the mysteriousness of television is a mystery of our own making. Television is a human construct. Much of the discussion of television talks about it, however, as if it were something natural, outside of human purview, as if it were as inevitable as it is inscrutable. There is an everyday version of this: you go to the store, buy a box, put it in your living room, and you have a television—a commodity, a technology, in any case, an object, a thing. But this objectification happens on an intellectual level as well. Academics are quick to suggest that television affects society, politics, psychology, but rarely remember that the medium is itself an effect of human actions. We explore how people do things *with* television—they "read" or interpret it, use it, manipulate it, find gratification in it—but rarely think of television as itself something that people *do.* We find it hard to remember that radio and television are not fixed objects to which people react; they are themselves collective human actions.

That we think of television as a thing instead of as a practice is reflected in the fact that we have no television equivalent to the film world's "Hollywood," understood as both a kind of film and the institutions that produce it.[3] So far I largely have been using the word "television" to describe my object of inquiry. In doing so, however, I have risked confusion, because my object of inquiry is really the system that was constructed and pioneered in the early days of radio, and then transferred to television in the late 1940s and early 1950s. Technically speaking, this book is about the historically embedded ensemble of social relations that make possible the production, distribution, and "consumption" of the majority of commercial American television programs in the United States. But our culture lacks a precise term for that ensemble, a condition that is in turn a part of what makes the ensemble the way it is.

So we speak and think of television as a thing, as if it were contained within that box in our living rooms, even though without the intricate and spectacularly collective set of activities that makes the box in our living rooms come alive as an integral part of our culture, that box wouldn't be much more than an oversized doorstop. "Television" includes the people in Hollywood and New York devoting their lives and careers to making programs, people in Washington making, changing,

3. Douglas Kellner, *Television and the Crisis of Democracy* (Boulder: Westview Press, 1990), 75.

and enforcing laws that enable and shape the institutions in which programs are produced and distributed, and elaborate international systems of manufacturing, marketing, and distribution that make the boxes available to audiences. And of course it involves the activities of audiences themselves: millions of people sitting down with millions of boxes all at the same time, and the cultures and patterns of daily life among those millions that provide the ability and motivation to buy boxes and tune in.

In a little noted but striking passage in his book on television, Raymond Williams argued that traditional research on the medium excluded questions of purpose. As he put it, what research has "excluded is *intention,* and therefore all real social and cultural process." A focus on the purposes of media would "direct our attention to the interests and agencies of communication."[4] Williams, the cultural neo-Marxist, was certainly not talking of authorial intention or of uncovering underlying "interests" of either the utilitarian or class-determinist variety. Rather, he was seeking to restore a broad sense of agency, a sense of collective human choice, to our understanding of television.

In this book I describe American broadcasting as "corporate liberal." The point of doing so is to provide a vocabulary that helps restore a sense of intentionality, of agency, of purpose, to discourse about the electronic media. If we are going to discuss broadcast structure, we need an effective way to grasp what the existing structure is. One of the principal impediments to public discussion of media structure is the belief that commercial broadcasting was born and is sustained by natural, impersonal forces, that it is something that happened, not something that is done. American broadcasting, it is said, is simply the product of the marketplace, or interest group pressures, or of a conspiracy on the part of the powers that be, or simply of greed run rampant.

My argument that broadcasting is corporate liberal, then, is intended to emphasize the ways in which the institution is the product of social and political choices, not of accident or impersonal economic or technological forces alone. "Corporate liberalism" is meant to breathe life into our vision of the electronic media, not to lock it up within a rigid framework. The concept is not meant to suggest underlying mechanical forces, elite conspiracies, or a mesmerizing false consciousness. Corporate liberalism is not so much a strict set of principles or formal ideology as it is an expression of values and hopes. It is a set of goals as well as a worldview; it expresses intention, agency—a policy. In this book I argue

4. Raymond Williams, *Television: Technology and Cultural Form* (New York: Schocken, 1977), 120.

that it is a deeply contradictory policy, but one need not be a critic of the policy to see the justice in making it explicit.

The epistemological principle at stake here is this: A distant planet or an exotic microorganism is indeed mysterious principally because of a lack of facts *about* it. But the electronic media did not fall from the sky or emerge fully formed from a test tube; they are the product of knowledgeable people doing things in a concerted, organized way, with certain purposes in mind. When people describe a distant planet as a wandering god, their guesses about the unknown object do not change the planet itself. But if people describe television alternately as an artwork or a commodity, in the right circumstances their talk can help shape it. The FCC chair who described television as a toaster with pictures, for example, did so as part of a successful effort to change the way television is regulated, which in turn noticeably changed the medium. And this is just a particularly obvious example; executives, employees, politicians, voters, audience members all have habitual ways of thinking about and acting toward the medium that together make the medium what it is. In at least one sense of the verb "to know," the people who collectively "do" television certainly know what they are doing. The activity of television is conditioned on certain kinds of knowledge, what sociologist Anthony Giddens calls "practical" knowledge.[5]

If there's something mysterious about the electronic media, therefore, it need not be the product of a simple lack of knowledge, a lack of "facts about"; it may very well be an aspect of the "fact of." The experience of incongruity we encounter in hearing television characterized alternately as a commodity, town meeting, and art form is itself part of the phenomenon in question. The fact is that television and radio have been constructed by people who talk about them in ways that don't seem to add up. The "mysteriousness" of the fact of television, in other words, is as much a product of the hopes we invest in it as it is a product of a lack of facts about the medium. The sense of worry, disappointment, and plain perplexity generated by television may tell us as much about ourselves as it does about television itself.

This book thus focuses on the electronic media as a set of imaginative activities, as something that people do out of hope and conviction. It starts from the simple premise that before radio and television can be businesses, public institutions, or technologies, people must have ideas and hopes about them and seek to implement those ideas and hopes.

5. Anthony Giddens, *Central Problems in Social Theory: Action, Structure, and Contradiction in Social Analysis* (Berkeley: University of California Press, 1979), 73.

And it looks at the incongruities, gaps, and blind spots in those works of imagination we call radio and television as historical encounters with the limits of our ideas and hopes. As a piece of scholarship, therefore, this book is as much an inquiry into our collective imagination as it is a study of a particular mass medium; it looks at the electronic media first and foremost as a kind of social philosophy in practice.

Law

Law is many things, but it is perhaps foremost a way of turning forms of knowledge into action, of making manifest collective ideas and hopes. One of the primary arguments of this book is that the American legal and political system is a principal but too often ignored arena for the practice of broadcasting, for "doing" the elaborate set of ongoing activities we call television and radio. Broadcasting, in other words, is to a large degree a legal activity. Although the importance of the constitutive character of law will be argued primarily by way of the history and social relations of broadcasting, it can also be defended on theoretical grounds.

In media, as in other fields, we tend to think of law as at once mechanical and arcane: its details are complex, and thus best left to experts, but its workings are straightforward and limited in scope, so the rest of us can trust the experts to tell us about the few details that are important, particularly those that constrain behaviors. It's helpful, after all, for the textual critic to know that stations are prohibited by law from broadcasting obscene programming, or for the economist to know that owners are prohibited from owning more than twelve broadcast stations. Law thus becomes simply a tidy subcategory of the collection of "facts about."

Law, however, even if arcane, is hardly mechanical, and its interpretation anything but straightforward. As this book will show, law is fluid both in meaning and in boundaries: its interpretation shifts dramatically from context to context, and its relevance flows in unexpected ways into areas normally thought of as remote from law. In the case of the electronic media, law flows into the "fact of" principally by virtue of its key role in the *creation* of radio and television. This book will show that law is not just an occasional constraint on the behavior of broadcasting, it *creates* broadcasting. It even creates broadcasters: to a large degree it defines who they are and what they do. Law, then, is a key to understanding the media as a product of meaningful habits of thought and action, as socially constructed.

This book will explore, for example, the ways that commercial broadcasting is a child of the collection of habits of thought some scholars call liberalism, understood not as a point on the political spectrum, but as a form of dominant social consciousness. On the one hand, liberalism involves ideas about markets, property, and private ownership; hence the idea of *commercial* broadcasting, the idea that broadcasting can and should be a process of buying and selling. But liberalism also involves ideas about freedom, communication, individuals, and democracy; in particular, it involves the hope that the process of buying and selling can complement or help create freedom and democracy for individuals, especially when integrated through the rule of law. Television as we know it is a product and expression of these ideas, of this hope.

Of course, though the electronic media are born of imagination, they are not simply blueprints come alive. Making broadcasting commercial, for example, involves taking the practice of broadcasting—the reproduction of disembodied sounds and pictures for dissemination to vast unseen audiences—and constituting it as something that can be bought, owned, and sold; it involves turning broadcasting into property. A large portion of this book is devoted to analyzing the tenuous and labyrinthine legal, political, and institutional processes by which this act of commodification is accomplished.

One reason for focusing on property is simply that it allocates control over the electronic media, and it does so in ways much more consequential than much-debated legal constructs such as free speech and the public interest. The creation and definition of property establishes the ground rules for market exchange in broadcasting, shapes who gets what and thus the distribution of power over the institution, and by framing broadcasting as a "free market" delimited more by rights than by privileges, helps legitimate the control by a few of an institution that affects the lives of millions.

But property also helps to underscore the profoundly imaginative character of the institutions of the electronic media. The creation of property in broadcasting is not simple. On the contrary, it involves a massive, tension-ridden effort of abstraction, an ongoing effort to make a kind of collective sense of broadcasting from within the liberal framework.

Property is just one of the liberal categories that illustrate a striking pattern in the encounter between broadcasting and liberal thought: although commercial broadcasting is in many ways a spectacular example of liberal principles put into practice, it often seems profoundly antiliberal. The encounter between liberal principles and broadcasting involves far more than a simple mapping of liberal concepts onto electronic technologies and practices. Our broadcast system is intended to foster a diverse "marketplace of ideas," for example, yet its programming is organized according to rigid formulas, and commercial broadcasters are notoriously unwilling to take political and aesthetic risks. Much of the political legitimacy of commercial broadcasting rests on the principle of free speech, yet its all-embracing dependence on entertainment values appears to enfeeble the political dialogue that free speech is supposed to foster. The commercial system is the historical product of a strenuous antistatism applied to radio and television, yet it is thoroughly dependent on regular and active forms of government intervention for its very existence.

In part as a result of the antiliberal effects of these liberal efforts, from the early days of broadcasting onward, there has been constant political and legal struggle over the proper place of broadcasting in our social and political systems. For reasons traceable to liberalism itself, the debate accompanying this struggle typically centers on the activities of government regulators, and is framed in such terms as freedom, fairness, and the public interest. In this book, my conclusions about this debate and its accompanying scholarly literature are skeptical: I argue that it can be usefully seen as an attempt, characteristic of twentieth-century liberalism, to regain the footing lost in the shifting sands of one set of liberal contradictions—the incoherence of atomistic individualism and of its industrial correlate, laissez-faire business principles—by shifting weight in the direction of another set of (equally contradictory) liberal principles—a faith in the power of expertise and objective scientific knowledge to make manifest a transcendent, reified "public interest." But the values and hopes to which the debate gives voice are nonetheless vital; it is one of my hopes that this book will help rescue these crucial issues from sterility by suggesting a way of reframing the debate in new terms.

If any conscious, significant changes are to come, however, they will come only by acknowledging the depth and breadth of the dilemmas. Solutions will not come from treating the dilemmas of broadcast law and policy as isolated problems amenable to solution by clever regulators. Broadcast law, however confused, is still a patterned confusion, shaped by the structures of history and contemporary social life, particularly those associated with liberalism. The contradictions of broadcast policy exemplify the tensions within our most fundamental beliefs and ways of acting; tensions revealed in the way we use terms like "individuals," "freedom," "fairness," and "public." No new law, policy, or bureaucratic structure can make those tensions disappear overnight. Precisely because the debated details of U.S. broadcast policy gain their meaning from the complex and varied framework of liberalism, most of the contemporary regulatory struggles and dilemmas must be understood as rooted in dilemmas within the larger liberal belief system.

Theory

This book is driven in the first instance by an intellectual encounter with the real historical experience of television and radio, not by a desire to prove one or another academic theory. Yet the idea that scholarly work can be intellectually neutral or theory-free is simply another theory; there's no escaping the fact that one comes to any inquiry already shaped by theoretical assumptions and habits.

Liberalism and Totality

It is fashionable these days to be suspicious of "totalizing" generalizations such as liberalism and corporate liberalism. Such generalizations, it is said, oversimplify and obscure contradictions, complexities, and resistance from the margins. True, the practice of imagining systems of thought as unified and coherent too often involves a simplifying projection on the part of the analyst, and can obscure important complexities. Although it is useful to speak of a totality called liberalism, it is dangerous to imagine that totality to be overly coherent, whole, and solid.

Yet, as Frederic Jameson puts it, "it is diagnostically more productive to have a totalizing concept than to try to make one's way without one."[6] The effort to identify and analyze general patterns in social life, moreover, is productive for more than analytic reasons. In the case of broadcasting, pointing to general patterns also has political value: too often, critics and apologists alike treat major historical decisions about media as if they were inevitable, a product of the inexorable workings of economic forces or struggles among interest groups. A concept like corporate liberalism helps keep in focus the fact that there is a general pattern of thought underlying the way the institution has been organized, that its character reflects collective human choices.

The idea of a broad pattern called liberalism, therefore, is a way into the changing complexity of social life, not a way to escape that complexity, and it is intended to call attention to human agency, not to obscure it. It is an empirical observation more than a philosophical one. The point is not to criticize liberalism as a philosophy in the abstract but to show how the sheer fact of broadcasting simultaneously brings attention both to liberalism's power as an imaginative system and to its contradictions. The key questions here, therefore, are not those of traditional philosophy or the history of ideas. They involve the interaction of ideas with social practices and structures, ideas in the trenches, so to speak.

6. Fredric Jameson, *Postmodernism; or, The Cultural Logic of Late Capitalism* (Durham, NC: Duke University Press, 1991), 212.

McLuhan, Postmodernism, and the Shock Effect of Media

There are some similarities, therefore, between this project and that of Marshall McLuhan and his successor Jean Baudrillard. Though McLuhan's optimistic, transcendental modernism ("The Global Village") is quite distinct from Baudrillard's darker postmodern denial of transcendentals, both scholars share an interest in the shock effect of the electronic media, in the ways that television and radio force us to reconsider some of our most basic assumptions about human life. To an extent, I agree with McLuhan that the medium of television—its organization, structure, and placement in contemporary social life—is "the message"; at least I believe that the medium is as interesting and perplexing as the particular programs the medium transmits. And like Baudrillard, I suspect that there's something important to the widespread feeling (common to both postmodernist scholars and my undergraduate students) that our electronically mediated world is one in which "all that is solid melts into air," a world in which life seems to be characterized by the dizzying manipulation of words, signs, and symbols, a world in which we no longer deal with things themselves, but with "simulations."[7]

Both McLuhan and Baudrillard, however, tend to speak of the media as primarily a technology, and thus obscure the legal and organizational formulas that clothe the technology.[8] And those two scholars tend to frame matters in millennial or apocalyptic terms. Postmodernists seem to speak of "simulation," for example, in terms of a nostalgic suggestion that we are at the end of an era or the "end of history," as if all signs had obvious meanings at some time in the past and only today have lost any connection to their referents.[9] Whether things were ever all that solid,

7. Within social and cultural theory, too much is made of the modernism/postmodernism distinction, which works best when applied to specific categories of art such as architecture. While there are important moral issues at stake in the distinction (particularly the value of authenticity), as a pattern of social life what most people mean by postmodernism seems to be merely a version or extension of the general trends discussed under the heading "modernism" by, for example, Marshall Berman in *All That Is Solid Melts into Air: The Experience of Modernity* (New York: Simon & Schuster, 1982). See also Jean-François Lyotard, *The Postmodern Condition: A Report on Knowledge* (Minneapolis: University of Minnesota Press, 1985); and Jean Baudrillard, "Simulacra and Simulations," in *Selected Writings,* ed. Mark Poster (Stanford: Stanford University Press, 1988), 166–84.

8. For a classic critique of the reduction of media to technology and technological determinism generally, see Williams, *Television,* 9–19.

9. There is a danger in interpretations of Baudrillard, if not in Baudrillard himself, that "simulation" is taken to mean that there was once a time when words and symbols all had solid references to things, whereas in our time they have come to refer just to each other. This interpretation is encouraged by Baudrillard by his use of certain metaphors, such as the map and the territory. ("Simulation," he writes in "Simulacra and Simulations," "is no longer

whether they are "melting into air" more now than before, is uncertain. In this book, radio, television, and simulation are interesting insofar as they embody specific historical configurations of events and trends. The concern here is with the specific historical circumstances that lead to a sense of certain "solid" things "melting into air" in certain conditions. The peculiarity of the electronic media, in other words, does not serve me as evidence for grand metaphysical (or antimetaphysical) claims, but as a way to explore the historical specificity of events we tend to experience as metaphysical.

Cultural Materialism, Bernard Edelman

If there is a predominant theoretical precedent or framework for this book, then, it is not so much the overly ahistorical McLuhan and Baudrillard, but a set of interrelated traditions that in different ways integrate the intellectual practices of critique, interpretation, and attention to historical complexity. These traditions sometimes have been divided into two competing camps, with the historically inclined poststructuralists such as Foucault or Gayatri Spivak on one side, and critical humanists and pragmatists such as Cornel West, E. P. Thompson, and Raymond Williams on the other. Yet it is possible to view the important differences between these traditions more as productive tensions than as competing positions. This is the strategy advocated, for example, by legal scholar Duncan Kennedy. Kennedy argues that a useful critique of legal practices should adopt a pragmatist or legal realist attention to the details of the ways that legal rules concretely operate to shape socioeconomic processes, but should combine that with a Foucauldian skepticism about categories like "interest" and the subject.[10] The combination of poststructuralism with a kind of pragmatist humanism is also characteristic of Stuart Hall.[11]

It is in this integrative sense, then, that I must mention the importance of the work of the French poststructuralist Bernard Edelman. After many years in relative obscurity, Edelman's book, *Ownership of the Image,* is now beginning to get the attention it deserves.[12] As the book's

---

that of the map . . . no longer that of a territory" [166]) The implicit idea here is that signs and symbols in general relate, or once related, to reality in the same way that maps relate to territory, by reference. Of course, as Baudrillard is certainly aware, it is a premise of most twentieth-century theories of language and signification, from Saussure onward, that signs and symbols *don't* work this way, and never have.

10. Duncan Kennedy, "The Stakes of Law, or Hale and Foucault!" in *Sexy Dressing Etc.* (Cambridge: Harvard University Press, 1993), 83–125.

11. Stuart Hall, "Cultural Studies: Two Paradigms," *Media, Culture, and Society* 2, no. 1 (1980): 57–72.

12. Bernard Edelman, *Ownership of the Image: Elements for a Marxist Theory of*

French title, *Le Droit saisi par la photographie,* suggests, Edelman is interested in how photography "seized" or surprised legal thinking in the nineteenth century by creating possibilities that did not readily fit into existing categories of property law, and to an extent threatened to undermine those categories. Is there property in the symbolic content of a mechanically produced photograph? More particularly, is there a liberal individual analogous to an author, a creative originator or "subject" entitled to ownership, of a photograph? If so, what exactly about a photograph is "original"? Who (or what) is the originator, the subject of a photograph, who or what its object?

By offering only nonobvious, arbitrary answers to these questions, Edelman suggests, the problem of photography threatened the underpinnings of the traditional law of property, which in various ways is premised on the belief that there must be something obvious, natural, and nonarbitrary to definitions of owner and owned, originator and originated. And this in turn touched on profound questions of what it means to be an individual, a legal and social subject. Hence, in the late nineteenth century, the technology of photography augured a reconsideration of basic social assumptions, in much the same way that McLuhan suggests television has done in the twentieth.

In sharp contrast to McLuhan, however, Edelman acknowledges that traditional thought is capable of responding to such profound challenges. He explores the mixture of intellectual, social, and political processes—ideological processes, in the Althusserian sense—by which the legal system was able to avoid that reconsideration and incorporate photography, successfully (if awkwardly) into its purview.

Though less ambitious and markedly different in tone, method, and emphasis from Edelman, this book nonetheless undertakes a parallel project: my interest, in a sense, is in how radio and television have "seized" or challenged American law and the liberal habits of thought that underpin it, and in how the legal system has responded to those challenges. Like Edelman, I believe that analysis of the encounter between media technology and legal thought helps reveal both moral and intellectual weaknesses and sociological strengths—that is, the resilience of structures of power—in contemporary American life.

Law, translated by Elizabeth Kingdom (London: Routledge & Kegan Paul, 1979). For remarks on the lack of attention to Edelman, see Jane M. Gaines, *Contested Culture: The Image, the Voice, and the Law* (Chapel Hill: University of North Carolina Press, 1991), 2–3.

Media, Culture, Text

Today it is rare to discuss the mass media in combination with Continental theorists like Edelman without also mentioning the word "culture." Under the rubric of "cultural studies," the concept of culture has become the focus of an interdisciplinary movement. One central theme of this diverse field is a reinvigoration of the interpretive sociological principle that human "reality" is socially constructed in processes of symbol use and interpretation. The logic of this well-known axiom encourages attention to the subtleties of interpretation, both as an aspect of social life to be analyzed—human action is fundamentally mediated by patterns and processes of interpretation—and as a central problem of inquiry; as Anthony Giddens puts it, inquiry into human life is conditioned by the "double hermeneutic," by the circumstance that scholarship is inevitably an interpretation of interpretations and just as inevitably an intervention into ongoing social processes.[13]

Law is a highly symbolic, interpretive activity; its raw materials are documents, rhetoric, and rituals. Law also shapes the distribution of resources and controls behavior; any discussion of it is necessarily political in that it involves us in debates and struggles over values and the distribution of power in society. Law thus forces us to look simultaneously at the textual quality of power and the powerful quality of texts.

Yet, for the most part, law does not happen in the sustained interaction between an individual and a distinct work that we think of when we think of literature or art. Law is a set of lived social relations; law happens when bargains are struck, hierarchies are enforced, and conflicts are initiated and resolved. Even law students diligently at work in libraries rarely read a book cover to cover the way one reads a novel; they concern themselves largely with the chains of cases, principles, and argumentative strategies of which casebooks and records contain only pieces. Intertextuality is no revelation to the law; in at least one sense of the word, it is one of the law's overt working principles.

Cultural studies is quick to assert the political character of scholarship, and is rife with discussions of symbols, intertextuality, power, and lived social relations. Yet cultural studies has devoted very little scholarly attention to law. In part this may be because the language and tone of the law might seem antithetical to the iconoclastic, mercurial, and populist spirit of cultural studies. The details of broadcast law and policy are matters of concern and fascination to those inside the corridors of power—

13. Anthony Giddens, *New Rules of Sociological Method* (London: Hutchinson, 1976), 155–69.

high-paid media executives, Washington insiders, and entertainment lawyers—whereas cultural studies is best known for calling attention to the importance of ephemeral, marginal, and informal phenomena in the lives of common people. The mass media may be important to cultural studies, but only because of their role as a key source of symbols and interpretations in the everyday life of audience members. It seems more in the spirit of contemporary cultural studies to analyze previously reviled cultural forms like television situation comedies than to dissect arcane legal terminology or behind-the-scenes machinations of industrial corporations and governments.[14]

Most practitioners of cultural studies give some credit to the argument that culture is embedded in social contexts that are shaped by structures of control and power; hence the grudging respect given to some forms of Marxism in a field that otherwise puts such importance on matters as ephemeral as symbols and ritual. Yet in cultural studies the question of control over production is traditionally segregated into the discipline of "political economy," and thus framed as a matter of industrial control and structure best subject to economic analysis. And once this act of segregation is accomplished, most students of cultural studies will be quick to assert, correctly, that a handful of executives in Hollywood and New York can hardly control how the many millions of audience members interpret and use the programs whose production the executives oversee. Power, particularly in matters of culture, is rarely if ever such a one-way, top-down affair; this is why, in most versions of cultural studies, political economy is considered a necessary but never suffi-

14. There are many exceptions to this trend in cultural studies, but a particularly articulate one can be found in the Australian "cultural policy debate." Beginning in the late 1980s, several major figures in cultural studies based in Australia focused their scholarly efforts toward influencing public policies, particularly in government agencies. Rather than merely criticizing culture from a safe, romantic distance, they argued, cultural studies should learn to deal more directly with policy-making apparatuses. This is in many ways in concert with the approach of this book. However, as will be discussed in chapter 4, in the U.S. context at least, the word "policy" carries the weight of specific technocratic connotations: "policy" is part of the set of practices by which government intervention on behalf of private corporations is reconciled with the liberal legal principle of the separation of public and private. In the United States, in other words, we have no generally accepted language for directly addressing "private" or corporate policies. In an effort to forge such a language, this book generally works with the terms "law" and "politics," emphasizing the material and legitimatory interconnectedness of the two terms. For a sample of the cultural policy studies argument, see Tony Bennett, "Putting Policy into Cultural Studies," in *Cultural Studies,* ed. Lawrence Grossberg, Paula Treichler, and Cary Nelson (New York: Routledge, 1992), 23–37; and Stuart Cunningham, *Framing Culture: Criticism and Policy in Australia* (Sydney: Allen & Unwin, 1992).

cient component of any full analysis.[15] Perhaps in reaction to the grim, reductive determinism of some forms of political economy, cultural studies has focused on the point where political economy is weakest: on the point of reception, on programs and the ways that audience members make sense of them. The problem is that, although television texts and television audiences are given the subtle attention they deserve, references to the media organizations themselves are fleeting—better to be brief and vague, it seems, than to be taken for a vulgar, economic reductionist—and thereby the reified monolith of economic structure is left intact, waiting in the wings.

Something is being missed here, in part because American cultural studies tends to rely on literary criticism as a model for understanding the process of interpretation. Stuart Hall has recently observed that "[o]ne of the problems just now is that everybody nowadays is, surprisingly after thirty years, a literary critic."[16] Cultural studies is tending to become a new brand of literary criticism: instead of writing about reading Dickens or James Joyce, one writes about "reading" television; instead of searching for eternal values in symbolic works, one looks for signs of social life. Interpretation thus tends to be understood in terms of an analogy with the literary model of a reader engaged in the interpretation of a novel or a poem. Mass media become understood principally as a kind of literature. Television is no longer an inert box, but a collection of symbolic works. From within this framework, law has little relevance. The broad power and effects of law are not constituted in isolated acts of reading of the kind we associate with works of literature.[17]

The goal of the interpretive tradition, however, is not simply finding social life in symbolic works, but finding the work of symbols in social life. As Grossberg puts it, "cultural studies does not need . . . theories of authors, texts, or audiences. Cultural studies needs theories of contexts

15. See, for example, Richard Johnson, "What Is Cultural Studies Anyway?" *Social Text* 16 (winter 1986/87): 38–80.

16. He continues: "We have made a surreptitious return to the undisciplined literary reading which this whole exercise [i.e., cultural studies] was designed to firm up. . . . in a funny kind of way, . . . we've gone back around to people trusting their intuitive understandings of the text and giving that a kind of authenticity, a kind of validity" ("Reflections upon the Encoding/Decoding Model: An Interview with Stuart Hall," in *Viewing, Reading, Listening: Audiences and Cultural Reception,* ed. Jon Cruz and Justin Lewis [Boulder: Westview Press, 1994], 273).

17. For recent critiques of the law/literature divide, see Stanford Levinson and Steven Mailloux, eds., *Interpreting Law and Literature: A Hermeneutic Reader* (Evanston, IL: Northwestern University Press, 1988); and Costas Douzinas and Ronnie Warrington with Shaun McVeigh, *Postmodern Jurisprudence: The Law of Texts in the Texts of Law* (New York: Routledge, 1991).

and of the complexity of cultural effects and relations of power."[18] If reality is indeed constructed in processes of symbol use and interpretation, then organizations, institutions, and social relations—such as those that bring the box in our living rooms to life—are themselves constituted in symbol use. The literary model draws our attention to one part of that life, the life of the stream of images on the screen, but at the same time it draws our attention away from the fact of the box itself; our attention stops at the boundaries of the moving image.[19]

Contemporary literary criticism itself offers a critique of this tendency to draw boundaries between social life and symbolic works: the call, first made by Roland Barthes, for a move from work to text.[20] A work, Barthes argued, is imagined as a finite object, delimited by, say, the obvious physicality of the book on the shelf (or, one might add, by the box that constitutes the borders of the television screen).[21] This delimitation, however, this drawing of boundaries, is arbitrary, and has the effect of obscuring the social context of the work's interpretation, which is constitutive of its meanings. As an alternative to the concept of the work, Barthes offered the concept of the text, which transgresses boundaries instead of creating them, whose boundaries in a sense extend into culture as far as the eye can see.

One need stretch the concept only a little bit to argue that the perspective of law leads to an understanding of television as text, as a process, as what Barthes called "an activity of production."[22] For relatively straightforward reasons, law cuts across or transgresses the imagined coherence of the boundary between the box and the images it displays. Television's structure and organization are as much a matter of symbolic process as its content. Television does not just provide symbols for the social construction of reality, it is itself socially constructed. And television, as text, is to a large degree constructed in the textual system of the

18. Lawrence Grossberg, "Can Cultural Studies Find True Happiness in Communication?" *Journal of Communication* 43 (autumn 1993): 93.

19. There are good reasons to borrow from literary criticism in cultural studies. If life is constructed by the interpretation of symbols, then widely interpreted symbolic works such as novels or television programs are likely to be of more importance than the traditional economist or positivist sociologist might expect. Literary theory, moreover, offers the most developed and nuanced sense of the complexities of the interpretive process. My point here is that, like most useful models of intellectual practice, literary criticism can conceal at the same time that it reveals.

20. Roland Barthes, "From Work to Text," in *Image—Music—Text,* translated by Stephen Heath (New York: Hill & Wang, 1977), 155-64.

21. A work can also be delimited by other devices of literary criticism, such as the imagined unity of the life of the author who wrote a work.

22. Barthes, "From Work to Text," 157.

law. Television as text is thus not constrained by the boundaries of the television screen; it extends into the box itself, and into the social relations that make it what it is.

Interpretation and the Construction of Subjectivities

This argument might not be so unusual to cultural studies if the critical community had taken to heart a point made many years ago by Edelman concerning the construction of the subject. Edelman at the time was responding to an early version of what has come to be called "*Screen* theory," which focuses on the ways that cinema operates by creating the spectator's subject position—in a sense, the ways that film defines the audience's sense of self.[23] A camera inherently constructs a point of view and in a sense puts the audience "inside the head" of an ideal single viewer. This, in combination with editing techniques and viewing practices, requires the audience to imagine themselves to be a particular kind of isolated individual—a subject—if they are to make sense of the film. From this imaginative process of viewing it is sometimes suggested that basic structures of film narrative have the profound effect of helping to create or reinforce the "bourgeois subject," the abstract, isolated sense of self characteristic of and necessary for contemporary capitalist social relations.

Edelman's criticism of this argument to a large degree anticipates later criticisms of *Screen* theory. *Screen* theory not only attributes improbably profound power to the "cinematic apparatus"—in its logical extreme, the theory suggests that people walk out of a film a different person than when they walked in—but it also assumes that the "bourgeois subject" is a monolithic, obvious, and predetermined construct; it unquestioningly assumes, in other words, the bourgeois definition of individuality that it purports to critique.[24] Unlike subsequent critics of *Screen* theory, however, Edelman made these criticisms in the context

23. See Edelman, *Ownership of the Image,* 62-67.

24. Edelman makes this criticism by arguing that the theory of bourgeois subject construction in the cinematic apparatus implies that a film of a workers' strike would have an antiworker effect simply because it reproduced the "humanist code of perspective" that favors the dominant order. This absurdity, he argues, is not simply the product of "determinism," but of a false understanding of determination. Being a Marxist, he puts this in terms of misdirected blame: the ideology of the individualist bourgeois subject—an effect of the capitalist system of social organization—is reified and thus mistaken for a cause, and the real culprit, capitalism, is let off the hook (ibid., 64-65). Being an Althusserian structuralist, however, he is not simply arguing the primacy of one linear cause against another, but using the word "capitalism" to stand for a social formation in which cause is not linear but structural. Edelman can be accused of reifying capitalism in the same way that his opponents reify the ideology of the subject, but it is certainly not his goal to do so.

of an alternative approach to the problem of subject construction, an approach centered on the problem of the subject in law.

On the one hand, there is a blunt materiality to law that makes the argument of subject construction more persuasive. In contrast to film, the construction of bourgeois subjects in law is both necessary and coercive. It is necessary because legal precedent and argument require it; when a judge is faced with a copyright dispute, he or she is bound by the system to settle the case in terms that at least give lip service to the notion that all copyrightable works are the unique creations of isolated individuals. And it is coercive for the obvious reason that law is enforced by the legitimized violence of the state; one need not believe in film or the law, but law, unlike film, coerces one to act according to its dictates. We all live our day-to-day lives within a coercively enforced web of legal constructs—contractual, financial, workplace, and family relations—that profoundly shape both our relations to others and, one suspects, our sense of who we are.[25]

On the other hand, Edelman's focus on law as a way into the problem of subject construction also helps point the way to an alternative to the literary model of interpretation in cultural studies. Novels, films, and laws all in their own ways contribute to the social construction of reality, to the collective enactment of values, ideas, hopes, and prejudices. But law illustrates the embeddedness of symbol use in ongoing social activities in a way that novels and films, considered in isolation, do not. Clearly, it was not the intention of the original *Screen* theorists to suggest that film viewing mechanically imprints a monolithic, undifferentiated bourgeois subject on viewers; yet the prevalence of a literary model tended to hypostatize the subject position suggested by the isolated individual interpreting a unified symbolic work, simply because, in practice if not in theory, it separated the moment of interpretation from the rest of social life. By bringing a critique of law into the equation, Edelman provides a model of analysis that addresses the relations among interpretation, media, the construction of subjectivities, and power in a way that cuts across symbolic works and their contexts. The problematic of the law, in other words, helps put the process of interpretation back into the stream of social life.

Analyzing broadcasting as a legal practice, then, is arguably consis-

25. Of course, Edelman's argument can be taken too far as well, and he has been criticized in terms similar to those he directs at the precursors of *Screen* theory. Law may be everywhere, but not everything is law; the law imagines a unified bourgeois subject, but in practice creates a bewildering variety of different, sometimes conflicting subject positions. Edelman certainly attempts to account for these facts; whether he succeeds is the subject for another essay.

tent with the theory of the text, even if it does not involve textual analysis in the conventional sense. Looking at law in broadcasting from a critical view cuts across the boundary between symbolic works and their social context; it is a transgression of boundaries, a questioning of conventional categories. Considering television as the product of a set of legal relationships, furthermore, offers a way to analyze it as a kind of social philosophy in practice, as a strategic enactment of ideals, hopes, and values. And this approach to television as a practice is consistent with the project of moving from work to text, of understanding the media as a process instead of an object, as value-laden instead of neutral.

Conclusion: Television as a Legal Inscription on Technology

To summarize, this book considers the activity of broadcasting as something not merely constrained by, but constituted in, a set of legal relationships. The tools of broadcasting, even the boxes in our living rooms, are to a large degree legal constructs. A television set itself is made practical, made into a practice, by its internal organization in concert with the elaborate social relations that make broadcasting possible, including everything from government regulation of the spectrum to a consumer economy. Those relations, in turn, centrally involve law and politics, that is, lawyers, judges, legislators, and a polity interpreting, making, changing, and enforcing laws and regulations that enable and shape both the equipment of broadcasting and the institutions that make the equipment come alive. So a television set is not just a technology; it is a collection of tubes, wires, and microchips whose organization is determined by, or inscribed with, law; it is a legal inscription on technology.

Part of the fact of television, however, is that its organization and social context obscure the process of inscription. We tend to see television sets, network structures, advertising, and all the other elements of the system as fixed in technological imperative, incontrovertible legal principle, and economic necessity; the fixity of broadcasting is part of its mystery. It is the hope of this book to show that fixity is historical, not inevitable, and thus, in the larger scope of things, subject to change.

Liberalism, Corporate Liberalism

[T]he Liberty of the man . . . consisteth in this, that he finds no stop, in doing what he has the will, desire, or inclination to doe.

THOMAS HOBBES, 1651

What is liberty? . . . Suppose that I were building a great piece of powerful machinery. . . . Liberty for the several parts would consist in the best possible assembling and adjusting of them all, would it not? . . . The piston of an engine [will] run with absolute freedom . . . not because it is left alone or isolated, but because it has been associated most skillfully and carefully with the other parts of the great structure.

WOODROW WILSON, 1912

Introduction: The Meaning of the "First Broadcast"

Media textbooks regularly treat college students across the United States to the following anecdote: On Christmas Eve of 1906, inventor and entrepreneur Reginald Fessenden astonished wireless telegraph operators scattered over the eastern seaboard of the United States by sending out voice and music over his experimental "wireless telephone."[1] Ship-

1. In an informal survey, the Fessenden "first broadcast" story turned up in four of seven undergraduate media textbooks: John R. Bittner, *Broadcasting and Telecommunication: An Introduction* (Englewood Cliffs, NJ: Prentice Hall, 1985), 65; Shirley Biagi, *Media/Impact: An Introduction to Mass Media,* 2d ed. (Belmont, CA: Wadsworth, 1992), 115; Melvin L. DeFleur and Everette E. Dennis, *Understanding Mass Communication* (Boston: Houghton Mifflin, 1981), 59; and Jay Black and Jennings Bryant, *Introduction to Mass Communication,* 3d ed. (Dubuque, IA: William C. Brown, 1992), 207. Erik Barnouw tells this story in *A Tower in Babel: A History of Broadcasting in the United States to 1933* (New York: Oxford University Press, 1966), 20. Christopher H. Sterling and John M. Kittross tell it in *Stay Tuned: A Concise History of American Broadcasting,* 2d ed. (Belmont, CA: Wadsworth, 1990), 28, and explicitly describe it as "the first broadcast." Susan J. Douglas repeats it in *Inventing American Broadcasting, 1899–1922* (Baltimore: Johns Hopkins University Press, 1987), 156, including the description of it as the first broadcast, though she adds evidence to the record that Fessenden was thinking more in terms of the telephone and in the rest of her book demonstrates a much more accurate and interesting way of understanding broadcasting and its origins. Hugh G. J. Aitken mentions the incident without committing to its characterization as the first broadcast in *The Con-*

board wireless operators called their officers to come listen as Fessenden played "O, Holy Night" on his violin, sang a few bars, and read some verses from Luke. This, it is often said, was the first broadcast.

As history, the importance and accuracy of this anecdote is debatable. Fessenden's 1906 transmission was merely one in the middle of a series of demonstrations of what he thought would be a kind of telephone, not a kind of broadcasting, and it involved a mechanical electrical oscillator that would turn out to be one of the many technological blind alleys that litter the history of invention. Fessenden, moreover, was not working alone. For example, the textbooks do not mention Fessenden's mentor and associate, Charles Steinmetz. Steinmetz not only cooperated with Fessenden in the construction of his transmitter, but was a leader in the development of the mathematical understanding of alternating current, a prerequisite to the science of radio.[2] Steinmetz, moreover, was an unusual character: a high official at General Electric and lifetime member of the Socialist Party, U.S.A., who believed that corporations were not essentially capitalist but instead embodied a rational transformation and centralization of production that prefigured socialism.[3]

The fact that Fessenden remains in the spotlight of textbook histories of broadcasting while Steinmetz remains in the shadows ultimately tells us less about history than about the present-day lenses through which history is viewed. The anecdote of Fessenden's "first broadcast" is enshrined in the textbooks because it appeals to familiar and popular themes: for example, the tale of entrepreneurial success. The story is much like other oft-told stories about American inventors from Thomas Edison to Steven

---

tinuous Wave: Technology and American Radio, 1900–1932 (Princeton: Princeton University Press, 1985), 74.

2. Steinmetz oversaw the manufacture of his former employee Fessenden's first high-frequency alternator, the central element in Fessenden's transmitting apparatus, which was a variation on Steinmetz's alternating-current electromagnetic generator (Douglas, *Inventing American Broadcasting,* 155; Aitken, *Continuous Wave,* 64–69). Alternating current was a product of a general understanding of electricity based, not on the crude mechanical metaphor of water in a tube, but as a matter of mathematically describable potentials, charges, and waves.

3. David F. Noble, *America by Design: Science, Technology, and the Rise of Corporate Capitalism* (New York: Knopf, 1977), xx. According to Charles Steinmetz, "financial consolidation is the first step of industrial cooperation. Administrative consolidation and reorganization must follow, and then technical or engineering reorganization, to reap the benefit of industrial cooperation. The technical side of the corporation is the purpose of its existence; manufacture, transportation, etc., are technical or engineering problems, and the administrative and financial activities, therefore, merely means to accomplish the legitimate object of the corporation—production" (*America and the New Epoch* [New York: Harper & Brothers, 1916], 37).

Jobs: an enterprising young man with a better idea and a desire for wealth strikes out on his own and through hard work and wits builds a better mousetrap that astonishes the world. Fessenden was indeed an entrepreneur, who rejected opportunities with both a corporation and the government,[4] and instead set off on his own to win his fortune.

Americans tend to tell the story of invention in terms of Fessendens instead of Steinmetzes because we like to think that inventions spring full blown from the mind of inventors working alone, from isolated individuals operating competitively—hence the often absurd search for "firsts" in textbook histories of technology: the first broadcast, the first radio station, the first television broadcast, and so forth. Steinmetz's role in the history of broadcasting, conversely, is omitted from the textbooks not just because he was a socialist (and a peculiar one at that) but because his contribution to the development of radio was not of a kind easily assimilable to the popular individualist vision of invention. The mainstream of our culture has difficulty acknowledging abstract and structural contributions like Steinmetz's, contributions that involve a community of shared understanding and goals where the relations between individuals are more important than the peculiarities of the individuals themselves.

This is not to say that our society *is* fundamentally entrepreneurial or individualist. Steinmetz's vision of a future of corporate-inspired socialism was wildly off the mark, of course, but what the textbooks don't make clear is that the same can be said for Fessenden. As an entrepreneur, Fessenden turned out to be a relative failure because he failed to anticipate the corporate institutional structures that would eventually dominate the medium. Fessenden was just one of many pioneering entrepreneurs who lost their patents to large corporations. Edwin Armstrong's story is more tragic than Fessenden's, and Lee De Forest's only slightly less so. They were all overwhelmed by the new corporate form of business that was pioneered in radio by Marconi—a lesser inventor than any of them, by many accounts, but a master of corporate organizational structure.

If there is less of the individualist, entrepreneurial form in broadcasting than we imagine there to be, moreover, there is more of Steinmetz's technocratic socialist utopianism. Many practices considered mainstream today, from bureaucratically centralized corporations themselves to a rigorous respect for free speech, were more valued among socialists

4. Fessenden, before forming his own business to develop radio, had previously worked for the Edison General Electric Company, a branch of Westinghouse, Purdue University, and the U.S. Weather Bureau (Douglas, *Inventing American Broadcasting,* 43–45).

than among the political mainstream in Fessenden's youth. Echoes of Steinmetz's social democratic principles are most evident in government regulation of broadcasting. Commercial broadcasting as we know it requires an elaborate bulwark of government-enforced pressures and restraints on the use of radio waves. In fact, if Fessenden were to duplicate his "first broadcast" today, it would be against the law.[5] But there is much of Steinmetz's vision on the business end of things as well. Trade associations, engineering schools, modern management techniques, and corporate research laboratories are contemporary institutions that at one time were either unknowns or anathemas to the majority of the business community.

This chapter presents a general framework for making sense of these contradictory themes: the belief system called liberalism and its twentieth-century variant, corporate liberalism. It outlines corporate liberalism as a way of thought, and makes the case for understanding U.S. broadcasting as an embodiment of that way of thought. As it has developed in the historical and sociological literature, the concept of corporate liberalism both describes the affirmative values that guide decision making about major institutions like broadcasting in the United States and calls attention to the tensions and contradictions within those values and the institutions they underwrite, such as tensions between liberal individualism and corporate collectivism. Applying the concept of corporate liberalism to the institution of broadcasting, then, provides a way of accounting for both the broad values that the institution embodies and the apparent contradictions in those values. The chapter begins with some of the conceptual themes of liberalism as a whole, and discusses corporate liberalism's social context, its relevance to broadcasting, and then its conceptual characteristics.

The Liberal Heritage: Rights, Individuals, Markets

Literatures of Liberalism

Although most would agree that there are systematic relations among concepts of individualism, rights, markets, property, politics, and law in American culture, there is little agreement about how best to character-

5. Our legal system would frown upon him for any number of reasons: broadcasting on an inappropriate frequency at an inappropriate bandwidth; operating unlicensed equipment; constructing and operating equipment that did not conform to appropriate technological specifications; broadcasting inappropriate content (music and entertainment to stations intended for ship-to-shore communications); and broadcasting without any approved and reliable means of making a profit.

ize those relations. Critical theorists often treat them as subcategories of "bourgeois" or "Western" consciousness, certain feminists speak of them as offshoots of patriarchy, Foucault at times referred to Western "technologies of the self," and humanist social critics speak variously of "possessive," "utilitarian," or "expressive" individualism, which are alternately contrasted with and lumped together with, for example, "republican" traditions in American culture. Conversely, both postmodernists and certain kinds of positivist historians (for very different reasons) like to suggest that all such generalizations are overconfident totalizing constructs that cannot hope to capture the impossibly complex, shifting contemporary world.

This is not the place to try to resolve the important questions these competing trends raise. For the purposes of this inquiry, what's important is that there *are* salient, general patterns of thought and practice that shape legal and political decision making in the United States. The term "liberalism," then, is in the first instance merely a shorthand term to refer to those admittedly complex and fluid general patterns. It is the liberalism that has shaped the culture of the legal, industrial, and political elites that built and continue to shape the institutions of the electronic media.

The use of "liberalism" to refer to broad currents underlying much of American legal and political decision making is probably most common among certain intellectual historians and political theorists. Important ideas are often found, however, not just in the writings of great minds, but in the details of institutional practices. This form of liberalism is a less self-conscious, messier, and perhaps cruder form of thought than that familiar to students of traditional intellectual history, with its list of great liberal political theorists from Hobbes to Mills and beyond. But, as this book goes to some lengths to show, it is patterned, shared, and learned: it is a form of thought. American broadcasting, for all its apparently illiberal characteristics, can only be explained as the outcome of a nearly century-long, deliberate social and political effort to put the liberal principles of the marketplace and private property into practice in the field of electronic mass communication.

To some degree, this sense of liberalism is analogous to what the Marxist tradition calls "bourgeois consciousness," with its emphasis on the emergence of ideas from material social conditions. Indeed, grand ideas may germinate in the rough and tumble of social life as much as they do in the writings of a few great men; famous philosophical statements may be better understood as distillations of aspects of common culture than as sources of it. Yet the term "bourgeois" carries with it connotations of an urban/rural dichotomy that does not fit well with American social experience, and can imply a base/superstructure model wherein matters of value and belief are reduced to mere epiphenomena of a narrowly construed economic structure or to legitimating alibis for class rule. Furthermore, the word "bourgeois" in American culture carries with it an implied sneer. It creates a distance between its user and the values being critiqued. "Liberalism," in contrast, helps maintain an awareness that the object of critique is one's own culture, and thus to some extent one's own ideas; it reminds one that one works from within culture, not outside it.

This view of liberalism, then, is closest to the use of the term in the early works of Roberto Unger.[6] Unger's work bridges two traditions of skeptical literature on liberalism: a humanist tradition represented by the likes of Alasdair MacIntyre and Thomas Spragens, and the critical legal studies movement in legal scholarship. This otherwise highly diverse collection of scholars share the observation that the career of liberalism in the modern era has followed a paradoxical path. Liberalism's problems stem as much from its successes as from its failures. Liberalism in practice comes into conflict with liberal aspirations; liberalism's deepest epistemological and social assumptions come into conflict with its ideals.

Roberto Unger calls liberalism "a kind of social life."[7] In this sense liberalism, as it has manifested itself in the culture, history, and social structure of the United States, cannot be understood as just a philosophy or an attitude; it has shaped an enormous variety of philosophies, frequently opposing ones. Liberalism is one of those things that seems to be "in the air" as people go about their everyday activities. It is evident in the fact that political movements from all points on the political spectrum are more likely to frame their claims in terms of the language of rights than any other political discourse (such as, say, appeals to religion or the social good). And it is also evident in the fact that American culture is still to a large degree, if not exclusively, a culture of business, a culture that believes in entrepreneurialism. Reginald Fessenden, in pursuing his radio telephony experiments for the purpose of making a profit, was acting from within liberal culture, and on the expectation that his efforts would be protected by liberal laws of ownership and property.

Liberalism is thus a set of habits of thought, talk, and action that can be used to construct a variety of positions. It is a highly dynamic and fundamentally historic pattern of thought and feeling, more a cultural

6. Roberto Unger, *Knowledge and Politics* (New York: Free Press, 1975); Thomas A. Spragens, Jr., *The Irony of Liberal Reason* (Chicago: University of Chicago Press, 1981).

7. Unger, *Knowledge and Politics,* 18.

form than a rigid set of abstract philosophical ideas or precisely defined legal and legislative principles. One could say of liberalism something similar to what Raymond Williams said of language and culture.[8] The fact that liberalism, like culture, is ephemeral, fluid, and often contradictory does not make it less important, less of a presence in social life. Liberalism provides some of the crucial habits of mind, some of the basic interpretive frameworks, upon which fundamental actions are based. It is thus as much a part of the base as it is of superstructure; even where it appears to contradict itself, it is productive of contemporary social relations, not just reflective of them.

Features of Liberalism

The dominant strains of liberalism that have shaped decisions about broadcasting in this century tend toward forms of utilitarianism, in which it is assumed that people the world over are in essence rational, self-interested individuals. The individual is an abstraction that is inherently active, desirous, and the source of difference and change. Differences between individuals, when not the result of exterior coercion or manipulation, are generally understood as the product of simple, inscrutable wants; beyond the assumption that those wants are the product of self-interest, little else can be said about them. What one means by the "individual" shifts in significant ways in discussions of broadcasting. For example, policy makers and executives frequently speak differently of corporate "stakeholders" and "consumers." Yet both stakeholders and consumers are imagined as self-evidently self-interested entities that are the source of change and agency.

In opposition to the individual is another abstraction. Hobbes called it Leviathan, today we call it society. Society is everything the individual is not: inherently static, and the source of continuity, stability, and domination. Typically, the individual is thought of as logically prior to, and radically autonomous from, society. Society's relation to the individual is thus one of constraint, and liberty is conceived as an absence of constraint (in Hobbes's words, the liberty of a person "consisteth in this, that he finds no stop, in doing what he has the will, desire, or inclination to doe").[9] Hence, the relation of the individual to society is one of activity, of resistance. Rights and freedoms are a crucial way of understanding the relation of individuals to society, and are thus in the first instance understood in the negative sense as a lack of social constraint on individual

8. Raymond Williams, "Base and Superstructure in Marxist Cultural Theory," in *Problems in Materialism and Culture* (London: New Left Books, 1980), 31–49.

9. Thomas Hobbes, *The Leviathan* (Buffalo: Prometheus, 1988), 110.

action ("Congress shall make no law . . ."). For example, one can describe an action or condition one dislikes as a social constraint imposed upon individual freedom, and an opponent might defend the action as a necessary constraint of unbridled private desire for the sake of the social good. Because both sides understand the situation through a dichotomy of an active individual set against social constraint, they are both liberal.

This central tension or "central contradiction" of liberalism between the individual and the social is better understood as historical and political, not formal, logical, or absolute. The point is that, in political discussions such as American broadcast regulation, the community of active participants tend to act on and justify their actions with a cluster of beliefs centered on an opposition between active individuals and static social structures. Subsequently, in trying to make sense of their actions and justifications, they repeatedly find themselves in conflict with themselves and each other, often involving conflicts about using political intervention to achieve the goal of limiting political intervention. The argument is *not* that this contradiction is true of all societies at all times.[10]

In any case, a result of this individual/social opposition is that debates about broadcast structure are regularly framed in terms of reconciling individual freedom with social constraint, of reconciling the competing self-interests of individuals (or stakeholders or consumers) with one another and with social stability and progress—the famous Hobbesian problem of order. Hobbes's model of individual and society was liberal, but his response appears to us as illiberal: individual freedom, Hobbes argued, cannot be reconciled with social order; if we are to have one, we must forgo the other. In a sense, most of liberal thought and social life can be seen as an attempt to provide a counterargument to Hobbes's pessimistic conclusion, to somehow construct a noncontradictory form of "ordered liberty."

The litany of proposed solutions to the Hobbesian problem of order is familiar: over the centuries both philosophers and rulers have explored

10. The dangers of interpreting a fundamental contradiction as formal and abstract instead of historical and experiential are illustrated in the career of the "fundamental contradiction" described by Duncan Kennedy in the opening pages of his essay on Blackstone's *Commentaries,* of which he subsequently said, "I renounce the fundamental contradiction [because] these things are absolutely classic examples of 'philosophical' abstractions." For the original formulation, see Duncan Kennedy, "The Structure of Blackstone's Commentaries," *Buffalo Law Review* 28 (1979): 205, 211–13. For the renunciation, see Peter Gabel and Duncan Kennedy, "Roll Over Beethoven," *Stanford Law Review* 36 (1984): 15–16.

the possibility that free individuals can be reconciled with the social good through some combination of the invisible hand of the market, democratic procedure, and a legal system based on impersonal, objective rules ("the rule of law, not of men"). In countless, often competing, variations, this set of ideas has shaped the core of social, political, and legal discourse in the United States for the last two hundred years. One thing they share is a vision of transcendence. Liberalism is not just a preference for things like individual freedom, a market economy, or the rule of law. It is the dream that such things can be happily integrated; that, for example, individual freedom can be reconciled with a market economy by recourse to formal procedures like the rule of law. This hope of transcending tensions between apparently opposed tendencies is what I take Unger to be describing when he writes that liberal consciousness "represents the religiosity of transcendence in secular garb."[11]

As Jennifer Nedelsky has shown, the idea of a natural right to private property, for all the permutations it has gone through over the years, remains at the center of U.S. liberalism's vision of justice and freedom.[12] The inviolateness of private property in early liberalism was a natural and absolute right, central to the moral and material progress of humankind, and serving as a clear legal limit to the power of democratically elected legislatures. Later, property may have been eclipsed by contract at the center of the legal imagination for a time, but in either case, the belief was that rights and freedoms could be reconciled with the social order, that politics could be transcended, by using law to uphold the inviolateness of relations of economic exchange.

Other ideas besides property have been added to the liberal pantheon over the years. The nineteenth century added electoral democracy to the system through the gradual expansion of suffrage. In this century freedom of speech has been elevated to central status, at times supplanting property as the symbolic centerpiece of the liberal liturgy. Yet these principles still play similar roles in liberal discourse. These various strategies for reconciling self-interested individuals with social order all share a certain trait: they are all imagined to be somehow neutral, objective, and irrefutable, and thus capable of transcending the subjective winds of politics and the arbitrary contingencies of subjective wants. Two constructs are central to this vision: the dichotomy between law and politics, and a metaphor of science.

Law in liberal thought is typically thought of as somehow the oppo-

11. Unger, *Knowledge and Politics,* 163.

12. Jennifer Nedelsky, *Private Property and the Limits of American Constitutionalism: The Madisonian Framework and Its Legacy* (Chicago: University of Chicago Press, 1990).

site of politics. In its simplest form, law is neutral and objective, politics subjective and partial. Ideally, lawyers and judges do not create rules and policies; they simply discover and enforce the laws that are already there in nature, reason, the Constitution, or common law; this is why it is considered appropriate to allocate certain kinds of decision making to judges instead of, say, subjecting those decisions to a vote. Politicians, in contrast, are thought to enact contingent, subjective choices in the name of the majority. Hence, law, as understood by an impartial judiciary, guarantees the existence of a realm of orderly social life protected from the buffets of the subjective, arbitrary whims of politics.

The liberal tradition is also constantly seeking strategies that are like science, particularly science in the model of Newtonian physics. It is hoped that if markets, government structures, and laws embody, or can be made to embody, the impersonal, objective, irrefutable clarity of scientific method and reasoning, then the arbitrary, coercive limitation of individual freedom for the sake of social order that Hobbes thought inevitable can be avoided. By offering a fixed truth or irrefutable certainty, they promise an order that prevents a slide into chaos, into Hobbes's "war of all against all," without interfering with the subjective desires of individuals. In different ways, science, the market, formal procedures, and the rule of law all suggest to the liberal mind a way to transcend the merely subjective and the contextual, and the chaos they threaten. They are all imagined, in other words, as tools for shielding us from the shifting, arbitrary winds of politics and social coercion.

Corporate Liberalism and Its Social Context

Introduction

Former FCC chair Mark Fowler once described his agency as a "New Deal dinosaur." Fowler was invoking the common belief that our existing framework of broadcast policy expresses the New Deal penchant for government interventionism and that his own scheme of "deregulation" marks a shift in the opposite direction. Typically, shifts such as the changes from the conservative probusiness republicanism of the 1920s to the liberal reformism of the New Deal in the 1930s are thought of as shifts of emphasis within the broad framework of "capitalism" or "free enterprise" that has characterized the U.S. political economy throughout its more-than-two-hundred-year history.

In recent years, scholarship from the fields of sociology, political economy, and business history have been providing an alternative view. According to Alfred Chandler,

> [t]he processes of production and distribution, the methods by which they were managed, the enterprises that administered them, and the resulting structure of industries and of the economy itself—all were, by World War I, much closer to the ways of the 1970s than they were to those of the 1850s or even of the 1870s. A businessman of today would find himself at home in the business world of 1910, but the business world of 1840 would be a strange, archaic, and arcane place. So, too, the American businessman of 1840 would find the environment of fifteenth-century Italy more familiar than that of his own nation seventy years later.[13]

Chandler has been a leading figure in the development of what is becoming a consensus view among historians and sociologists of U.S. political economy: roughly between 1880 and 1920, the American economic, political, and social order was transformed.[14] The words "rights," "freedom," "markets," and "property" remained part of the standard political and legal vocabulary, but their meanings shifted dramatically. Business, politics, law, and social consciousness were all changed. New structures emerged, structures that are for the most part still with us today: industry trade organizations, administrative law, federal regulatory agencies, policy research institutes, and oligopoly corporations. Commercial broadcasting in the United States is a creature of the changed environment.

The scholarship on the new corporate form of liberalism that emerged from the 1880-1920 period is diverse; it includes scholars from all points on the political spectrum, and even the term "corporate liberalism" is not universal to the literature.[15] The literature seems to have two poles, with many scholars working somewhere between them. On one end, there is the "managerial school" of business history led by Alfred Chandler, which takes an apologetic stance toward the new order. Chandler calls the new economic system "managerial capitalism," believes it is caused largely by changing technology and managerial techniques, and believes it to be inherently efficient and productive; his

13. Alfred D. Chandler, Jr., *The Visible Hand: The Managerial Revolution in American Business* (Cambridge: Harvard University Press, 1977), 455.

14. For an overview of the literature, see Ellis W. Hawley, "The Discovery and Study of a 'Corporate Liberalism,'" *Business History Review* 52 (autumn 1978): 309-20.

15. For a sense of the broad range of political views contributing to corporate liberal theory, see the editors' preface to *A New History of Leviathan: Essays on the Rise of the American Corporate State* (New York: E. P. Dutton & Co., 1972); the book is a collaboration between editors Murray N. Rothbard, a right-wing free market economist, and Ronald Radosh, an associate editor of *Studies on the Left.*

works are standard reading at the Harvard Business School.[16] On the other end is a body of scholarship known as the revisionist school, based in the new left and led by scholars such as William Appleman Williams, Martin Sklar, James Weinstein, Gabriel Kolko, and David Noble. Over the last thirty years, they have produced a series of rich historical studies of the efforts of various groups and individuals during the tumultuous and crucial period in U.S. history roughly between 1880 and 1920—the four decades immediately preceding the birth of commercial broadcasting. They see in that period the origins of the corporate liberal political and economic environment that has characterized the political economy of the United States ever since. They originally set out in part to use historical evidence to cut through the sanguine pieties of the "end of ideology" and related "liberal pluralist" views of U.S. society that appeared in the 1950s and 1960s. It was not a simple absence of ideology, the revisionists argued, that made U.S. politics appear to lack class and ideological warfare, that made it appear a relatively open and innocuous pattern of struggles between interest groups. Rather, the United States has been dominated by a very real but largely taken-for-granted ideological and political economic framework described alternately as corporate capitalism or corporate liberalism. This framework does not rigidly dictate economic and political behavior, but it sets the terms and broad boundaries of acceptable action within which interest group struggles can take place.[17]

16. See Chandler, *Visible Hand;* Alfred D. Chandler, Jr., with Takashi Hikino, *Scale and Scope: The Dynamics of Industrial Capitalism* (Cambridge, MA: Belknap Press, 1990); and Otto Mayr and Robert C. Post, eds., *Yankee Enterprise: The Rise of the American System of Manufacturers* (Washington, DC: Smithsonian Institution Press, 1981).

17. The revisionist tradition probably began with William Appleman Williams's *Contours of American History* (Chicago: Quadrangle Books, 1961). The term "corporate liberalism" first appeared in print in an essay by Martin J. Sklar ("Woodrow Wilson and the Political Economy of Modern United States Liberalism," *Studies on the Left* 1, no. 3 [1960]: 17-47), a leading proponent of the revisionist school. See also Robert H. Wiebe, *The Search for Order, 1877-1920* (New York: Hill & Wang, 1967); Radosh and Rothbard, *New History of Leviathan;* Ellis W. Hawley, "Herbert Hoover, the Commerce Secretariat, and the Vision of an 'Associative State,' 1921-1928," *Journal of American History* 61 (June 1974): 116-40; Noble, *America by Design;* Martin J. Sklar, *The Corporate Reconstruction of American Capitalism, 1890-1916: The Market, the Law, and Politics* (Cambridge: Cambridge University Press, 1988); R. Jeffrey Lustig, *Corporate Liberalism: The Origins of Modern American Political Theory, 1890-1920* (Berkeley: University of California Press, 1982); and Olivier Zunz, *Making America Corporate, 1870-1920* (Chicago: University of Chicago Press, 1990). For a debate over the value of "corporate liberalism," see Alan L. Seltzer, "Woodrow Wilson as 'Corporate-Liberal': Toward a Reconsideration of Left Revisionist Historiography," *Western Political Quarterly* 30 (June 1977): 183-212.

For all its diversity, scholarship on corporate liberalism has come to generally agree on several tenets. First, history belies the popular vision of an immutable American political and economic framework, unaltered since it was constructed by the revolutionary founding fathers and enshrined in an unchanging Constitution and state of nature. On the contrary, we live in a world that is organized along dramatically different lines than that of the founding fathers. Second, the old story of an ongoing struggle between a business elite defending a market system and reformists seeking to limit the power of business and the market is inadequate. History shows the story to be a good deal more complex than the traditional government/business dichotomy suggests. Much of the impetus for "big government" in this century, for example, has come from business leaders, and much of the opposition to big government from the grass roots. Third, the new political economy has been accompanied and fostered by the emergence of an "organizational sector" that cuts across government/business boundaries, a sector inhabited by professionals and bureaucrats of various sorts that operate according to new, "technocratic" logics. Fourth, the historical details of the 1880-1920 period clearly illustrate the extent to which corporate liberal social structures are the product of struggles among thinking human beings, not simply the outcome of implacable economic or technological forces (hence the preference for the term "corporate liberalism" over "corporate capitalism"). "[C]orporate reorganization," argues Sklar, is "better understood not simply as an 'external force' or an 'objective' economic or organizational phenomenon but as a social movement, no less than populism, trade unionism, feminism, Afro-American equalitarianism, or socialism."[18]

Like any historical movement, corporate liberalism is full of nuance, conflict, and compromise, and thus cannot be reduced to either a few

18. Sklar, *Corporate Reconstruction of American Capitalism,* 13. Perhaps because the turn-of-the-century leaders found it necessary to discuss, debate, and struggle over many basic principles of social and political organization, their words and deeds reveal in bold relief the principles of twentieth-century American social organization that are today generally hidden behind a haze of platitudes and business-as-usual. Reading revisionist histories often brings a shock of recognition: the arguments of a Woodrow Wilson or Herbert Hoover, or the tales of struggles between newly formed industry trade associations, sound strikingly contemporary and unusually explicit. Martin Sklar nicely captured the sense one gets from the events and characters of the period with a quote from F. Scott Fitzgerald: "They were making the first tentative combinations of the ideas and materials they found ready at their hands—ideas destined to become, in future years, first articulate, then startling and finally commonplace. At the moment . . . they were sitting with disarming quiet upon the still unhatched eggs of the mid-twentieth century" ("The Scandal Detectives," in *Taps at Reveille,* quoted in Sklar, *Corporate Reconstruction of American Capitalism,* 431).

propositions as if it were a simple political platform or to a narrow ruling class whose interest it mechanically serves. One can point to certain central tendencies in terms of both ideas and membership, however. Corporate liberals were originally centered in the leadership of big business and, in the words of David Noble,

> aimed at absorbing moderate reform movements, anticipating or redirecting them, while at the same time isolating the proponents of more radical change. Espousing a general theme of cooperation, social harmony, and economic and political order, they stood in opposition to socialism, on the one hand, and the anarchy of unrestricted competition, on the other. [The corporate liberals] sought above all to reconcile traditional liberal democratic notions of individualism, self-reliance, free enterprise, and anti-statism with corporate-capitalist and scientific-technological demands for order, stability, and social efficiency. Emphasizing first one, then the other, they worked to regulate the corporate economy through the agencies of government, through private associations like chambers of commerce and trade organizations, and through such research agencies as the National Bureau of Economic Research, the Brookings Institution, and the National Industrial Conference Board. Through reform bodies like the National Civic Federation, they promoted social-welfare legislation in order to reduce the burdens and antagonisms of working people, and strove to enlist the labor unions as voluntary partners in the corporate industrial system, thereby hoping to substitute orderly and predictable negotiation for industrial warfare.[19]

The Problem of the Corporation

At the center of the social changes that accompanied corporate liberalism was the new economic form of the large multiunit industrial corporation, which arose first in railroads, and then in oil, steel, and banking. The corporation brought with it several ground shifts in the shape of economic life. Most agree that the rise of corporations was accompanied by a decline in classical, free market competition. In the twentieth-century corporate economy, instead of numerous small players in the marketplace, in many cases we now have only a handful, each heavily insulated from traditional competitive forces. Business within and between corporations has become heavily inflected by rigid bureaucratic structures and quasi-cooperative relationships. And control has

19. Noble, *America by Design*, 61.

shifted in the direction of salaried managers, away from owner-entrepreneurs.

This is not to suggest that classical markets no longer exist in many parts of the economy. In broadcasting, for example, relatively limited and oligopolistic conditions tend to dominate in areas like program supply, internetwork competition, and the consumer-product industry. By most accounts, however, the generally volatile and freewheeling market for advertising time more closely conforms to the free, open, and unfettered marketplace of Adam Smith's ideal.[20] Nor is it to suggest that corporations no longer struggle mightily with one another. Rather, their struggles are no longer easily characterized by the mathematical certainty that both Adam Smith and Marx believed accompanied an open market.[21] In an open marketplace, if your goods were inferior or priced too high you would inevitably be forced out of business and die an economic death; no noble status, political connections, or government patents could save you. What seems to have faded from sight, therefore, is not the general struggle for power and profit, but the ability to easily characterize that struggle in terms of open markets and the corollary economic laws. In such circumstances, what's to distinguish corporate rivalries from, say, noncapitalist struggles between feudal lords?

One common answer is that we have a new mode of production, that our economy is simply *not* essentially competitive and that as a consequence new noneconomic forces such as technology and bureaucracy have asserted themselves alongside traditional economic forces. Marx himself, in an aside on the formation of the then-embryonic "stock companies," remarked, "This is the abolition of the capitalist mode within capitalist production itself, a self-destructive contradiction, which represents on its face a mere phase of transition to a new form of production. . . . It is private production without the control of private property."[22] Many others have countered with new theories and definitions of competition that try to locate economic laws operating in changed but nonetheless primary ways within the corporate economy. While some, such as Milton Friedman, have denied the decline of competition outright, many others have devised various theories of

20. Willard G. Manning and Bruce M. Owen, "Television Rivalry and Network Power," *Public Policy* 24 (winter 1976): 36.

21. "For [classical economists], and for Marx most of all, competition was an elemental force, somewhat comparable to the force of gravity, which keeps the parts of the system in place and interacting with each other in intelligible ways" (Paul M. Sweezy, *Four Lectures on Marxism* [New York: Monthly Review Press, 1981], 57).

22. Karl Marx, *Capital,* Kerr ed., chap. 27, 3:519, quoted in Sweezy, *Four Lectures on Marxism,* 58.

"altered" competition—from the liberal camp, for example, theories of "imperfect" competition, and from the Marxists, theories of "monopoly competition." The point is simply that competition of the *classical* sort is no longer dominant.

The corporate form of organization, then, tends toward vertical integration, bureaucratic rigidity and hierarchical organization, oligopolistic market behavior, large size, and in general replacement of open, entrepreneurial competition with what Chandler calls the "visible hand of management." The key players in the broadcasting business—electronics manufacturers and networks, for example—are and have always been organized along corporate lines, and those that are organized more entrepreneurially—such as independent stations or some program producers—have to deal with corporate structures as part of their everyday business. The peculiarities of corporate, managerial business organization thus permeate the institution of commercial broadcasting at numerous levels.

It is often said, with some alarm, that media corporations tend toward monopoly.[23] There is a kind of truth to this: they are for the most part not entrepreneurial, and over time corporate logic and economics encourage vertical integration and consolidation so that larger firms tend to merge with smaller ones, thereby reducing the number of firms in a market. But accusations of "media monopoly" can oversimplify. Corporate structures do concentrate control over cultural production in the hands of a few, but describing that concentration as economic monopoly does not fully capture the character of the situation. The target of criticism ought to be the corporate system of organization as a whole, not individual corporations within that system.

First, the trend toward concentration, although constant and predictable, is often counterbalanced in the long term by occasional episodes of competition, typically not within an established industry, but from new institutional structures, frequently associated with new technologies. Point-to-point radio was faced with competition from broadcasting, network radio was overwhelmed by network television, network television has been challenged by cable television, and these may both be challenged in the near future by new players that add computers to the mass media mix. Lists of recent corporate buyouts and mergers accompanied by charts showing shrinking numbers of major industry players, then, frequently tell only part of the story.

Second, the term "monopoly" can oversimplify the corporate pattern, which tends to be a center-periphery system with a handful of

23. Ben H. Bagdikian, *The Media Monopoly* (Boston: Beacon Press, 1983).

large, dominant corporations at the center surrounded by a periphery of smaller, more entrepreneurial firms that, though clearly less powerful, are an integral part of the industry. The major broadcast networks are a classic example: each network corporation is legally and technologically linked to hundreds of largely independently owned affiliates, each of whom is dependent on the networks for much, though not all, of their programming. There is a fairly lively market for affiliate stations themselves, a complex and highly competitive market for advertising time between affiliates, but each affiliate's link to its network is stable; affiliates hardly ever "jump ship" to go to another network or become independent. In such a system, then, there is always some competition present, increasing in degree as one moves away from the center, just as there are always monopoly, or more precisely oligopoly, conditions operating at the core; those who speak as though the system can only be *either* monopolistic *or* competitive will always be able to find some of what they are looking for regardless of what side of the debate they are on.

Third, the concern for monopoly often implicitly conflates ownership with management, and presumes that the central constraint on media product is the intentions of the wealthiest owners. While this is certainly a concern, it eclipses the constraints on media exercised by managers and managerial consciousness. The networks do have censors—the program practices departments—but they censor according to the dictates of bureaucratic formulas, rarely if ever according to explicit political agendas. The results are different, if equally inane. For every case of a media owner killing a controversial story, or of a William Randolph Hearst or a Rupert Murdoch trying to inflame public opinion with his own agenda, there are thousands of cases of managers, directors, editors, and other career professionals constraining media content in the name of "professionalism," "neutrality," or some abstract formula.

Besides the corporate form itself, another trend characteristic of corporate liberalism concerns the relation of corporations to the state and to one another. If, in the nineteenth century, business organization was coordinated by courts of law under the banner of property and contract, in the twentieth century, that function has been largely taken over by administrative regulatory agencies: both public administrative entities like federal regulatory agencies, and private entities like trade associations and professional schools now serve as key support structures of corporations, and are integral to the day-to-day operation of commercial broadcasting. The FCC, for example, enforces a series of regulations that shape and constrain the networks' relations to both affiliates and to independent program producers, regulations that are less constraints on business overall than they are devices for maintaining managerial order within the industry.

A final, and sociologically profound and complex, trend involves a shift from viewing the mass of the population as simply workers or customers to viewing them as consumers. Raymond Williams has pointed out that it wasn't until the middle of this century that the word "consumer" passed from specialized use in economics to general and popular use; its predecessor, "customer," implied some kind of local, ongoing relationship to a supplier, whereas "consumer" suggests a generalized and abstracted sense of something going on in society as a whole.[24] This terminological shift reflects a development often described as "Fordism": in the early part of this century many corporate and government leaders came to agree with Henry Ford that higher wages and shorter hours for workers can create new markets for mass-produced goods purchased by workers in their leisure time.[25] The result has been the rise to new legitimacy of the advertising industry, and the rise of marketing as a central component in industrial strategy—developments that have also proven central to, and in some ways constitutive of, the institution of commercial broadcasting.

The Political Character of Corporate Liberalism

The 1880–1920 shift need not be seen simply as an economic change with social and political side effects. On the contrary, corporate liberalism properly understood involves political and social components that not only reflect but are *necessary* to the creation of the corporate economy. The construction of corporate America, in other words, was a political, not just economic, achievement.

Corporate liberal social organization does not simply mean control by private corporations. It involves a complex, dynamic pattern of interaction among corporations, small businesses, the state, and an electoral polity. In general, the pattern involves a hierarchical distribution of power, with a core dominated by an alliance of corporate and government elites, orbited by less powerful—but not powerless—peripheries: an economic periphery of smaller enterprises and a political periphery of electoral politics.

Nor is corporate liberalism a "dominant ideology" in the simple

24. Raymond Williams, *Keywords: A Vocabulary of Culture and Society,* rev. ed. (New York: Oxford University Press, 1983), 79.

25. For an insightful analysis of the importance of communications to this "Fordist" industrial strategy, see Kevin Robins and Frank Webster, "Cybernetic Capitalism: Information, Technology, Everyday Life," in *The Political Economy of Information,* ed. Vincent Mosco and Janet Wasko (Madison: University of Wisconsin Press, 1988), 44–75.

sense of that term; it is not a self-explanatory, monolithic framework, imposed on a hapless populace by the ruling elite. Corporate liberalism has always been profoundly shaped by a pattern of constant and flexible response, accommodation, and reaction to a broad variety of dissident and resistant forces ranging from socialists to small capitalists to feminists. Corporate liberalism must regularly respond to, and is in some cases dependent on, a variety of noncorporate social groups and structures. Corporate liberalism, in other words, is a dynamic response to complex social contradictions and conditions, conditions that include various forms of resistance to corporate control.

One reason that the study of corporate liberalism has been centered in the field of history is that its political character was most strikingly evident in its formative years. The political turmoil surrounding the unprecedented concentration of power embodied in the giant "trusts" in the late 1800s dramatically underscored the contradiction between the principle of democracy and the conditions of work and economic inequalities experienced by the bulk of the population—a problem complexly expressed in numerous ways throughout society.

Some reactions came from within the business world itself. Smaller businesses threatened by the corporations sought to check their power. The new industrial giants themselves worried about the high economic risks associated with such capital-intensive, large-scale enterprises, and about the sometimes highly destructive, cutthroat competition within their ranks.

But the concerns were expressed outside the business world as well, particularly in the populist Progressive movements around the turn of the century. Organizations of farmers, labor unions, socialists, journalists, urban liberal intellectuals, and others all rocked the political landscape with bitter denunciations of the excesses of big business.

While the chorus of complaints seemed to speak in one voice, however, that unanimity covered up a considerable ambiguity concerning what was to be done about the problem. Socialism, of course, was in the air, and many key figures in the Progressive movement, such as Upton Sinclair, professed to varieties of socialism. And as the case of Steinmetz illustrates, there were even socialists within the corporate world.

The opposition to socialism, however, was fierce. Many found it more comfortable to think that the excesses of big business were simply that—excesses, not fundamental flaws. For these individuals, the goal was somehow containing, guiding, or controlling capitalism to eliminate its most serious problems, not replacing it altogether. Yet even within these reformist groups, there were widely variant interpretations of both the purpose and nature of reform. For smaller businesses, reform meant checking the power of corporations that threatened their well-being and very existence. For labor and many other groups, reform meant restraining the power of business in American life altogether, so as to further noneconomic aspects of the public welfare. For the corporations, reform eventually took on the shape of both a necessity and an opportunity: on the one hand, moderate reforms seemed necessary as a way to ward off the threat of more drastic changes, especially socialism, while on the other, those same reforms offered the possibility of dampening competition and otherwise introducing stability and order that would provide a more congenial and protected environment for the vertically integrated, complex and bureaucratic industries characteristic of the corporate system.

Corporate liberalism is thus in the first instance a set of values and forms of social life that helped resolve conflicts and knit together the diverse interests and points of view of the Progressive Era into a relatively stable social formation. As an underlying framework for understanding and legitimating the U.S. political economy, corporate liberalism has persisted, with variations, for the rest of this century, providing a set of shared values and assumptions to the mainstreams of the business community and the Republican and Democratic parties.

Corporate Liberalism and Radio: Finding a Role—and a Meaning—for the Public

Herbert Hoover and the New Public Interest

One of the principal players in the formation of radio was also, and not coincidentally, a principal player in the formation of corporate liberalism itself: Herbert Hoover. The life and thought of Hoover nicely illustrates the character of the social reorganization that occurred at the beginning of this century. His career and the vision of society that motivated it consequently has become the subject of an illuminating body of historical research.[26]

For a while known as the hapless president who reacted pathetically to the disaster of the Great Depression, Herbert Hoover has been in a sense rehabilitated by the revisionist literature into one of the most important figures of the twentieth century and the quintessential corporate liberal. Trained as an engineer, Hoover made his fortune in mining, and then became an industrial statesman, moving easily between the

26. For a characteristic example, see Hawley, "Vision of an 'Associative State,'" 116–40.

worlds of business and government. Courted by both the Democratic and Republican parties, his career in government before the presidency, particularly as secretary of commerce through the 1920s, was long and fruitful, and helped establish many of the institutions and patterns of government-business interaction that are still common today.

This revisionist view of Hoover began, interestingly enough, largely with the work of Marxist historian William Appleman Williams, for whom Hoover "was the crucial figure in the evolution of the [corporate liberal] approach."[27] The understanding of some of the specific details of Hoover's thought has been refined by recent scholarship, but in its basic outlines Williams's analysis of Hoover is still pertinent.

Hoover's basic faith in capitalism and the liberal philosophy of minimalist government was clear. For example, he was a leading opponent of the antitrust movement on the grounds that it was an unwarranted government intrusion into private business affairs. Nonetheless, Hoover saw limits to the business individualism that dominated the nineteenth century. Hoover, according to Williams, "was quite aware that the American economic system was not functioning satisfactorily." Quoting Hoover, Williams argues that the secretary of commerce was convinced that "any return to 'individualism run riot' would only increase 'social ferment and class consciousness' among the lower classes and thereby accelerate the 'drift toward socialism'" (427). Indeed, in 1922 Hoover announced, "We are passing from a period of extremely individualistic action into a period of associational activities" (413).

Hoover thus set out to analyze the corporate economy, and concluded that "it was composed of three basic functional and syndicalist elements: capital . . . , labor, and the public at large, represented institutionally by the government" (427). The crucial task, he thought, was to struggle "to balance and control the units so that they would not drive the system toward fascism (business control), socialism (labor dominance), or the tyranny of bureaucratic government" (385). Toward this goal, Hoover advocated and did much to put into practice a number of measures that are familiar to us today. He was enthusiastic, for example, about the formation of self-regulating industry trade associations as an alternative way to settle industrial disputes, and was happy to use the offices of government, particularly the Department of Commerce, as a facilitator in the formation of these organizations. In general, he saw the task of coordinating the economy as a balancing act, requiring flexible and nuanced cooperation and compromise between the sometimes antagonistic three elements of society.

27. Williams, *Contours of American History*, 385 (hereafter pages given in text).

As the man who introduced the phrase "the public interest" to broadcast regulation, Hoover's concept of the "public" is worth exploring in some detail. Three things can be said about Hoover's understanding of the "public." First, Hoover's vision of the "public interest" has a subordinate place in relation to the "free enterprise" system as a whole; the "public" is not everyone all the time, but a unit within a larger social system. Hoover viewed the public interest as a necessary element of a system whose proper goal is continued development and growth on a capitalist basis; in the broadest sense, the public is not opposed to capitalist interests, but is a part of those interests. Second, the "public" in Hoover's vision is not the dominant interest, but simply one among three major forces that have to be balanced against each other; the public interest can be paramount only in certain circumstances, not as a general rule. Third, Hoover's "public" is a constant presence on the stage of human affairs, requiring constant attention and involvement.[28]

Hoover's vision of society as a system made of interacting functional units of capital, labor, and the "public" provided both an analysis of discord—labor unrest, for example, was a product of dysfunctional relations within the system, not any fundamental antagonism between capital and labor—and solutions to that discord—capital and labor need to be brought into a more harmonious, functional relationship, perhaps by better attention to the "public" on the part of capital. "There are great areas of mutual interest between employee and employer which must be discovered and cultivated," he argued.[29]

Hoover was just one of many leading industrialists who saw things this way and set about establishing harmony between competing "sectors" of society. Many interrelated strategies emerged from the effort to establish that harmony. One involved the Fordist idea of shorter hours and higher wages to create new consumer markets while quelling industrial unrest. "The very essence of great production," Hoover argued, "is high wages and low prices, because it depends upon a widening range of consumption only to be obtained from the purchasing power of high real wages and increasing standards of living."[30] Another policy involved improved housing and health care to reduce the stress of urban concentration, and the cultivation of new domestic relations centered on the

28. For a related analysis of the meaning of the public interest, see Willard D. Rowland, Jr., "The Meaning of 'the Public Interest' in Communications Policy: Part 1, Its Origins in State and Federal Regulation," paper presented to the Mass Communication Division of the International Communication Association, San Francisco, May 28, 1989.

29. Herbert Hoover, *The Memoirs of Herbert Hoover: The Cabinet and the Presidency, 1920–1933* (New York: Macmillan, 1952), 101.

30. Ibid., 108.

nuclear family, wage-earning husbands and unpaid, consuming, homemaking wives. Along these lines, Hoover created a volunteer organization called Better Homes in America, with a largely female membership to encourage the development of better housing for workers.[31]

The Public and the Consumer Society

Alongside the formulas of shorter hours/higher wages, industry self-regulation, and corporate/government associationalism, another corporate liberal industrial strategy was discovered and adopted during Hoover's day: the coordinated advertising of name brand, mass-produced consumer goods for a mass public. Most accounts suggest that, as an industrial strategy, consumerism came into full flower in the 1920s, the period when Hoover was most actively working to bring to life his vision of a harmonious corporate industrial system. It is telling that Hoover himself was a bit uneasy with some of the by-products of consumerism: his statements about radio advertising during the early 1920s suggest that advertising's culture of frenetic hedonism conflicted with the values of rationality and social responsibility that he associated with a corporate liberal social order.[32] Yet in the minds of corporate leadership, consumerism bore all the marks of corporate liberal thought: it was a kind of social engineering in the name of liberal goals.

Stuart Ewen's classic *Captains of Consciousness* clearly reveals how modern advertising and consumer production were, in the minds of at least some members of the industrial elite early in this century, elements of the same corporate liberal social project championed by Hoover. The system of consumer advertising, Ewen convincingly demonstrates, was imagined as a way to bind the various aspects of this social project together: while reducing the economic trauma of capitalist overproduction by increasing and regularizing consumption (in Vance Packard's words, "to end glut by producing gluttons"), advertising was also seen as an advocate for the new way of life, educating the public in the new habits of thought and action associated with domesticity, consumer spending, and corollary apolitical visions of the self.[33]

Ewen's book, like some other products of the revisionist historical

31. Ibid., 92.

32. In 1924 Hoover said, "It is inconceivable that we should allow so great a possibility for service to be drowned in advertising chatter" (Barnouw, *Tower in Babel*, 96).

33. Stuart Ewen, *Captains of Consciousness: Advertising and the Social Roots of the Consumer Culture* (New York: McGraw-Hill, 1976). See also Roland Marchand, *Advertising the American Dream: Making Way for Modernity, 1920–1940* (Berkeley: University of California Press, 1985).

tradition, has been criticized as narrowly conspiratorial.[34] There is some merit to this criticism. By focusing almost entirely on the ideas and actions of industrial elites, Ewen gives the impression that consumerism was a kind of plot foisted on a hapless public by a homogeneous ruling class. Subsequent historical work on advertising and consumerism has added nuance and complexity to the subject by emphasizing the role of communities of consumers themselves, the myriad ways in which the new world of consumerism addressed modern discontents, and the often confused and misguided actions of corporate management in consumerism's development. As one scholar has put it, Ewen's "captains of consciousness" were less captains of the consumer society than they were shipwrecked on its shores.[35]

What Ewen did convincingly demonstrate, however, is that whatever its sociological roots and character, in the early decades of this century, consumerism provided a new mode of *conceptualizing social relations* in an industrial society. Whether corporate executives shrewdly led the way into a consumer society or were haphazardly thrown up on its banks, the fact remains that consumerism eventually became available to managers and consumers alike as a way of imagining the relations between corporations and the rest of society, between industrial production and everyday life. And in the manager's mind this relation was imagined functionally, as a system that integrated democracy and oligopolistic capitalist industrialism by constructing the bulk of the population primarily as a body of potential consumers.

What was emerging in Hoover's day, therefore, were two interrelated ideas: a particular vision of the public as a social force in need of harmonious integration into the larger political economy, and the belief that the consumer system would facilitate that integration. The public, in other words, was a body of potential consumers, and the public inter-

34. T. J. Jackson Lears criticizes Stuart Ewen for using a conspiracy theory to account for the consumer culture in "The Concept of Cultural Hegemony: Problems and Possibilities," *American Historical Review* 90 (June 1985): 587. Robert Britt Horwitz similarly characterizes revisionist historians in general, particularly Gabriel Kolko and James Weinstein, as working from an elaborate conspiracy theory in which regulation is understood as a tool created by capitalists in their own interest (*The Irony of Regulatory Reform: The Deregulation of American Telecommunications* [New York: Oxford University Press, 1989], 34–35). Without denying that Kolko and Weinstein tend in this direction, Sklar's characterization of corporate reorganization as a social movement (*Corporate Reconstruction of American Capitalism,* 13) provides a more full and nuanced understanding of revisionist theory.

35. Francis Couvares, response to the panel "Broadcasting, Mass Culture, and Audiences" at the Conference on Culture and Communication, Philadelphia, October 6, 1989.

est lay in the cultivation of a consumer society. These ideas would eventually provide the guiding spirit of commercial broadcasting.[36]

Radio, the Public, and Corporate Capitalism

As we will see in the next chapter, broadcasting did not crystallize as an institution until the early 1920s. Yet corporate leaders were already formulating the general principles they would use to organize broadcasting between 1910 and 1920. During that time, ways of thinking about the political, economic, and social importance of the mass public to the corporate system were being formulated. Hence, although broadcasting itself took corporate leadership by surprise in the 1920s, the ideas used to respond to the surprise were already in place.

AT&T is a case in point. Under the leadership of Theodore Vail—another quintessential corporate liberal—AT&T between 1910 and 1920 was in the process of building a monopoly nationwide telephone network. Vail's telephone network involved his company in new ways with a mass consuming public. On the one hand, Vail was discovering that government regulation was helpful in establishing monopolies, and in any case was politically necessary in the face of the antitrust movement. This made him concerned about politics in a way most businesses were not. On the other hand, telephones were being installed and used in private homes; AT&T was providing an ongoing service, not just selling mass-produced objects. For both these reasons, Vail faced the necessity of developing consistent corporate strategies for dealing with and communicating to the public at large.

Those strategies were exemplified when AT&T began experiments with radio voice transmission in 1915. In that year, AT&T conducted a carefully orchestrated public demonstration of transatlantic "wireless telephone," gripping the public imagination with voice transmissions from New York to Paris. The company took advantage of the resulting press coverage to promulgate its vision of the relationship of radio to the public at large.[37]

Predictably, there was much that was self-serving, even misleading, in AT&T's statements. AT&T, by seizing control of a series of radio patents, had just eliminated the possibility of radio becoming a competitor to the telephone (thereby cutting off the interesting social possibility

36. For a sophisticated and elegant, if also functionalist, account of the role of broadcasting in a "Fordist" political economy, see Kevin Robins and Frank Webster, "Broadcasting Politics: Communications and Consumption," *Screen* 27 (May–August 1986): 30–51.

37. Douglas, *Inventing American Broadcasting,* 246.

of widespread popular communication by two-way radio outside of corporate control). Of course, the public statements made no mention of this, and in fact implied that the technologies involved were AT&T inventions, not implementations of the work of others.[38] Yet the AT&T statements reflect more than mere cynical manipulation, even when they were that as well.

The company eschewed the individualist language of competition and self-interest. AT&T's chief engineer in charge of the project, for example, told reporters that AT&T's wireless telephone was a "humanitarian rather than [a] commercial venture." Similarly, Theodore Vail assured a reporter that the new technology was not important because of profits but because "tens of thousands, hundreds of thousands, millions, the whole race, will draw from it a profit more desirable than dollars."[39] AT&T acted for the good of all, not out of greed.

Vail also articulated a particular understanding of just what constituted the "good of all." He spoke not in the manner of the nineteenth-century elite, not about the abstract individual or moral values, but of the problems of what he called the "unassimilated mass," which was less in need of rights than of an "interchange of ideas and thought" that would do away with "prejudice" and help lead to "nonpartisanship" and a "common course of action."[40] By offering, via the wireless telephone, "free communication between people," Vail thus imagined a public, not so much free to do what it willed, not so much free to go its own different ways, but free to become part of a consensual, homogeneous, integrated social system, with corporations like AT&T paternalistically leading the way.

Corporate Liberal Consciousness: From Rights to Functions

Classical Liberalism, Formalism, and the Bright Line

In order to appreciate the magnitude of the intellectual change associated with corporate liberalism, it is helpful to look at the state of liberal

38. Vail on occasion denied outright AT&T's reliance on patent purchases. For example, an AT&T demonstration of cross- country wireless telephone with a transmission from Arlington, Virginia, to Mare Island, California, was made possible largely by De Forest's Audion tube and his discovery of the possibility of "cascading" tubes for amplification. Vail told reporters, however, "As far as Mr. De Forest's lamp goes, if it played any part in the wireless conversation with Mare Island it is news to me" (ibid., 246-47, quoting the *New York Times,* October 22, 1915, 3).

39. Douglas, *Inventing American Broadcasting,* 248.

40. Ibid.

culture and thought in the late nineteenth century. At this time (Fessenden's formative years), American legal culture was rigorously liberal, but it was a kind of liberalism that seems strange and cruel to us today. As late as 1906, the time of Fessenden's "first broadcast," considerably more than half of the adult population was still prohibited from voting by law; Jim Crow was in full flower, and women's suffrage was more than a decade away. The U.S. Supreme Court had recently scoffed at the notion that individuals had a constitutional right to express their views in public parks.[41] Individuals recently had been, or would soon be, arrested or fined simply for advocating birth control, unions, pacifism, and socialism.[42] And it was in living memory that the legal system had viewed the Constitution with reverence yet looked on slavery with approval or at best indifference.

Much of the difference between then and now must be attributed to factors outside liberalism proper, to the social and cultural patterns in which liberalism was couched. For example, strict, unquestioned European middle-class social norms and moral codes still held sway over much of the population, particularly when bolstered by religion. The fact that the freedom for the privileged few was made possible by the unfree conditions of the many—the unpaid labor of women, for example, and severe legal restraints of minorities and laboring classes—was considered as it should be. Nonetheless, the United States was understood in the late nineteenth century to be a nation of freedom, democracy, and rights, at least in the realm of legal thought.

Part of the difference between nineteenth-century liberalism and corporate liberalism concerned characteristic modes of thought and legal argument. Under classical liberal thought, particularly in the second half of the nineteenth century, law claimed for itself the status of apolitical neutrality by way of a geometric model of science: legal thought relied on an image of itself as resting on a rigid, formal model, based on an ideal of axiomatic deduction from rules and unequivocal, "bright-line" legal distinctions. The role of law, then, was to locate and uphold clear boundaries—"bright lines"—between the rights of individuals and between individuals and the state. Nineteenth-century industrial disputes, for example, were most often treated as a matter of

41. *Commonwealth v. Davis,* 162 Mass. 510 (1895), 511. Reverend William F. Davis was arrested and fined for preaching the gospel on the Boston Common. He appealed to the Massachusetts and eventually the U.S. Supreme Court on the ground that he had a constitutional right to such preaching; each court turned him down (David Kairys, "Freedom of Speech," in *The Politics of Law: A Progressive Critique,* ed. David Kairys, rev. ed. [New York: Pantheon, 1990], 238–39).

42. Kairys, "Freedom of Speech," passim.

locating the formal boundary between the property rights of the parties involved. If the effluent from a coal mine spilled into a neighboring farmer's field, for example, the courts would set out to derive the line between the farmer's and the mine's property rights from the proper (and preferably Latin) legal axioms. Were the farmer's property rights being violated by the spill or would forcing the mine to limit operations violate its owner's property rights? Locating such boundaries was often tricky. But to the nineteenth-century legal mind it was obvious that the boundaries were there, even if difficult to discern.

In part because of the concern with bright lines, the nineteenth-century legal mind was heavily concerned with crystalline systems of property and contract. In one of the first American texts of legal theory, James Kent asserted as the central principle of law that "[e]very person is entitled to be protected in the enjoyment of his property, not only from invasions of it by individuals, but from all unequal and undue assessment on the part of government."[43]

The nineteenth-century legal imagination did nonetheless have room for the occasional exception to the rule; in practice the nineteenth-century legal system was perhaps not as pure as its legal theorists imagined. Exceptions to the inviolateness of property were in fact quietly acknowledged as part of the system. In Kent's words,

> there are many cases in which the rights of property must be made subservient to the public welfare. The maxim of law is that a private mischief is to be endured rather than a public inconvenience. On this ground rest the rights of public necessity. . . . it is lawful to raze houses to the ground to prevent the spreading of a conflagration. . . . the legislature [may] control private property for public uses, and for public uses only. Roads may be cut through the cultivated lands of individuals without their consent, provided it be done by [elected officials] and amount of the damages must be . . . paid to the owner. . . . In these and other instances which might be enumerated, the interest of the public is deemed paramount to that of any private individual.[44]

Here, in a legal tract written in 1827, are foreshadowings of the language familiar to broadcast law of today: "public inconvenience," "public

43. James Kent, *Commentaries on American Law* (New York: O. Halsted, 1827), 2:268.

44. Kent continued, "[T]his principle in American constitutional jurisprudence, is founded in natural equity, and is laid down by jurists as an acknowledged principle of universal law" (ibid., 274-76.)

necessity," "the interest of the public is . . . paramount." While this is not as expansive a use of the term as Hoover's "public interest," it demonstrates a discursive continuity between the nineteenth- and twentieth-century patterns of legal argument. Significantly, Kent turned to the idea of the public welfare in cases where the traditional understanding of property rights as inviolate lead to obvious threats to the larger economic system: emergencies, the construction and maintenance of roads, and the like. In fact, nineteenth-century uses of the phrase in law were most often used to help ensure effective use of bridges, canals, railroads, and other infrastructural links in the economic system. The "public interest" here was not being understood as a limit to the market or as a constraint on commercial interests; it was not a limit to the economic system that would eventually be called capitalism. On the contrary, it was typically used to untie perplexing knots in economic systems so that the market system as a whole would benefit; it was a necessary element to the nineteenth-century vision of laissez-faire. Although Hoover's use of the "public interest" would be different from Kent's, it would retain this basic orientation.

Corporate Liberalism, Science, and Bureaucracy

In 1912 Woodrow Wilson said the following in a campaign speech: "What is liberty? . . . Suppose that I were building a great piece of powerful machinery. . . . Liberty for the several parts would consist in the best possible assembling and adjusting of them all, would it not? . . . The piston of an engine [will] run with absolute freedom . . . not because it is left alone or isolated, but because it has been associated most skillfully and carefully with the other parts of the great structure."[45] Wilson's statement nicely captures the changed spirit of liberalism in the corporate era. If classical liberalism relied on formalist bright lines, twentieth-century corporate liberalism has come to increasingly rely heavily on images of technology, administration, bureaucracy, and functionalist metaphors of order. That the new way of thinking would lead to a restructuring of legal fundamentals was already becoming evident in 1896, when Oliver Wendell Holmes shocked his contemporaries by advocating such heresies as a right to strike for workers and the belief that "the absolute protection of property . . . is hardly consistent with the requirements of modern business."[46]

45. Quoted in Lustig, *Corporate Liberalism,* 29–30.

46. Oliver Wendell Holmes, Jr., *The Common Law* (1881; reprint, Boston: Little, Brown & Co., 1963), 100. Holmes had argued for the right to strike in a dissent to *Vegelahn v. Guntner,* 167 Mass. 92 (1896), 104.

Keeping in mind the limits of abstraction from context, it is possible to discuss some theoretical principles of corporate liberalism, that is, to analyze it as a problematic, as a characteristic set of concerns, questions, and problems. The quandaries created for liberal thought by the rise of the giant business corporation were many, but most of them can be seen as versions of one central dilemma: if the legitimacy of a market society rests on its control by individuals, how can one justify a capitalism dominated by the giant impersonal collectivities we call corporations? If the magic of the marketplace and private property in liberal theory is that it neatly draws a line between the realm of freely acting private individuals and the realm of collective, political constraints, then the corporation threatened to blur that line beyond recognition. The theoretical and institutional ideas of corporate liberalism can be usefully understood as an interrelated set of strategies for overcoming these difficulties.

One set of strategies concerns science and expertise. Around 1875 Charles Francis Adams (the brother of Henry) read an essay by John Stuart Mill on Auguste Comte. The essay, Adams said, "revolutionized in a single morning my whole mental attitude. I emerged from the theological stage in which I had been nurtured and passed into the scientific."[47] Comte, following Saint-Simon, was one of the originators of the idea that society could be studied scientifically, in a manner analogous to the scientific study of the natural world. Like many of the American elite of his day, Adams was struck by this new "scientific" outlook on society emerging from Europe, and saw in this outlook solutions to contemporary problems. Adams went on, in his own fashion, to put this new wisdom into practice. He is best remembered today as the father of the independent regulatory commission, a centerpiece of corporate liberal political organization.

Adams envisioned regulatory commissions as expert, apolitical bodies for overseeing industry that would overcome the political and economic turmoil surrounding industry in the late 1800s. "Commissions," he argued, "might scientifically study and disclose to an astonished community the shallows, the eddies, and the currents of business . . . the remedies no less than the causes of obstructions."[48]

What Adams was borrowing from Comte was not the idea of science per se, but a particular idea of science as applied to society. Adams was a harbinger of a new concern for what historian Robert Wiebe calls the

47. Charles Francis Adams, Jr., *Autobiography, 1835-1915* (Boston: Houghton Mifflin, 1916), 179, quoted in Lustig, *Corporate Liberalism*, 83.

48. Thomas K. McCraw, *Prophets of Regulation: Charles Francis Adams, Louis D. Brandeis, James M. Landis, Alfred E. Kahn* (Cambridge: Harvard University Press, 1984), 15.

new values of "continuity and regularity, functionality and rationality, administration and management."[49] Adams helped bring into being a new industrial logic that views the business world as a system to be efficiently maintained and a conception of the manager as that system's engineer, armed with the tools of administration, technology, and science. The ideas that came to dominate American political and social thought in the early part of this century, Wiebe writes, "were bureaucratic ones, peculiarly suited to the fluidity and impersonality of an urban-industrial world. They pictured a society of ceaselessly interacting members and concentrated upon adjustments within it. . . . predictability really meant probability. Thus the rules, resembling orientations much more than laws, stressed techniques of constant watchfulness and mechanisms of continuous management" (145).

The turn to bureaucracy in the name of liberal goals is a characteristic trend of the last hundred years. As Wiebe puts it, "[b]ureaucratic thought filled the interior" of our dominant social consciousness beginning in the early decades of this century (163). As bureaucratic terms and procedures repeatedly have been invoked in the service of classical values, bureaucracy has come to fill a shell of traditional liberal ideals.

It was Weber, of course, who observed that, contra Marx, capitalism was becoming characterized less by an "anarchy of production" than by an *increase* in bureaucratic organization.[50] Since Weber, many have observed that, although liberalism as a whole promises to enable individual freedoms, instead it often produces bureaucracies, with all their associated petty tyrannies and restrictions. Weber himself discussed how democracy, in spite of its opposition to bureaucracy, nonetheless tends to unintentionally promote bureaucratization.[51] Unger describes bureaucracy "as the characteristic institution that is the visible face of liberalism's hidden modes of consciousness and order."[52] Largely because of liberalism's search for neutrality in formal rules and procedures, the effort to reconcile disparate goals in legal and political structures seems over time to breed burgeoning bureaucratic institutions and logics. The turn to bureaucracy is not, strictly speaking, then, a clear-cut departure from liberal principles so much as it is a predictable if unintended outcome of the effort to enact liberal hopes.

49. Wiebe, *The Search for Order*, viii.

50. Max Weber, *From Max Weber: Essays in Sociology*, ed. and translated by H. H. Gerth and C. Wright Mills (New York: Oxford University Press, 1946), 49.

51. Ibid., 231.

52. Unger, *Knowledge and Politics*, 20. See also Gerald E. Frug, "The Ideology of Bureaucracy in American Law," *Harvard Law Review* 97 (1984): 1276–1388.

A key question here is whether or not bureaucracy is truly, as Weber suggested, always the most efficient way to do things. "The decisive reason for the advance of bureaucratic organization," Weber wrote, "has always been its purely technical superiority over any other form of organization. The fully developed bureaucratic mechanism compares with other organizations exactly as does the machine with the nonmechanical modes of production."[53] This view of bureaucracy is widespread; it is how bureaucrats typically justify themselves. Chandler's work, for example, elegantly explains the logic and history of the extension of bureaucracy (or "management") into business life, but he generally operates from the assumption that the simple success of the "managerial revolution" in American business proves its inherent productivity, rationality, and superiority. This assumption is bolstered by his use of case studies such as the railroads, where consolidation and introduction of administrative logic in the nineteenth century allowed for a much more efficient and effective coordination of scheduling and pricing.

The fact that bureaucratic methods help make the trains run on time need not, however, stand as proof that administrative practice is more efficient in all its applications, such as broadcasting. It may be "efficient" from the point of view of television program schedulers to force all programming into a rigid schedule of half-hour time blocks, but from the point of view of both program producers and the audience the effects of this practice (predictable plots, scripts mangled to fit the schedule) may be highly undesirable. Since Weber, therefore, a variety of critics from Jacques Ellul to the Frankfurt school have approached bureaucratic logic, not just with Weber's "nostalgic liberalism" which mourns the inevitable passing of the autonomous individual in the face of efficient bureaucracies, but with skepticism toward bureaucracy's own self-definition as inherently efficient. Central to this strain of thought is the argument that the values of administration and bureaucracy are simply that: *values,* a particular and in some ways limited vision of how human life ought to be, that tends to be enacted for its own sake, not because of some inexorable force of history. Bureaucratic consciousness, in other words, is not so much a matter of the application of rationality *as opposed* to myth, tradition, or religion as it is *another kind* of myth or tradition.

Significantly, technology has provided both a social context and metaphor for the new mode of thought; it is not coincidental that Weber said bureaucracy "compares with other organizations exactly as does the

53. Weber, *Essays,* 214.

machine with the non-mechanical modes of production." Technologies need not be imagined this way, in terms of rigid coordinated systems. In a countertradition ranging from the earliest radio amateurs through avant-garde electronic composers to today's computer hackers, a minority has associated technology, not with the values of "impersonality, regularity, efficiency, and uniformity," but with the values of "heterogeneity, randomness, and plenitude."[54]

Yet in our corporate liberal society, this countertradition has been for the most part safely marginalized. A mystified, overly uniform vision of technology has become dominant. As a result, political and social values are regularly hidden behind the supposed neutrality of technology and technological progress. David Noble has put it well. "[A]s technology has increasingly placed the world at people's fingertips," he writes,

> those people have become less able to put their finger on precisely what technology is. A general mystification evolved just as modern technology was becoming a dominant aspect of social life. . . . The development of technology, and thus the social development it implies, is as much determined by the breadth of vision that informs it, and the particular notions of social order to which it is bound, as by mechanical relations between things and the physical laws of nature. Like all others, this historical enterprise always contains a range of possibilities as well as necessities, possibilities seized upon by particular people, for particular purposes, according to particular conceptions of social destiny.[55]

What underlies much of the "technocratic" tone of corporate liberalism, then, is not so much technology itself, but a particular way of *imagining* technology within a larger social and political framework.

Hence, science-based industries' needs for technological standards and educational structures for engineers and managers provided a corporate motivation for the construction of corporate liberal institutions. Engineering societies, trade associations, universities like CalTech and MIT, and government bodies for coordinating corporate businesses all were fostered by corporate leaders for these purposes. At the same time, the wonders of new technologies helped justify corporate liberalism to the polity, and enabled corporate liberal decision makers to occasionally

54. Kathleen Woodward, "Art and Technics: John Cage, Electronics, and World Improvement," in *The Myths of Information: Technology and Postindustrial Culture,* ed. Kathleen Woodward (Madison, WI: Coda Press, 1980), 176.

55. Noble, *America by Design,* xxvi, xxii.

imagine themselves to be the political and social counterparts of scientists and engineers, and thus possessed of the same scientific neutrality.

Needs, Interests, Systems: Popular Functionalism

One of the more striking patterns in the history of decision making about broadcast institutions, we will see, is the vision of broadcasting and society at large as integrated, dynamic systems. The idea of a machine as a *system,* in fact, captures much of the way that metaphors of technology operate in corporate liberal decision making. This is symptomatic of a commonsense form of functionalism (to be distinguished from scholarly variants), which is a characteristic habit of thought and action central to corporate liberalism. The problems of functionalism point to key problems of corporate liberalism as a form of social life.[56] Generally speaking, functionalism is a pattern of explaining and attempting to order social life as if society were a self-regulating system or collection of systems that possessed needs. Explanation then tends to work from phenomena to the needs those phenomena satisfy; social institutions exist because they satisfy systemic social needs.

Woodrow Wilson's definition of a free society as "a great piece of powerful machinery" with all the parts "associated most skillfully and carefully with the other parts of the great structure" is an arresting example of functionalist logic put into service for liberal ideals. So is Hoover's understanding of society as a system made up of three interdependent units, and, in a less formalized way, Vail's idea of the ideal public as a component of a consensual, homogeneous, integrated social system. In each case, society is envisioned as a system with needs that can be satisfied by smoothly integrating the differing parts.

But functionalist system logic is not limited to moments of broad social speculation. It has become part of contemporary common sense, a reflex regularly used to make sense of and organize contemporary social institutions, particularly in the worlds of business and government. Why, for example, do we have federal administrative agencies even though they are not mentioned in the Constitution? They arose in the late nineteenth century, it is said, to serve the organizational needs of an increasingly complex economy. The economy—a system—had a

56. The term "functionalist" here is not meant to invoke the traditions of functionalist sociology or other scholarly traditions that use the word. Rather, it is useful because it nicely captures certain gestures and habits of thought characteristic of corporate liberalism. My use of the word here is in many ways analogous to, and is informed by, the concept of "evolutionary functionalism" discussed in Robert W. Gordon, "Critical Legal Histories," *Stanford Law Review* 36 (1984): 57–125.

need for integration which administrative agencies filled.[57] That the administrative agency also embodies vision and values is eclipsed.

Numerous other peculiarly twentieth-century American institutions are justified along similar lines. The belief that advertising supports the corporate system by promoting a consumer-oriented, quiescent workforce, for example, is a quintessentially functionalist vision. Whether or not it is accurate, it has been regularly used by industrial leaders throughout this century as an organizing principle for institutions like commercial broadcasting. Copyright collectives and centralized systems of distribution like broadcast networks—to mention just two examples that will be discussed in following chapters—are similarly justified in functionalist terms. In each case, these institutions are said to exist and be legitimate because they fill a need of the system, they serve a function.

Functionalist logic crops up in justifications of legal and organizational practices, as well. The legal fiction of the corporate individual, the "business judgment rule," and other legal devices used to create and grant power to corporations and their management are regularly said to be legitimate because they are necessary: they fill a need of advanced industrial societies. Within the broadcasting industry, the acceptable accuracy level of audience ratings is determined largely by reference to the system: the (surprisingly low) level of accuracy is adequate, industry executives argue, because it serves the needs of the advertising and programming system.

Functionalism can often serve as a powerful tool of critique, particularly, as Giddens points out, because it can point to unintended consequences of social actions.[58] As we will see, beginning early in this century, functionalist arguments have regularly worked in small ways to justify the abandonment of classical liberal formalist arguments. For

57. Ibid., 65.

58. Anthony Giddens, *Studies in Social and Political Theory* (New York: Basic Books, 1977), 96–129. The most powerful of these forms of critique familiar to many readers are in scholarly traditions that have at various points advanced important critiques of social institutions by locating systematic patterns of interaction whose consequence is other than that which on the surface seems to be the intention. Structural-functionalist sociology, for example, showed that certain kinds of zoning policies may be intended to aid the real estate industry, but they have the consequence of systematically supporting racial divisions in society. In law, legal realism showed that formal legal distinctions such as that between a right and a privilege are indistinguishable when considered in terms of their effect on social relations as a whole. Neo-Marxist theories of the state have demonstrated how government regulation of business may be advanced as a check on business abuses, but has the consequence of supporting the interests of capital overall.

example, Hoover overcame claims for formal property rights in the broadcast spectrum by justifying the administrative criteria of "the public interest" functionally, in terms of the necessity of such a criteria for the industrial and technological system.

Functionalism is better known for its conservative tendencies than for its radical ones, however. The dangers of functionalism, most would say, lie in the tendency toward tautologous argument that, in the guise of explanation, provides support for one or another status quo. (Even the best orthodox structural functionalists acknowledge this as "an inherent rhetorical opportunity" in functionalism, though they would deny that it is necessary to functionalism as such.)[59] At its crudest, the argument from system to presupposed need is simply circular: the institution or practice is explained in terms of the needs of a system, and the system is explained in terms of the institutions that supposedly serve it. Needs are presupposed, and then offered as causes, even when their presence and character is known only through the institutions being explained. Hence, the imagined "system," though presented as empirical necessity, in fact rests on normative presuppositions.

The problem is not so much that systems and functions don't exist as it is that their existence is not in itself an explanation of anything. Nietzsche said it well:

> But purposes and utilities are only *signs* that a will to power has become master of something less powerful and imposed upon it the character of a function; and the entire history of a "thing," an organ, a custom, can in this way be a continuous sign-chain of ever new interpretations and adoptions whose causes do not even have to be related to one another but, on the contrary, in some cases succeed and alternate with one another in a purely chance fashion. The "evolution" of a thing, a custom, an organ is thus by no means its *progressus* toward a goal, even less a logical *progressus* by the shortest route and with the smallest expenditure of force—but a succession of more or less profound, more or less mutually independent processes of subduing, plus the resistances they encounter, the attempts at transformation for the purposes of defense and reaction, and the results of successful counteractions. The form is fluid, but the "meaning" is even more so.[60]

59. Arthur Stinchcombe, *Constructing Social Theories* (New York: Harcourt, Brace, & World, 1968), 91, quoted in Giddens, *Studies in Social and Political Theory,* 103.

60. Friedrich Nietzsche, *On the Genealogy of Morals,* ed. Walter Kaufmann, translated by Walter Kaufmann and R. J. Hollingdale (New York: Vintage Books, 1969), 77–78.

Systems logic, in other words, obscures the "will to power," the desires, choices, passions, struggles, and moral aspirations embedded in the "systems" that are taken-for-granted constituents of contemporary institutions like radio and television. As Gordon puts it, "the inevitable ambiguities of legislative command, prior case law, custom or constitutional text need never force a legal system to the pain of political choice because its managers can always claim to be serving the logic of an historical process or immanent social consensus that exists beyond and prior to politics."[61]

Conclusion

Corporate liberal patterns of thought and practice are ultimately neither rational extensions of classic liberal legal principles, nor clear-cut departures from those principles. There is a tendency in contemporary critical theories—Foucauldian discourse theory, for example, or critical legal studies—to look at such matters with deep skepticism: corporate liberalism is a maneuver to rescue liberalism from itself, an attempt to regain the footing lost in the shifting sands of one set of liberal contradictions—the incoherence of atomistic individualism and of its industrial correlate, laissez-faire business principles—by shifting weight in the direction of another set of (also contradictory) liberal principles—a faith in the power of expertise, systems, and objective scientific knowledge to make manifest a transcendent, reified "public interest."

There is much to this view of corporate liberalism; it profoundly informs this book. But if left on a purely philosophical level, this kind of caustic, dismissive critique begins to lose its bite. For the disappointing quality of contemporary political and social discourse is first of all historical, a condition of our times, not simply a matter of incorrect thinking. Abstract discussions of principles, logical entailments, and contradictions can point out the contours of social life, their tendencies and weak points, but to get at the substance of problems one must turn to history and social life itself. This is the task of the next chapter.

61. Gordon, "Critical Legal Histories," 68.

A Revisionist History of Broadcasting, 1900–1934

What is found at the historical beginning of things is not the inviolable identity of their origin; it is the dissension of other things.
MICHEL FOUCAULT

Introduction

The early history of radio in the United States offers a useful lens for exploring just what it means to organize broadcasting on a corporate liberal basis. While radio did not emerge from a social and political vacuum, the range of organizational options facing radio before 1912 was broader than it has been any time since. Implicit in the struggles of the period were not just questions familiar to us today—for example, Should radio be regulated more by government or by business?—but also questions that today seem almost unthinkable: Does radio need to be controlled at all? Can it be organized on a purely voluntary, informal basis? Should the military be granted control over any part of the spectrum? Should nongovernment radio be developed in a for-profit or nonprofit framework? By large organizations or small ones? For which purposes? Should radio, for example, replace or compete with the telephone? How should the resulting organizational arrangements be justified?

The American answers to these questions were already taking clear shape by 1912, were firmly established by 1920, and by 1934 had been elaborated into a form close to what they still take today. The history of these developments reveals the broad contours, the patterns of pressure and constraint, that corporate liberal organization implies.

This chapter will show how it was decided that radio would be controlled by a coalition of large bureaucratic organizations, principally the military and corporations; small businesses would be allowed significant but nondominant roles at the industry's peripheries, whereas nonprofits would be aggressively marginalized. Radio would be developed, manufactured, and sold on a for-profit basis, but principally in a corporate oligopolistic mode in cooperation with government, rather than an entrepreneurial, fully competitive mode. Its most important uses would be strategic point-to-point communication for the purpose of controlling large, dispersed bureaucratic organizations like the navy and transnational corporations. These arrangements would be broadly justified in liberal terms: radio was to be free because it would be operated on a private basis. Yet the specifics would be justified in the characteristically *corporate* liberal terms of technical necessity, administrative expertise, and a functionalist vision of the public good. Formal legal categories such as property and contract would be hardly discussed.

The story of how these principles were arrived at is a classic and illustrative example of corporate liberalism in practice; it closely parallels the stories of the telegraph, steel, railroad, and chemical industries.[1] Yet the story of radio has its own unique contribution to make to the historical understanding of corporate liberalism. And broadcasting, the use of radio as a popular means of communication, stands at the center of radio's uniqueness.

As major industrial technologies go, radio tends to be relatively small, lightweight, inexpensive, and flexible. In contrast with many of the other technologies around which corporate liberal institutions and practices were forged, radio was relatively easy to experiment with even in its early days. Unlike steel, railroads, electric power, the telegraph, and the telephone, radio required no massive manufacturing plants or capital-intensive overland constructions. It could be assembled and experimented with by small entrepreneurs and hobbyists working in attics and backyard shacks. After the discovery of the crystal detector in 1906, moreover, radio receivers became positively cheap, bringing a new technology into easy reach of thousands when it was still in its earliest stages of development and as yet only dimly understood.

As a result, the domination of radio by giant bureaucratic organizations is less easily attributed to technological necessity or capital intensiveness than is the case with most other technology-based industries of this period. Much of the development of radio technology, and, more importantly, the exploration of its uses, originally occurred outside corporate walls. As this chapter will show, it took nearly thirty years and a complex series of political and institutional developments to bring radio firmly under the corporate umbrella.

The story of radio broadcasting's hybrid beginnings thus provides a backdrop of alternative visions for radio, against which the corporate liberal choices that eventually prevailed stand out in relief. Because the groups and individuals involved were operating from within a variety of

1. Gabriel Kolko, *The Triumph of Conservatism: A Reinterpretation of American History* (New York: Free Press, 1963); James Weinstein, *The Corporate Ideal in the Liberal State, 1900–1918* (Boston: Beacon Press, 1968).

different frameworks, one can see the possibilities that were abandoned as well as those that were pursued. Against the backdrop of those abandoned possibilities, the corporate liberal choices can be seen, not as technical necessities or simple common sense, but as the political choices that they are.

The discovery and development of broadcasting is at the crux of radio's uniqueness in the history of corporate liberalism. Among communications technologies, radio transmissions are unique in that they travel in all directions. The varied reactions to radio's omnidirectionality serve as a measure of social purpose and vision. As a general rule, governments, militaries, and large corporations struggled mightily against omnidirectionality. They were most of all interested in using radio to exert control at a distance, and thus focused on point-to-point uses of radio. When purely technological means to eliminate omnidirectionality failed, they turned to legal and institutional measures to overcome the problem. The assertion of legal control over the spectrum was largely a by-product of the desire to overcome omnidirectionality. It was less a technological necessity than an attempt to limit a technological potentiality.

Radio hobbyists and some radio entrepreneurs, conversely, took to omnidirectionality with enthusiasm, and were consequently resistant to legal controls of any sort; they, not corporations, discovered the extraordinary potential of radio as a means of popular communication. As a social institution, one-way mass distribution of electronic signals does not require radio waves (witness cable television). But it was in the early exploration of radio's omnidirectionality by amateurs and entrepreneurs that the social possibilities of broadcasting, of deliberately sending signals to numerous unseen listeners, were discovered.

It is telling that the corporate liberal elite remained blind to the potential of broadcasting for nearly twenty years. Yet it is equally telling that they eventually came to adopt broadcasting as their own and define it in their terms. Around 1920, as broadcasting began to become a popular craze, the corporate liberal establishment began to discover the value of broadcasting as a publicity and advertising medium. Within a few years, they successfully shaped the law and institutional structures so as to turn broadcasting into a linchpin of the consumer economy, while aggressively eliminating or marginalizing all other potential popular uses of radio. The success of these efforts testifies to corporate liberalism's extraordinary social and political powers of accommodation.

This chapter explains how all this was accomplished. It begins with a look at the early days of radio before and during World War I, when, in the face of several different visions of radio and its uses, a corporate liberal framework came to dominate. The next section focuses on the emergence of broadcasting in the 1920s, emphasizing the extent to which the crucial events of that period were a working out of principles already arrived at instead of a new departure. The final section quickly sketches the subsequent development of American broadcasting up to 1934, illustrating the continuity of the corporate liberal policies crystallized in the 1920s with the practices since.

A brief historiographical comment is warranted here. This chapter, like any serious discussion of American broadcast history, depends heavily on Barnouw's classic three-volume work on the topic. Besides making a vast amount of historical detail available, Barnouw showed how major structural decisions about broadcasting were the product, not of technological or economic inevitabilities, but of complex social interactions ranging from backroom bargaining among power elites to cultural trends from 1910 to 1940. In particular, though I take issue with some of his characterizations of legal and legislative events, Barnouw's discussion of corporate-government interactions in the 1920s is fundamental to the analysis presented here.[2]

While there are many other historical works important to this discussion—Danielian, Aitken, and Sterling and Kittross all have been particularly useful—Susan Douglas's *Inventing American Broadcasting* has been the most central. Douglas's underappreciated history of the pre-1920 events that led to the formation of broadcasting not only adds much to the historical record, but provides a crucial reinterpretation of broadcasting's origins. Both broadcasting and its commercial form, her book convincingly demonstrates, are not best understood as natural phenomena or processes that were "discovered" around 1920. Rather, they were collectively *invented* during the preceding two decades by diverse groups of people working in specific social contexts, people operating with particular visions, not just of technology, but of social life and culture.[3]

2. The most relevant volume to this chapter is the first: Erik Barnouw, *A Tower in Babel: A History of Broadcasting in the United States to 1933* (New York: Oxford University Press, 1966).

3. Susan J. Douglas, *Inventing American Broadcasting, 1899–1922* (Baltimore: Johns Hopkins University Press, 1987); N. R. Danielian, *AT&T: The Story of Industrial Conquest* (New York: Vanguard, 1939); Hugh G. J. Aitken, *Syntony and Spark: The Origins of Radio* (New York: John Wiley & Sons, 1976); Hugh G. J. Aitken, *The Continuous Wave: Technology and American Radio, 1900–1932* (Princeton: Princeton University Press, 1985); Christopher H. Sterling and John M. Kittross, *Stay Tuned: a Concise History of American Broadcasting*, 2d ed. (Belmont, CA: Wadsworth, 1990). Susan Smulyan's *Selling Radio: The Commercialization of American Broadcasting, 1920–1934* (Washington, DC: Smithsonian Institution Press, 1994) became available only as this book was in the

Like this book, Robert McChesney's important study of the politics of broadcasting in the 1920s and 1930s calls into question the sense of inevitability that permeates so much of the discussion of the commercial system, and his book has proven invaluable. McChesney, however, takes issue with Douglas, taking the more conventional position that questions of broadcast structure were undecided up until the late-1920s, when behind-the-scenes maneuvering led to the triumph of corporate commercialism. While McChesney offers powerful evidence for his view, here I side with Douglas: the corporate liberal interpretive frameworks that were in place by 1920, I argue, created a context that enabled and legitimated the specific policy decisions McChesney describes.[4]

Finding an Organizing Framework for a New Technology: Radio, 1900–1919

If there ever was a period when radio was truly free and unfettered, when it resembled the utopia suggested by the rhetoric of today's free marketeers, it was during radio's first decade. Before 1912, radio developed outside the control of law, government, or large bureaucratic corporations. Yet that period and the decade that followed were not so much shaped by pristine market forces—there was very little buying and selling of radios before World War I—as they were shaped by competing social visions. The first decade of radio, and to some extent the decade following, was a time of intense social experimentation, during which institutional arrangements and social uses for the technology were tinkered with as much as the technology itself.

Competing Visions

Four groups played key roles in the struggles during the formative period of radio and broadcasting, and each group had a distinct way of envisioning the medium. Radio hobbyists or amateurs discovered in the spectrum a playful democratic forum for leisure-time exploration. Inventor-entrepreneurs approached radio in a nineteenth-century competitive style, seeking to gain wealth by building a better mousetrap in the form of better radio equipment. The leading Western nation-states

---

final stages of completion, but it has nonetheless proven useful and in general is compatible with my position.

4. Robert McChesney, *Telecommunications, Mass Media, and Democracy: The Battle for the Control of U.S. Broadcasting, 1928–1935* (New York: Oxford University Press, 1993).

and their militaries began to see the radio spectrum as something analogous to a territory with strategic implications for imperialist expansion. And managerial businesses in the corporate mold sought to integrate radio into existing corporate structures that would complement, but not compete with, the telegraph and telephone. By 1910, within a decade of radio's first practical applications, all four visions of the spectrum were being actively pursued by their supporters.

Amateurs

While the exact numbers are difficult to determine, the evidence suggests that by 1914, and perhaps earlier, the largest system of communication by radio in the United States may not have been the product of corporations, the military, or inventor-entrepreneurs. Rather, it may have been an ad hoc, nonprofit network run by young radio hobbyists.[5]

In a culture that reveres professionalism, it is easy to dismiss or trivialize the role of nonprofessionals in social and technological innovation. Yet radio amateurs must be credited with numerous technical and social innovations that paved the way for broadcasting; arguably,they deserve credit for the discovery of broadcasting itself.[6] They discovered the long-distance propagation characteristics of shortwaves, for example, and pioneered the instant distribution of news, informing listeners of events such as the outbreak of World War I hours before newspapers.[7] They were one of the first social groups to engage in leisure-time activity in the home using electronic technology, and played a crucial role in popularizing and democratizing radio, bringing large parts of the public into contact with it for the first time. And they were also probably the first

5. According to Douglas, navy and commercial high-power transmitting stations made up only 15 or 20 percent of the total number of stations by 1910 (*Inventing American Broadcasting,* 207). Many popular accounts asserted at least the quantitative dominance of the amateur early in the second decade of this century; the *New York Times,* for example, guessed that amateurs numbered in the hundreds of thousands in 1912 (*Inventing American Broadcasting,* 195, 198). The American Radio Relay League was formed in 1914, and within four months had two hundred official relay stations in the United States. In some famous cases, amateur relays became the only source of communications for communities isolated by natural disasters (206).

6. The amateurs' remarkable success in the face of their lack of support and funds when compared to their institutional counterparts had a number of causes. Besides sheer enthusiasm and imagination, their tinkerer's approach allowed them to experiment and fine-tune apparatus without any concern for the bureaucratic constraints often faced by military and commercial operators, and they felt no compulsion to honor patents, such as that for Marconi's crucial tuner (ibid., 197–98, 207). As a result their technology, often assembled from Quaker Oats boxes, pieces of brass bedsteads, and telephones stolen from public booths, was sometimes more effective than professional equipment.

7. Ibid., 203.

"mass" audience of a simultaneous communication; estimates of the time suggested that as many as one hundred thousand amateurs could be reached in an evening.[8] The amateurs' impact on the development of radio and its uses is certainly equal to, and perhaps surpasses, the impact of the modern-day "hackers" who fomented the microcomputer revolution.

The early history of amateur radio points to organizational possibilities that are neither commercial nor corporate nor governmental. In response to the formation of an official, amateur radio relay network spanning the United States in 1914, *Popular Mechanics* exclaimed that the coming of wireless telegraphy "has made it possible for the private citizen to communicate across great distances without the aid of either the government or a corporation."[9] The amateurs were a grassroots, voluntarist group. They envisioned radio as anarchically democratic—in its own way, a traditional American vision.

For the amateurs, the radio spectrum was desirable largely because it was a realm free of hierarchy. They saw the formless, wide-open character of the spectrum as a fascinating and enjoyable potentiality, as something to be played with and explored and as the source of an alternative community. The elements of surprise and openness that came with late-night radio listening and signaling—those very elements that were an anathema to military and corporate radio users—were central to the pleasures of being a radio hobbyist. To amateurs, the omnidirectional and public character of radio was an asset, not the liability it was to corporations and the military. They discovered in the spectrum, not just ways to perform existing tasks quickly, but the opportunity to playfully discover new social possibilities. They used the spectrum, not as a means to preestablished ends, but as a source of amusement, new experiences, and new social contacts. In the process of doing so, they helped lay the foundation for modern broadcasting.

Entrepreneurs

Like many radio pioneers, Fessenden had worked at various times for universities, the government, and both General Electric and Westinghouse.[10] It is telling, then, that Fessenden eventually chose to operate as a classic entrepreneur: he found a few financial backers, formed a company, and set off to build his better mousetrap on his own, shirking universities, corporations, and government alike. It is perhaps equally telling that, as an entrepreneur, he failed.

8. Ibid., 205.
9. Ibid., 206.
10. Ibid., 42–45.

Fessenden was one of many of the first inventor-entrepreneurs that sought to commercialize the new technology of radio at the turn of the century. The entrepreneurs in many ways embodied the spirit and business approach of laissez-faire competitive capitalism of popular mythology. They also believed in the myth, as their actions and blunders reveal.

For the entrepreneurs, to commercialize radio meant to sell objects, physical things, on an open market; the route to success was thus building better radios, and they devoted their efforts to constructing radio sets that could transmit and receive farther, more clearly, or more reliably than those of their competitors. With varying degrees of scientific understanding, they tinkered with their equipment and hawked their wares, always with enthusiasm if only sometimes with success. Their eagerness to explore any and all alternatives made them more technologically flexible than the electrical corporations, with the result that their contributions have become the stuff of legend in American textbooks: Fessenden's continuous-wave alternator, Lee De Forest's "Audion" vacuum tube, Edwin Armstrong's superheterodyne and frequency-modulation circuits.

The idea of open competition is more central to the entrepreneurial than any other vision, and the radio pioneers behaved accordingly: they approached technological development like a sports event. Getting one's signal to travel longer distances, with greater reliability, was the inventors' principal challenge. Success was measured in miles covered and words transmitted per minute. As one might expect of a sporting event where the rules are as yet unsettled, the competitive spirit sometimes erupted into chaos. Many of the early cases of radio interference were the result, not of simple overcrowding, but of deliberate attempts on the part of entrepreneurial radio operators to drown out their competition. Perhaps the most famous early case of interference occurred in 1901, one of the first highly publicized demonstrations of radio. After Marconi announced to the press that he would use his radio apparatus to report the results of yacht races off Newport, Rhode Island, De Forest decided to get in on the publicity stunt by setting up his competing equipment alongside. Marconi—whom, as we shall see, was less the entrepreneur and more the managerial statesman—struck an agreement with De Forest to eliminate the resulting interference, but a third, little-known company with hopes of increasing its stock sales stepped in and began transmitting randomly in order to drown out and thus embarrass the two leaders in the radio field.[11]

The behavior of the radio entrepreneurs, in keeping with their

11. Ibid., 56–57.

image, was sometimes brash, risky, cutthroat, and self-aggrandizing. De Forest, for example, was notorious for claiming others' inventions as his own, and he was not alone in this practice. And De Forest was only one of many who became famous for exaggerating his achievements for the purpose of garnering stock sales.[12]

Importantly, the inventors were less programmatically hostile to the omnidirectionality of radio waves than were the corporations. They were interested in exploiting any competitive advantage they could find, and thus were not constrained by the corporate tendency to stick to known markets and applications. Many of the early protobroadcasts, such as Fessenden's famous Christmas Eve broadcast in 1906, were a mixture of experimentation and self-promotion. Radio operators accustomed only to the dots and dashes of Morse code were reportedly amazed to suddenly hear the sound of a human voice or music in their earphones; Fessenden must have taken a certain pride and pleasure in that effect. De Forest engaged in similar activities in the early years. The entrepreneurs were thus the first to exploit the omnidirectionality of radio waves for promotional purposes associated with a business. They, unlike large corporations, had to struggle to make themselves known and had the flexibility of vision to explore new ways of doing so.

Susan Douglas, in a touching and brilliant discussion of Lee De Forest's role in the early history of radio, has pointed to a more subtle but perhaps just as important contribution of the entrepreneurial perspective to the social construction of broadcasting. By the winter of 1906, at a time when most efforts were still devoted to telegraphic, point-to-point uses of radio waves, Lee De Forest envisioned and pursued the idea of using radio to transmit news and entertainment, particularly opera, into the homes of common people.[13] Like many of the radio entrepreneurs, his fortunes waxed and waned dramatically throughout most of

12. De Forest claimed to have invented the electrolytic detector and the "oscillation valve," technologies actually developed and patented first by Fessenden and Fleming, respectively. His famous addition of the "third element" to the vacuum tube came only *after* he had announced Fleming's "Audion" as his own (ibid., 169–70; Aitken, *Continuous Wave,* 220–22). And in a case that is still notorious among radio engineers, he successfully claimed patent rights to the pathbreaking oscillating circuit, an invention today generally attributed to Edwin Armstrong (Aitken, *Continuous Wave,* 239–42). De Forest's first company sent out fraudulent press releases to encourage stock sales. It is perhaps not surprising that his career oscillated wildly: in 1905 he was a wealthy and famous inventor; in 1906 he had to flee to Canada to avoid arrest for patent violations and was left nearly penniless. Between 1907 and 1910 he repeated his journey from rags to riches to rags, founding a new company, selling questionable stock, and ending up in bankruptcy.

13. Douglas, *Inventing American Broadcasting,* 172.

his life, leaving him wealthy one minute and nearly penniless the next. De Forest, an opera fan, first conceived of broadcasting during a time of poverty, when he could hardly afford to buy opera tickets. He envisioned radio, according to Douglas,

> as a way to serve the culturally and economically excluded—and as a way to make money. Having been in his life, by turns, the ridiculed outcast and the exploiter of the gullible public, De Forest carried with him two very distinct impulses that guided the development of radio. For De Forest, radio broadcasting blended his altruistic and self-serving impulses. It resolved his internal contradictions just as it would later straddle, and mask, contradictions in the culture at large.[14]

Whether or not De Forest was really the first to "invent" broadcasting, the important point is that his position as an entrepreneur, as an outsider struggling on his own to rise above his modest roots via market success, led him to envision a social use for radio and the spectrum to which others, particularly corporations, were blind. It was De Forest's entrepreneurial position that allowed him to fuse the amateur's sense of the spectrum as a source of pleasure and popular community with the businessman's desire for profit.[15] De Forest's heirs can be found scattered throughout today's commercial broadcast system.

Admirals

Accounts of the early development of radio in the United States frequently mention the navy. The dramatic strategic value of radio for ship-to-ship and ship-to-shore communication was obvious from the beginning. Consequently, the navy advocated a military monopoly of radio as early as 1904, and continued to promote the idea well past World War I; it held a seat on RCA's board of trustees into the 1920s. The U.S. government still controls roughly half of the radio spectrum, and most of that is in the hands of the military.

In many accounts of early radio, the navy stands for the possibility of "government monopoly," which is juxtaposed to "private" or business

14. Ibid.

15. Douglas's discussion of De Forest's early visions of broadcasting should finally put to rest the sense of importance given to David Sarnoff's oft-reprinted "music box memo," which is supposed to have "predicted" broadcasting in 1916, ten years after De Forest publicly proposed the same idea. The memo is reproduced by Barnouw, *Tower in Babel,* 78; Sterling and Kittross, *Stay Tuned,* 43; and Frank J. Kahn, ed., *Documents of American Broadcasting,* 4th ed. (Englewood Cliffs, NJ: Prentice Hall, 1984), 23.

control. The navy's position is thus sometimes represented as "the road not taken" toward government monopoly of radio, as if the U.S. Navy had an interest in state-controlled mass media.[16]

Interpreting the navy's role through the lens of a simple government/business dichotomy—a hallmark of contemporary liberal discourse—vastly oversimplifies what was in fact a fundamentally cooperative, if not always harmonious, relationship between the state and business interests in the early part of this century. By 1900, overseas military conquest had become national policy, as evidenced in the Spanish American War and the subsequent takeovers of Hawaii, Puerto Rico, and the Philippines. Popular support for these expansionist activities may have rested on simple chauvinism, but among the leadership of the country the interest was explicitly economic. With the closing of the western frontier, secure overseas markets were understood as necessary to maintaining a growth economy.[17] The navy's goal of extending military power across the oceans, and its interest in using radio to achieve that goal, was part and parcel of a larger economic vision.

The navy's interest in radio, therefore, had nothing to do with such state-aligned projects as government-controlled media services. The navy was acting in accordance with a plan at the center of which were the interests of American business; that was the national policy. Such tensions as did exist involved differences in process, in style, not in ultimate goals. The military preference for top-down command and orderliness at the expense of innovation, for example, threatened businessmen's abilities to move into and compete in new, developing markets. In particular, radio's omnidirectionality, lack of secrecy, and easy accessibility to tinkerers was threatening to the military. The navy's interest in government control, therefore, had more to do with radio's indiscriminate propagation characteristics than any kind of resistance to business domination of the economy and culture.

Managers

In retrospect, it seems that major corporations were remarkably reluctant to invest at first in radio, and similarly shortsighted about the institution of broadcasting. AT&T, for example, considered but then dismissed as impractical Fessenden's technology a few months after his Christmas

16. Sydney W. Head and Christopher H. Sterling, *Broadcasting in America: A Survey of Electronic Media*, 6th ed. (Boston: Houghton Mifflin, 1990), 39.

17. Martin J. Sklar, *The Corporate Reconstruction of American Capitalism, 1890–1916: The Market, the Law, and Politics* (Cambridge: Cambridge University Press, 1988), 78–85.

Eve broadcast.[18] This apparent shortsightedness, however, should not be attributed to simple bad management or bureaucratic incompetence. It was a direct product of corporate structure and corresponding corporate principles of operation—the same principles that eventually would lead to successful corporate dominance of radio and broadcasting.

Before the appearance of radio, the corporate world had already been introduced to privately owned, monopolized communications systems in the United States, first by Western Union's telegraph and then by AT&T's telephone.[19] Learning from the late-nineteenth-century experience of the railroads, chemical companies, and Edison's General Electric, AT&T in particular had adopted many of the strategies for industrial dominance that are the trademarks of twentieth-century technology-based corporations: vertical integration, the cultivation of patent libraries in industrial research labs, and efforts to limit competition (or enhance "market power") through domination of distribution networks. Under the leadership of Theodore Vail, AT&T made a few contributions to corporate strategy of its own, such as the institutionalization of "public relations" as a means to overcome the robber-baron image of corporations, and the discovery that in the right circumstances government regulation could enhance corporate power instead of diminish it.

Underlying many of these innovations is a particular vision and organizational principle: an understanding of corporate enterprise as an elaborate, integrated, bureaucratically organized system. Corporate enterprises are not single factories or simple aggregates of factories; they are not organized as if they were individuated atoms in a Newtonian social universe. Modern corporations, as Chandler has pointed out, are better understood as administrative systems that coordinate and rationalize the activities of numerous units of production and distribution; this is as much a way of thinking as it is an organizational form.

This fact is central to the corporate approach to new technologies and technological innovation. Corporations are always on the lookout for new technologies. But an individual corporation will generally view

18. According to Douglas (*Inventing American Broadcasting,* 159–60), in 1907, when Theodore Vail took over AT&T, he asked the company's chief patent attorney to assess the value of a proposal to invest in Fessenden's apparatus; the attorney recommended against it because "wireless competition was too great . . . commercial outlets too unpromising" and the technology too primitive.

19. Western Union, formed in 1866, had an effective monopoly of the telegraph by 1878, and AT&T achieved dominance of the telephone industry, particularly long-distance service, by 1910 (Alfred D. Chandler, Jr., *The Visible Hand: The Managerial Revolution in American Business* [Cambridge: Harvard University Press, 1977], 197–203).

those technologies strictly in terms of their potential for integration into the corporation's existing "system." If a technology can be profitably integrated or used to enhance a corporation's existing practices, therefore, it is aggressively pursued, both within the corporation's own laboratories and, if necessary, through the purchase of patents. But if a new technology cannot be clearly tied into a corporate system, it is, if considered innocuous, overlooked or ignored. And if it is considered a potential alternative or competitor to corporate activities, efforts are more often than not undertaken to squelch its development.

The corporate approach is thus not really conservative in the sense of simply striving to maintain the status quo; technological change is a given of corporate planning. But the corporate approach is blinkered. It is shaped and limited by a narrow range of corporate plans and preference for order. As CBS founder William Paley once said: "[S]udden revolutionary twists and turns in our planning for the future must be avoided. Capital can adjust itself to orderly progress. It always does. But it retreats in the face of chaos."[20]

It should thus not be surprising that radio was at first either ignored or actively resisted by the major electrical and communications corporations such as General Electric, Western Union, and AT&T. The turn-of-the-century giants were jealously committed to existing electrical technologies, which at that time were all wired: the telegraph, telephone, and traditional electrical power. The new, imperfect, and still experimental wireless technology of radio was both too unpredictable and too different from existing technologies to fit easily into entrenched institutional structures. Moreover, the very omnidirectionality and lack of privacy of radio waves that would make broadcasting possible appeared to the established corporations only as flaws, as annoying obstacles in the path toward effective use of radio for point-to-point communication. The wide-open accessibility of radio waves conflicted both with established technological practice—point-to-point communication—and with corporate principles of organization, which were generally based on linear, hierarchical chains of command and the strict control of access to services (and markets) that wired networks enabled.

The task of building a bridge between the peculiar technology of radio and the world of corporate structure was thus left to an entrepreneur with a corporate approach, Guglielmo Marconi. Marconi began in the radio business much like other entrepreneurs: in 1898 the newly

20. Statement of William S. Paley, FCC, *Informal Engineering Conference,* June 16, 1936, 2:252-53, quoted in Frank C. Waldrop and Joseph Borkin, *Television: A Struggle for Power* (1938; reprint, New York: Arno Press, 1971), 72-73.

founded Marconi Company set out to make a profit by selling radio equipment. But two years later, he adopted a new strategy that set him apart from all others in the business. Instead of selling equipment, he undertook to sell a service. Largely in response to the relatively closed conditions he faced in his home base of Britain, he began to lease his equipment exclusively, provide his own radio operators, and prohibit them from communicating with non-Marconi radio systems.[21] Marconi, in other words, sold a service instead of a device and secured that service by limiting access to it.

Marconi's "exclusivity policy" was not, as some of his rivals suggested, based solely on a fear of competition. It sprang from an alternative understanding of the nature of the business. Marconi envisioned radio as an integrated system, not as a technological apparatus for sale to individuals who would then be free to do with the apparatus as they wished. His policy thus included within its scope specific plans for distribution and consumption as well as production. It reflected a vision of radio that extended far beyond the technology itself to specific uses, users, and modes of use.

Marconi thus came to stand alone in the way he *conceptualized* radio, structured his company, and pursued his business strategy, and he thereby brought the corporate, managerial perspective to the field of radio. His behavior became more and more corporate as the years went by. Like large corporations, he engaged, not in raw salesmanship, but in public relations, presenting himself to the press as an industrial statesman enveloped in the dignifying aura of science. His attention was directed more toward long-term capital gains than toward short-term profits.[22] Instead of competing head to head, he sought to eliminate or limit competition by conquering and securing the spectrum, thus establishing a near-exclusive monopoly for himself. Marconi was the first to pursue the radio business with a service-based, hierarchically structured organization that did not so much compete as it *limited* competition by controlling access to the spectrum and the system of communication it made possible. Marconi's most historically significant legacy, in sum, was the practice of extracting profit from radio by treating it as a service instead of as a manufactured product, and by controlling that service

21. Marconi's strategy was adopted in response to problems encountered by his British-based company in dealing with the British Post Office's monopoly of electrical communications. The British Telegraph Acts of 1868 and 1869 seemed to prohibit any private sale of radio for land communications or ship-to-shore communications, unless a company was sending the messages for its own use or—and this was Marconi's loophole—providing that service to others (Aitken, *Syntony and Spark*, 232–35).

22. Ibid., 230, 228–29.

through the control of critical technology and through policies restricting how that technology should be used.[23]

In keeping with his corporate approach, Marconi pursued radio technology, not playfully, not idiosyncratically, but with a clear-cut plan that was easily formalizable and communicated to subordinates through a bureaucratic hierarchy. Radio was to be used to fulfill the already existing, well-defined need of large businesses and governments for long-distance telegraphic point-to-point communication in areas where there were no alternatives: over large bodies of water.[24] This adherence to specific purposes and established plans helps explain why Marconi worked almost exclusively with ship-to-ship, ship-to-shore, and transatlantic communications, and why he directed so much effort toward overcoming the omnidirectionality of radio waves, principally through "tuning": omnidirectionality conflicted with the established goal of strategic point-to-point communication.

Marconi's corporate approach did lead to gradual improvements in the effectiveness and range of his equipment, and in one case it actually lead to a breakthrough: although there was no scientific reason at the time for thinking that radio waves could travel over the horizon, Marconi's hazy scientific understanding combined with the enormous potential value of communicating beyond line of sight led him to try it and, to the surprise of the scientific community, succeed. But that same approach also led to some serious failures. He was blind to the potential of voice communication and omnidirectionality long after Fessenden and De Forest conducted the first experimental voice broadcasts in 1906. His efforts to develop "tuning" were doomed by his exclusive focus on gradual refinements of the original, inherently untuned spark-gap technology modeled on Heinrich Hertz's apparatus, allowing Fessenden's vastly superior "continuous-wave" technique to slip beyond his grasp. And Marconi's dogged adherence to long waves as the best means to achieve distance sent the entire industry down a technological blind alley until, many years later, amateurs stumbled upon the superior long-distance capabilities of shortwave transmission.[25]

Nonetheless, within little more than a decade of its foundation, in both the United States and Britain the Marconi Company completely dominated radio.[26] Within two decades, Marconi's American subsidiary

23. Douglas, *Inventing American Broadcasting,* 101.

24. See Aitken's comparison of Marconi to Lodge, *Syntony and Spark,* 161–62.

25. Ibid., 179–297.

26. In 1912 the Commerce Department concluded, "[T]he supply of apparatus and operators for radio communication in the United States is now in the hands of the Marconi company of America" (Department of Commerce and Labor, *Annual Report* [1912], 777).

would become the core of RCA, and one of Marconi's protégés, David Sarnoff, would go on to lead RCA and profoundly shape the development of both radio and television broadcasting. Entrepreneurs like Fessenden, De Forest, and Edwin Armstrong, meanwhile, would be sidelined.

The Creation of Corporate Liberal Radio, 1906–1912

Spectrum Chaos or Organizational Conflict?

As one might expect of an embryonic technology and institution, the world of radio was faced with considerable confusion and disarray in its first two decades. It is easy, however, to both exaggerate and misunderstand those conflicts. The difficulties were of a piece with the enthusiastic, exploratory spirit of the time. The problems were in many ways analogous to the problems of incompatible hardware, operating systems, and software frequently encountered with today's desktop computers. Phrases like "spectrum chaos" and "crisis of the airwaves" that are so frequently used to describe the early days today—particularly in legal explanations of the origins of federal broadcast regulation—obscure the sporting quality of early radio.

Some of the confusion undoubtedly came from straightforward interference of the kind that results when two individuals inadvertently transmit on the same frequency at the same time, rendering each others' signals unintelligible. But to a degree rarely acknowledged, the individual groups involved with radio were often quite successful at developing ways of overcoming interference without legal intervention. Within their spheres of influence, managers and military leaders could deal with problems of interference simply by administrative fiat. And the amateurs often worked out informal time-sharing arrangements and codes of conduct among themselves, sometimes in cooperation with entrepreneurs.[27]

Much of the confusion was thus less a purely technological problem than it was a product of the organizational *visions,* and the tensions between visions, of the different groups interested in radio. The no-holds-barred competitive approach of the entrepreneurs alone was enough to inspire occasional deliberate interference, such as occurred

27. The famous and most spectacular example of extralegal radio "regulation" was the American Radio Relay League, which organized a coast-to-coast network of radio amateurs beginning in 1914 and which has successors in today's ham radio and packet radio operators. But there were many smaller examples as well. For example, an amateurs' group in Chicago with one hundred members worked out a spectrum-sharing arrangement with local radio entrepreneurs in 1910 to the mutual satisfaction of both groups (Douglas, *Inventing American Broadcasting,* 209).

during the Newport yacht races of 1901. But more typically, conflicts arose *between* groups, when their different purposes and senses of order brought them into conflict. The confusions of radio's first decades, in sum, were less a product of "chaos," of a simple lack of order, than they were a product of different *ideas about* order.

Initial Conflicts

The two groups that overlapped the least in terms of purpose and organizational style were the anarchic amateurs, interested in maximum openness, and the bureaucratic navy, interested in maximum secrecy and restraint. The conflict between them was consequently the most overt, and most clearly illustrates the conflict's origins in competing visions. Annoyed by the amateurs using the airwaves to discuss everything from sports scores to school, navy officials used examples of amateur pranks to bolster their argument that the airwaves should be brought firmly under government control in the name of national security and the safety of ships at sea. Amateurs, according to an official navy report, were often "seemingly semi-intelligent and wholly irresponsible operators," who "at any time through carelessness or stupidity may render hopeless the case of a shipwreck" by interfering with maritime transmissions. The amateurs responded by publicizing examples of navy radio operators' frequent incompetence, refusing to yield to navy operators over the air, and on a few occasions generating radio messages from fictitious admirals that sent navy ships steaming off on spurious missions.[28]

The internal machinery of corporate liberal broadcasting, however, did not originate for the most part in skirmishes between officious naval officers and mischievous amateurs. Rather, the corporate liberal framework for radio was forged for the most part in the interaction of Western nation-states, corporations, and militaries. Radio, particularly Marconi's construction of it, aroused important territorial concerns at a moment when the European imperialist frenzy had reached its peak and the United States was enthusiastically setting out to join in the fray.

Legend has it that the first international convention on radio regulation, convened in 1903 by the German government in Berlin, was inspired by the experience of a German prince when a Marconi operator, following the company's exclusivity policy, refused to relay signals from the prince's yacht.[29] Whether or not this is true, it nicely characterizes the nature of the initial concern about Marconi's policy: the power elites of the first decade of this century saw his restrictive policy not so much as a threat to freedom as a threat to their power.

28. Ibid., 208–12.
29. Sterling and Kittross, *Stay Tuned,* 37.

Marconi was certainly not opposed to the interests of navies and nation-states. He saw them as his principal market. He was acutely aware of the strategic value of the telegraph and undersea cables in creating and maintaining the far-flung British colonial empire. Part of his corporate policy was called the Imperial Wireless Scheme because he meant to connect the British empire together with wireless, and he cultivated contacts with governments in America and Europe as a central part of his corporate strategy.[30]

But Marconi's approach, besides providing a tool for the administrative and military aspects of empire building, was also "a political time bomb."[31] Among governments and their military organizations, Marconi's approach to radio occasioned great interest, hope, and anxiety. The newly discovered "ether" under Marconi's hands looked like territory with strategic value. It was neither the entrepreneurs' competitive playing field nor the amateurs' anarchic forum. It was an as yet unconquered expanse analogous to the lands of Africa and Asia that the European powers were racing to claim as their own. By treating the spectrum as something to be cordoned off and controlled, in other words, Marconi raised the possibility of envisioning the spectrum as something to be imperialized; his business strategy, seen through the eyes of national governments, made the spectrum appear as another territory to be conquered in the struggle for global supremacy. Radio, from the perspective of the nation-state, was both a tool of empire building and itself a territory open to imperialist expansion.

As a result, Marconi faced powerful opposition from many Western nation-states. The navy, already the chief vehicle for the United States' nascent imperial project, approached other groups of American radio users in much the same way it approached foreign powers: the navy believed it was its duty to keep others from usurping its, and thus the nation's, control of the spectrum. It thus came to view the Marconi Company as a threat. The Marconi Company not only was foreign (British) and the dominant commercial interest in radio at the time, but also pursued a monopolistic strategy that seemed to assert exactly the kind of territorial control that the governments of nation-states viewed as their exclusive prerogative.

The U.S. Navy, in concert with the German government, thus initiated the assault on Marconi's transatlantic monopoly at the first international conventions on radio regulation in 1903 and 1906. The primary practical concern at these conventions was maritime uses of radio,

30. Aitken, *Continuous Wave,* 87, 356–60.

31. Douglas, *Inventing American Broadcasting,* 106.

and the rhetorical focus was public safety. But the principal source of controversy was Marconi and his prohibitions on communication with non-Marconi radios. Germany and the United States became the leading advocates of international rules directed precisely against Marconi's exclusivity policy: the rules, among other things, required maritime users of radio to communicate with all other users, regardless of the systems being used.[32] Their efforts were successful, leading to the passage of the first international radio regulation, the Berlin treaty of 1906.

This opening episode in the development of corporate-government relationships became an object lesson in the necessity of cooperation for both sides. It is significant that Marconi's major opposition in the 1903 and 1906 conferences came from Germany and the United States, whereas the government of Britain—the home base of Marconi's operations—at first sought to give Marconi qualified support. The lesson of this experience for Marconi and others interested in the exploitation of radio in the corporate mold was twofold. On the one hand, a monopoly company could face serious, perhaps devastating opposition in a head-to-head conflict with the interests of nation-states. On the other, by *allying* itself with a particular nation-state, a corporation could find support in the international arena. In the process of establishing control of the spectrum, in other words, cooperation with national governments was beginning to be seen as perhaps beneficial and in any case necessary.

Initial Resolution: The 1912 Radio Act

Marconi learned his lesson well. He abandoned his exclusivity policy in 1908, and quietly joined forces with his former opponent, the navy, in advocating government regulation of radio.[33] Other entrepreneurial wireless companies and groups of amateurs resisted such regulatory efforts, and successfully lobbied against attempts to bring U.S. law into line with the 1906 treaty for four years.[34] By 1910, however, pressure from international forces, from the navy, and from concerns about maritime safety combined to goad Congress into passing the Wireless Ship Act of 1910. Two years later similar pressures prompted passage of the Radio Act of 1912, which created full compliance with the 1906 treaty.[35]

The immediate concern motivating both pieces of legislation was public safety at sea—the 1912 act was passed less than six months after

32. Aitken, *Continuous Wave,* 255.

33. Aitken, in *Syntony and Spark,* 236–37, discusses the end of the Marconi monopoly policy in 1908.

34. These groups argued that the 1906 international rules were restrictive, premature, and technically naive (Douglas, *Inventing American Broadcasting,* 216).

35. Wireless Ship Act, 36 Stat. 629 (1910); Radio Act, 37 Stat. 302 (1912).

the *Titanic* disaster. Often treated as a mere footnote in the history of spectrum regulation, however, the 1912 act in particular asserted several basic principles upon which U.S. regulation of the spectrum has been based ever since. Its passage is properly described as a watershed.[36]

First, the 1912 act clearly asserted the principle of legally sanctioned limitations on spectrum access. It specified different portions of the spectrum for different types of service: in a compromise between advocates of government and corporate control over the spectrum, useful portions of the spectrum were divided between the navy and commercial operators. While the act had the effect of making illegal Marconi's exclusivity policy, it also required all ships to carry radio apparatus, thus expanding the market for commercial operators as a whole, and Marconi in particular. Because Marconi already dominated the market, he fared well in the ensuing expansion. The amateurs, in contrast, were given only token consideration: they were relegated to shortwave, a realm then thought to be of no practical value.[37] Regardless of its specifics, however, by asserting a government power to make such spectrum assignments, the 1912 act eliminated the possibility of a spectrum without boundaries; all further disputes would involve the legal technicalities of the system, but not the question of legal restraint itself.

Second, the restraints would be enforced, not by the courts in the name of common law property rights, but by agencies of the federal government, in the name of the public good; access would be characterized more as a privilege than a right. All radio operators were required to obtain licenses from the secretary of commerce and labor. Third, those given full access to the spectrum would not be simply "private" interests—the amateurs were certainly private, but were banished to a spectrum wasteland—but would be large, bureaucratic institutions, in this case the navy and the Marconi Company, by then a burgeoning transnational corporation. The navy and the Marconi Company were now working in concert. The navy saw this as a chance to assert certain, if not absolute, control over the spectrum, and because Marconi enjoyed overwhelming dominance of the field of radio in 1912, the act offered to secure for him a close approximation of the monopoly that he had forsaken when he ended his exclusivity policy.

Beginning in 1912, therefore, the force of law was brought to bear

36. The following analysis of the importance of the 1912 act is greatly indebted to Douglas, *Inventing American Broadcasting,* 234–37.

37. The navy was given the 600–1,600 meter range, commercial operators above 1,600 and between 200 and 600 meters, and amateurs were relegated to below 200 meters.

on a new communications medium and effectively set the terms for resolving the tensions between the groups interested in its use. The law had the effect of marginalizing the mixed group of small entrepreneurs and hobbyists who had played a major role in radio's creation, and eliminating their vision of an unconstrained spectrum. The law also elevated a coalition dominated by the navy, the Marconi Company, and the U.S. government into a commanding position of leadership over the organization and use of radio.

These events represented a political accomplishment. They were neither the natural working out of economic forces nor a simple triumph of big organizations over individuals. They reflected the triumph of a particular configuration of business organization, technology, and state action, a configuration characteristic of corporate liberalism: corporate private-sector cooperation with the public sector, small businesses relegated to a secondary role, and grassroots nonprofit activities pushed to the fringes.

Making Sense of Corporate Liberal Control

Similar actions were taken in Europe. What was peculiar about the U.S. context was not the fact of cooperative corporate-government relations but the fact that these relations were made politically acceptable to a liberal polity accustomed to individualism, rights, and free enterprise. After all, in 1912 private individuals—the amateurs—were forcibly ejected from their place in the spectrum without compensation, while others, notably the Marconi Company, were granted a place of privilege by what amounted to a government bequest. Was this not massive government intervention in the service of impersonal bureaucracies and at the expense of private individuals and their rights?

Corporate liberal faith in expertise and a functionalist social vision helped make sense of the situation by couching actions as a matter of neutral, technological necessity in service of the social system. The consolidation of control over radio by a coalition of oligopoly corporations and government was rendered legitimate, and perhaps even enabled, by the new thinking. The new logic was not formalist: there was little talk of absolute rights, or of legal categories like property and contract. Rather, the logic was functionalist and systemic: radio was not a realm of autonomous individuals, it was a system that if properly organized could fulfill beneficial social functions such as public safety, the national interest, and the furtherance of technological and economic progress. To a large degree, the apparently illiberal outcomes of the act were thus reconciled with liberal goals by framing the issue as a matter of system maintenance; maintaining the system was less a matter of rights than one of neutral, technological necessity and overriding public purpose—all in the service of broadly liberal goals.

Technology played a dual role in the new logic. On the one hand, radio itself was still an exotic and wondrous technology; its mysteriousness lent heightened authority to those who claimed to be its masters. On the other hand, the aura of technology provided a metaphor for social organization. If the self-regulating steam engine served as an initial inspiration for systems logic, radio provided a metaphor that could take systems logic to new heights. Now social relations could be imagined to be like, and conflated with, the dispersed and mysterious radio devices interconnected by invisible webs of radiation. Proper relations among the military, private corporations, and private individuals were not so much a matter of formal legal boundaries as they were technical relations among different parts of an integrated whole, parts in need of careful synchronization, of tuning, for optimal efficiency. Just as massive steamships needed orderly radio systems to operate safely and efficiently, corporations and government needed orderly relations among themselves to successfully develop the technology. This magical product of engineers and engineering, in other words, lent credence to the idea of *social* engineering as a way to organize the technology's use.

Of course, this was just one of several possible visions of technology. Radio technology in corporate liberal discourse was neither the heroic, individualist technology of the entrepreneurs nor the unconstrained, spontaneous, anarchic technology of the amateurs. It was a technology imagined in tight association with predictability, the absolute truths of science, and the orderly march of progress. The press-savvy and publicly visible "experts" such as Marconi and navy officials better fit the corollary stereotypes of scientific authority than ragtag bands of amateurs or quarrelsome and eccentric inventor-entrepreneurs, and were given consequent respect and authority in the halls of Congress and in the press.

In the discourse surrounding the passage of the 1912 act as well as in the act itself, then, the basic corporate liberal themes are clearly, if not yet emphatically, visible. A series of accidents at sea culminating in the *Titanic* disaster, complete with sensationalist press coverage, had associated the interests of the radio industry with the public safety. The act was thus understood for the most part as a response to complex technical problems amenable to solution by experts, such as safety of ships at sea, the needs of the navy, effective coordination and maintenance of radio service, and so forth.

By 1912, then, the foundations were laid. It was becoming certain that radio was going to be developed under some mixture of government and corporate auspices; both sides had learned the value of close cooperation, and a shared conceptualization of radio as both a commercial industry and a social-technological system was emerging. Only a few questions remained: Which corporations would dominate? What were going to be the terms of the corporate-government relationship?

Consolidation, 1912–1919: Patent Pools, War, and the National Interest

The story of how these questions were resolved, though it has been told many times, remains a classic example of corporate liberal industrial behavior. It is a story of technological innovation, quasi-feudal struggles among giant organizations, and world war.[38]

One key part of the story involved the classic corporate strategy of defensive patent acquisition. With radio's growing financial and technical success, the leaders in electrical technology, AT&T and General Electric, finally recognized its importance, and its threat to their command of electrical communications. For AT&T in particular, there remained the possibility that two-way radio could become an alternative and thus competitor to the telephone, perhaps even among ordinary citizens.[39] They quickly made up for their shortsightedness by using a mixture of legal cunning and financial lures to persuade Fessenden, De Forest, and others to part with their patents, effectively eliminating the entrepreneurs from the field.[40] Once this was accomplished, the American electrical giants squared off with Marconi in a struggle for dominance.

They found themselves, however, in a stalemate: the best radios worked only with a combination of the technologies, control over which was distributed among the corporate giants.[41] As a result, no corporation

38. The best telling of the tale remains Barnouw, *Tower in Babel,* though Aitken's *Continuous Wave* has also made a major contribution.

39. The question of why two-way radio has remained a specialized medium and has never emerged as an alternative to the telephone has yet to be thoroughly investigated. Certainly, AT&T's patent acquisitions and industrial strategies during the teens were designed to prevent such a possibility, and the careful legal and institutional partition of broadcasting from amateur two-way radio in 1920 (see below) also worked to shut off exploration along these lines. It is difficult to say whether, absent these actions, some form of broad-based popular use of two-way radio, perhaps one that mixed point-to-point with broadcast uses as the amateurs did in the teens, might be practical; but it is certainly worth looking into.

40. Douglas, *Inventing American Broadcasting,* 240–50.

41. The problem centered on the vacuum tube, which in 1916 the courts had ruled belonged in part to De Forest and in part to Marconi interests. Armstrong's invention of the feedback circuit further complicated matters, as did the use of the vacuum tube to generate radio waves, which several parties claimed as their own. Each of these inventors gradually sold or lost control of their patents to corporations during this period, but no single

could achieve dominance alone—and a new window of opportunity opened for amateurs, whose willingness to combine technologies with impunity allowed their homebuilt equipment to sometimes surpass some of the better commercial devices.[42]

The other key element in the story came with war. James Weinstein has remarked: "The entrance of the United States into the First World War in April of 1917 provided a full-scale testing ground for the new liberalism and the new liberals. Out of the war came striking proof that the ideas and institutional reforms developed in the prewar days . . . served the interests of the new corporate giants and their political economy of corporate liberalism."[43]

Radio is a case in point. With the onset of American involvement in the war, using a clause in the 1912 act, the navy established legal control over all of radio in the name of national security.[44] Amateurs were banished from the airwaves altogether, and their equipment seized; many of them were then recruited into the military as wireless operators. Corporations, conversely, were enlisted in the war effort, and the navy overcame the patents stalemate by legally protecting corporations from responsibility for patent violations, and thereby "created from quarrelsome enterprises a coordinated industry."[45] During the war, radio technology was greatly advanced, and thousands of military radio operators honed their knowledge and skills. Government intervention, in sum, was once again the corporate solution, and this time war provided the perfect justification.

The months immediately following the war's end provide a breathtaking example of just how certain of their power and importance the corporate liberal elite had become by 1918. Without consulting Congress or the courts, and with clear indifference to the idea of a boundary between private and public sectors, a coalition of corporate managers, the military, and a few representatives of the Wilson administration quietly forged an institutional system of control for radio.[46] The ban on ama-

corporation was able to gain control of all of the relevant technologies (Barnouw, *Tower in Babel,* 47).

42. Douglas, *Inventing American Broadcasting,* 197.

43. Weinstein, *Corporate Ideal in the Liberal State,* 214.

44. Section 2 of the 1912 act states that "[e]very such license shall provide that the President of the United States in time of war or public peril or disaster may cause the closing of any station for radio communication and the removal therefrom of all radio apparatus, or may authorize the use or control of any such station or apparatus by any department of the Government, upon just compensation to the owners."

45. Barnouw, *Tower in Babel,* 52.

46. The most exhaustive discussion of these events to date is Aitken, *Continuous Wave,* passim.

teurs was continued while negotiations went on that eventually led to the creation of Radio Corporation of America (RCA) to control the pooled radio patents and thereby dominate the industry.

It was at that point that Marconi lost control of his American subsidiary: the "national interest" that justified the already intimate relations between the corporations and the navy was hard to reconcile with the delegation of so much power over a crucial technology to a British corporation. Faced with an insurmountable political obstacle, Marconi sold his American holdings to General Electric, which then, under a joint patent-pooling agreement with AT&T and several other corporations, created RCA. The navy seems to have played nursemaid to this arrangement, and received a position on RCA's board of directors for its efforts.[47]

There were conflicts between the government and corporate leadership during this period. The navy clearly would have preferred having radio to itself, and it was only opposition from the corporations, with some help from Congress, that prevented a navy monopoly.[48] And the threat posed by the U.S. government to the Marconi Company's ambitions proved devastating to Marconi's designs for an international radio monopoly. But this was a defeat for Marconi, not for the corporate strategy and structure he had pioneered.

It should be emphasized that the debate in the teens about the relative merits of "private" versus "government" control rested on the underlying corporate liberal consensus that had been established in 1912. In defending legislation that would have created a navy monopoly of radio, Navy Secretary Josephus Daniels argued that radio "must be a monopoly. It is up to Congress to say whether it is a monopoly for the government or a monopoly for a company." Navy commander S. C. Hooper argued similarly that radio "is a natural monopoly; either the government must exercise that monopoly by owning the stations or it must place the ownership of these stations in the hands of some one commercial concern and let the government keep out of it."[49] Both sides thus agreed that radio was to be cordoned off and controlled by an alliance of corporations and the government. The only concern was the relative degree of control to be delegated to each side.

47. Ibid., 415. See also James Schwoch, *The American Radio Industry and Its Latin American Activities, 1900–1939* (Urbana: University of Illinois Press, 1990), 57–60.

48. Bills to create a navy radio monopoly were introduced but defeated by Congress in 1916 and 1918. See House Committee on Merchant Marine and Fisheries, *Government Control of Radio Communication: Hearings before the Committee on Merchant Marine and Fisheries,* 65th Cong., 3d sess., December 12–19, 1918.

49. Quoted in Barnouw, *Tower in Babel,* 53.

The larger relation between state and business that evolved around the question of radio, therefore, was more a kind of tense fusion of interests and perspectives than it was one of a fundamental struggle between the opposite forces of public and private power. The resulting joint vision did not by any means resolve all differences or tensions; the question of whether and to what degree broadcasting should be subject to public or private controls has remained a matter of dispute ever since. But it did become a practical organizing framework that set clear boundaries for acceptable action and dispute, boundaries that have hardly changed to this day.

The Rise of Broadcasting, 1919–1926

Introduction: KDKA

Throughout the teens, the amateurs had been exploring the possibilities of broadcasting. Before the war, they had steadily expanded their ranks and activities, and discovered that the realm of shortwave to which they had been banished was more a paradise than a wasteland. Numerous cases of scheduled broadcasts of music and news exist from the teens (inspiring fruitless debates about which station should get credit for being the first broadcaster).[50] But it is probably just as important that, even without scheduled, deliberate broadcasts, more and more individuals were being introduced to the pleasures of listening to radio sets, not for any specific purpose, but just to see who was there; and more and more amateurs were becoming accustomed to sending signals to just such listeners.[51] As a social practice, then, broadcasting existed in the teens, freely intermixed with amateur point-to-point communications. And after the war, thousands of amateurs returned home from positions as military radio operators, now with more sophisticated knowledge of the technology, ready to take up where they had left off. Only when broadcasting was brought to the attention of corporate managers, the press, and the government would it be partitioned off from amateur activities.

In September 1920 Westinghouse vice president Harry P. Davis noticed a department store newspaper ad for radio sets that described evening radio "concerts" transmitted by one of those returned amateurs: Frank Conrad, an amateur who by day was one of Davis's own employees. Davis suddenly realized that radio could be profitably used

50. For example, Joseph E. Baudino and John M. Kittross, "Broadcasting's Oldest Stations: An Examination of Four Claimants," *Journal of Broadcasting* 21 (winter 1977): 61–83; and Sterling and Kittross, *Stay Tuned*, 58.

51. Barnouw, *Tower in Babel*, 33–37.

for publicity, as a mass medium; it was more than just a point-to-point device. In his words, "Here was an idea of limitless opportunity."[52] Davis persuaded Westinghouse to bring Conrad's activities inside corporate walls. On the theory that regular broadcasts would stimulate sales of Westinghouse radio receivers, a new, more powerful transmitting station was built at the Westinghouse plant, and a license obtained from the secretary of commerce. On November 2, 1920, station KDKA was inaugurated with a broadcast of the election results, and the legendary broadcast boom of the early twenties began.[53]

The lightening bolt that struck Davis that fall did not come out of the blue. The conditions were all in place; if not Davis, someone else would have married the amateurs' habit of broadcasting with corporate designs. Inside managerial circles, the belief that corporations should look beyond the factory door into the habits, hearts, and minds of the mass consuming public, both for enhancing corporate legitimacy and for stimulating the consumption of corporate products, was by then becoming dominant. Broadcasting, when it appeared on the corporate doorstep, was an obvious means for pursuing those goals.

Up to the moment of Harry Davis's epiphany, however, the corporate world generally had remained oblivious to the amateurs' explorations of radio's omnidirectionality (though it had taken note of the amateurs' technical discoveries during the teens, such as the long-range capabilities of shortwave). The established corporate liberal institutions thus were taken by surprise by this unforeseen use of a technology they thought they thoroughly commanded and understood; the new situation presented a serious legal and organizational challenge. But, as the events of the 1920s demonstrate, it was a challenge successfully met: over the next decade or so, existing institutional and legal arrangements would be elaborated and modified, and their justifications refined to bring the practice of broadcasting into the corporate fold.

Hooverian Associationalism and Broadcast Policy, 1920–1926

The Broadcast Boom as an Organizational Problem

Between 1920 and 1922 the broadcast boom swept the country, and for a brief moment amateurs and a new crop of small entrepreneurs seemed to return from their position on the margins to become major players in the field of radio. Amateurs that were not already doing so began regular

52. Harry P. Davis, "The Early History of Broadcasting in the United States," in *The Radio Industry* (Chicago: A. W. Shaw, 1928), 194, quoted in Barnouw, *Tower in Babel,* 68.

53. The best description of the boom remains Barnouw, *Tower in Babel,* 68–69.

broadcasting, and numerous others sought to get in on the new phenomenon. Reasons for broadcasting were various and often vague. Department stores often set up stations in the hopes of drawing in customers. Newspapers began broadcasting on the theory that it might become an extension of the newspaper.[54] Schools and universities set up stations, sometimes with an eye toward simple publicity, sometimes for the purpose of offering courses over the air. Churches and religious groups began broadcasting the good word over the airwaves. Unions and socialist organizations set up stations to advocate their cause. Meanwhile, as congestion in the airwaves grew, broadcasters experimented with various cooperative arrangements to prevent interference, particularly the sharing of frequencies by transmitting at different times of day.[55]

From the point of view of corporate managers and corporate liberal public servants, the new situation was simultaneously tantalizing and daunting. It was tantalizing because the possibility of a radio set in every home presented new opportunities for integrating everyday life with the corporate order. There was the potential for direct profits from the sale of radio sets and other corporate products, and the no less important if indirect profits to be gained by cultivating a positive image of corporations and the corporate order in the public mind. Broadcasting was too important to be ignored or suppressed.

It was daunting, on the other hand, because of the problems broadcasting presented to a corporate way of doing things: because broadcasting was largely an amateur invention, control and order of a kind compatible with existing interests would have to be somehow asserted, and any new structures would have to be compatible with the delicately balanced arrangements between corporations and the military arrived at in 1919. The difficulty of asserting control was both organizational—how could such a dispersed, anarchic activity be made compatible with rigid corporate and military structures and goals?—and political—how could the necessary constraints be justified to Congress, the courts, and the population at large?

Corralling Broadcasting: Patent Pools and the Interdepartmental Radio Advisory Committee

The swiftness with which the corporate liberal establishment asserted control over broadcasting is testimony to the organizational maturation that had gone on between 1910 and 1920. Fresh from the corporate liberal organizational advances of World War I and its aftermath, government and managerial elites were intimately familiar with the methods

54. Ibid., 61–64.
55. Ibid., 97–99.

and value of cooperative behavior. As a result, one thing was clear: solutions to the problems of broadcasting were not going to be arrived at in the turmoil of either an open marketplace or open political debate.

Much of the decision making occurred out of the public eye. The RCA-centered patent pool, barely solidified in 1920, threatened to unravel under the uncertainty of the new situation. In a series of thoroughly secret negotiations,[56] AT&T, General Electric, RCA, Westinghouse, and the other patent-pool members divided up the broadcast pie among themselves, creating an arrangement that would survive for more than fifty years: the centerpiece of the agreement was that AT&T, in exchange for an exclusive right to interconnect broadcast networks, would leave the field of broadcasting to the other members, particularly RCA, which then created NBC.[57] In 1922, while the patent-pool negotiations were ongoing, the federal government quietly formed the Interdepartmental Radio Advisory Committee (IRAC) to organize spectrum allocation among government agencies, thus ensuring government, especially military, primacy in roughly half of the spectrum. To this day, IRAC's direct descendant remains the primary institution for allocating spectrum in the United States.[58] Protected operating terrain for the established point-to-point uses of radio was thereby secured, and the tumult of the broadcast boom safely corralled.

Channel Assignments and the Radio Conferences

Even with core issues concerning military control, patents, and corporate turf resolved, however, major problems *within* the sphere of broadcasting remained. The key moves would be definitional. The first would be to partition amateur activities and broadcasting, thus allowing broadcasting to be defined in strictly business terms and separating it from the voluntarist grassroots organizational precepts of the amateur community. This crucial action was largely accomplished in the months

56. Knowledge of the existence of negotiations between the patent-pool members began to leak to the press only in mid-July 1926, seven months after they were completed, and the details were not made public until their publication by Danielian in 1939. See also ibid., 184–86.

57. Ibid., 184–88.

58. After a period in the 1970s under the now-defunct Office of Telecommunications Policy, IRAC's functions are now under the National Telecommunications and Information Administration. To this date a full history of IRAC with a sense of its political economic implications has yet to be written. Discussions of the FCC, particularly those concerning public ownership of the airwaves, frequently appear to assume that the FCC has ultimate control over the airwaves, whereas in fact its control is secondary to that of IRAC and its successors, whose powers do not rest on any legislation or even clear-cut legal precedent. See Ronald H. Coase, "The Interdepartment Radio Advisory Committee," *Journal of Law and Economics* 5 (October 1962): 17–47.

immediately after the creation of Westinghouse's KDKA in the fall of 1920, with little fanfare. On January 11, 1921, Secretary of Commerce Hoover prohibited amateurs from "broadcast[ing] weather reports, market reports, music, concerts, speeches, news or similar information" and on September 15 of that year, Commerce began licensing broadcast stations as "limited commercial stations" on a wavelength designated for such purpose.[59] The policy was clear: amateurs were forced to choose between abandoning broadcasting or abandoning the amateur community by turning themselves into "limited commercial stations." As the Frank Conrads of the nation moved their operations from their attics into institutions, Hoover made sure there would be no going back. The amateurs and the organizational possibilities they represented were returned to the periphery of radio.

Some of the remaining organizational and political issues would be hammered out in the recently invented institution of trade associations; broadcasters created the National Association of Broadcasters (NAB) in 1923.[60] But broadcasting was an inherently public practice, and one that happened to be to a considerable degree in the hands of groups outside the corporate liberal establishment. Neither behind-the-scenes secret negotiations nor industry-centered private trade associations would be politically acceptable means for addressing all the issues facing such a public and popular phenomenon.

The public side of the basic legal foundations for broadcasting, therefore, would be laid in a series of four Radio Conferences organized by the Department of Commerce that met every year from 1922 to 1925. Appropriately enough for a gathering that embodied quintessential corporate liberal behavior, the Radio Conferences were organized by the quintessential corporate liberal, Herbert Hoover, the newly appointed secretary of commerce.

In a sense, Hoover's decision to organize the conferences simply reflected the principle articulated by Woodrow Wilson several years before: "the truth [is] that, in the new order, government and business must be associated."[61] The conferences brought government and business together in a cooperative context in order to resolve mutual problems. But the conferences reflected a particular style of government-

59. Marvin R. Bensman, "Regulation of Broadcasting by the Department of Commerce, 1921–1927," in *American Broadcasting: A Sourcebook for the History of Radio and Television,* ed. Lawrence W. Lichty and Malachi C. Topping (New York: Hastings House, 1975), 547–48.

60. Barnouw, *Tower in Babel,* 120.

61. William Appleman Williams, *The Contours of American History* (Chicago: Quadrangle Books, 1961), 410.

business cooperation, a mode of interaction of which Hoover was such an enthusiast and a skilled practitioner that it has since been dubbed "Hooverian associationalism."[62]

Although publicly announced and reported in the press, the conferences were not official public forums along the lines of a Senate hearing or a political convention. They had no clear legal status. They were not a response to a congressional inquiry. The conferences had an aura of sophisticated practicality about them; they seemed to be simply gatherings of interested experts, brought together to solve mutual problems by way of a gentlemanly process of right reason.

Creating this atmosphere was an artful accomplishment, for which Hoover deserves much of the credit. Part of the success was due to skillful co-optation of potential opponents. Amateurs, for example, were allowed a representative (Hiram Percy Maxim, the founder of the amateurs' American Radio Relay League), and their accomplishments lauded, yet their opposition to regulation and corporate dominance was politely ignored.[63] Representatives of large corporations, the professional engineering community, the government, and the military, on the other hand, were well represented and provided with clear leading roles.

But much of what made the conferences successful is properly described as legitimatory or ideological. On the one hand, the discussions of the conferences were framed so that certain things were taken for granted and thus not open for discussion. The initial separation of broadcasting from amateur activities was not discussed, nor was the necessity of military control of half the spectrum. The functional separation and coordination of different elements of society was treated as a given.

Also taken as given were the liberal values of business, private property, and the market. The legislative proposal of the final Radio Conference was quite frank: it suggested "[t]hat in order to insure financial stability to radio enterprises, capital now invested must receive reasonable protection."[64] Hoover put the matter in more exalted terms: he proudly drew a contrast between the emerging private American system and the government-controlled systems of Europe such as the BBC. "[W]e should not," he argued, "imitate some of our foreign colleagues

62. Ellis W. Hawley, "Herbert Hoover, the Commerce Secretariat, and the Vision of an 'Associative State,' 1921–1928," *Journal of American History* 61 (June 1974): 116–40.

63. Amateurs had voiced objections to the impending first conference in their own gathering a few weeks before (Sterling and Kittross, *Stay Tuned,* 84).

64. Senate Committee on Interstate Commerce, *Radio Control: Hearings before the Committee on Interstate Commerce,* 69th Cong., 1st sess., 1926, p. 42 (hearings on S. 1 and S. 1754), January 8 and 9, 26.

with governmentally controlled broadcasting supported by a tax upon the listener . . . [private broadcasting] has secured for us a far greater variety of programs and excellence in service free of cost to the listener. This decision has . . . preserved free speech to this medium."[65] On a broad level, therefore, the framework was orthodoxly liberal. The general purpose was to encourage the development of broadcasting on a for-profit basis with maximum autonomy for capital and minimal governmental interference.

In the context of the early 1920s, however, "reasonable protection" to "capital now invested" could not have been provided simply by the classical liberal technique of restraining the government's ability to interfere with business activities. The major threat to corporate investments at the time was not government but a community of nonprofit organizations and small entrepreneurs, many of whom were still infected with the democratic amateur spirit. Broadcasting itself, by putting radio technology in the hands of throngs of ordinary citizens, threatened to undermine the system (the "investments") that had been established in the patent-pooling agreements that created RCA. Even after the amateurs were eliminated from broadcasting, there were growing numbers of small entrepreneurs and various nonprofit organizations to be dealt with.

The principle mechanism for steering the development of broadcasting in a direction that provided protection for "capital now invested" was spectrum allocation and licensing by the Commerce Department. Hoover took care to avoid the appearance of autocratic decision making; his actions generally followed guidelines laid down by the Radio Conferences, and care was taken to give everyone, even the amateurs, at least token recognition. But as he established different classes of stations and assigned them to different frequencies, he was not acting simply as a traffic cop of the airwaves; he was building the streets and deciding where they were going to go. And some were getting superhighways while others were getting narrow and congested back roads.

In August 1922 Commerce created a distinction between classes of broadcast licenses, where stations at the original broadcast wavelength of 360 meters had less power, and stations with more power were allowed to move to the less crowded 400 meter wavelength. Broadcast content was already being specified: the privileged 400 meter stations were expected to broadcast original programming, whereas the poorer stations were allowed to continue broadcasting recordings.[66] In March

65. Herbert Hoover, opening address to the Fourth Radio Conference, reprinted in ibid., 50.

66. Bensman, "Regulation of Broadcasting," 550; Senate Committee, *Radio Control,* 42.

1923 Commerce elaborated on this trend by creating a third category, a C class of stations, which were under five hundred watts of power and were assigned to, and thus forced to share, a single frequency. Class A and B stations, on the other hand, began to receive exclusive frequencies, giving them prominent places in the broadcast world. That same year, the Westinghouse Corporation proposed and received its own exclusive, interference-free classification, class D, for its now famous KDKA.[67]

Neither economic nor technological necessity drove the specifics of these channel assignments. The Commerce Department could have done things differently, or it could have done nothing, without threatening already-established principles such as public safety or the national interest. It was creating policy that drove the development of new institutions, a policy that favored particular interests and particular organizational patterns. It was not favoring free enterprise in the simplest sense: many small, entrepreneurial stations were being sequestered on the same crowded 360 meter wavelength as struggling nonprofits. It was favoring large, well-funded enterprise, particularly of a corporate sort.

The Legitimating Framework

Any political action needs a legitimating framework, and the policies of the early 1920s were no exception. The language and procedures of the 1920s were a refinement of the terms that had dominated the discussion in 1912: technical necessity and the criteria of the public good.

Significantly, the strategy of licensing classes of stations conveniently allowed matters to be couched in technical language: assignments were presented in terms of wavelengths and wattage, not by naming the individual beneficiaries of the assignments. This was not cynical. When Hoover handed Westinghouse, at its own request, the extraordinary privilege of an exclusive channel on the principle that class D channels were "for developmental work,"[68] he was acting on a sincere faith that corporations were society's best hope for achieving that enlightened fusion of social and technological expertise that created "progress." Yet the allocation and licensing process nonetheless allowed the Commerce Department to pursue a procorporate policy without simply declaring unqualified favoritism for Westinghouse and its corporate cousins.

The policy pattern being established in the first half of the 1920s, therefore, was all the more influential because of its implicit character. The larger, well-funded corporations were clearly the favored entities in

67. Bensman, "Regulation of Broadcasting," 551-52.

68. Ibid.; Sterling and Kittross, *Stay Tuned,* 86.

this policy, but that fact was the necessary outcome—so it seemed—of largely technological matters.

The aura of mysterious technical complexity that surrounded radio technology in the 1920s extended well beyond the licensing process itself, and probably did much to smooth over any potential discord concerning government intervention and corporate favoritism. Speaking to the First Radio Conference, Hoover described the problem of regulating the spectrum as, not political, but "one of most intensely technical character."[69] Finding solutions to "the ever new problems which have developed in the growth of this astonishing industry," he suggested, was intimately linked to technological progress. "[W]e still have difficulties to face and overcome," he said,

> but before I come to a discussion of them it seems proper to describe some of the progress in the various branches of radio during the 12 months past. . . . There has been gratifying improvement in the character of equipment. . . . Increase in the frequency range of receiving sets is making the shorter wave lengths of the broadcasting band more available. Improvement in sets has given far greater perfection in tone and quality. Experimental work in the high frequencies is giving encouragement to the further development of the art. The most profound change during the year, however, has been the tremendous increase in power and the rapid multiplication of powerful stations. . . . The vast expenditure of money and skill in our great industrial laboratories is not only advancing the application of the art but has been conceived in a fine sense of contribution to fundamental science itself.[70]

During the Radio Conferences, then, matters of ownership, control, legality, and political legitimacy were regularly sandwiched in between discussions of questions of transmitter power, effects of atmospheric conditions, and tests of the effective range of various transmitting techniques. The implication was that new and important technologies and institutions like radio involved foremost issues of science and technology, not formal problems of law like private property or political problems of justice and democracy.

The aura of technology, with its connotations of scientific neutrality and objectivity, lent weight to the second key element in the legitimating framework of broadcast regulation: the notion, carried over from

69. Herbert Hoover, *The Memoirs of Herbert Hoover: The Cabinet and the Presidency, 1920–1933* (New York: Macmillan, 1952), 141.

70. Senate Committee, *Radio Control*, 50–53.

1912, that radio involved unique matters of public purpose. In a crucial articulation, Hoover argued: "The ether is a public medium, and its use must be for public benefit. The use of a radio channel is justified only if there is public benefit. The dominant element for consideration in the radio field is, and always will be, the great body of the listening public. . . . Whatever other motive may exist for broadcasting, the pleasing of the listeners is always the primary purpose. . . . public good must overbalance private desire."[71] Hoover was by no means alone in advocating the public-interest principle in broadcasting, though he seems to have played a key role in introducing the term to the conferences.[72] All four conferences produced recommendations in favor of public-interest principles. Nor was the phrase unheard of in the larger sphere of American politics, having been written into law in the Transportation Act of 1920. And the more savvy corporate executives had long been familiar with the advantages of references to the "public": in 1903, for example, Marconi rewrote a corporate director's report by substituting the phrase "commercial purposes" with the phrases "the general public" and "a public service."[73]

The meaning of the public interest as used in the conferences was clearly corporate liberal. In suggesting the public good should be the dominant criteria in broadcasting, the conferences were not trying to remove it from private influence. The public interest was part of a legal and rhetorical strategy for organizing broadcasting's further development as a commercial, for-profit institution. The "public interest" was

71. Ibid., 56–57. Hoover had invoked the language of the public interest before. In addressing the Third Radio Conference in 1924, Hoover said: "Radio has passed from the field of an adventure to that of a public utility. Nor among the utilities is there one whose activities may yet come more closely to the life of each and every one of our citizens, nor which holds out greater possibilities of future influence, nor which is of more potential public concern. Here is an agency that has reached deep into family life. . . . Radio must now be considered as a great agency of public service, and it is from that viewpoint that I hope the difficult problems coming up before this conference will be discussed and solved" (Stephen Davis, *The Law of Radio Communication* [New York: McGraw-Hill, 1927], 57).

72. A Hoover protégé and solicitor of the Department of Commerce heavily involved with radio regulation at the time, Stephen Davis, remarked that, before the Third Radio Conference, "no one had suggested that broadcasting was anything but a strictly private enterprise of the station owner or that any public element was present in it" (*Law of Radio Communication,* 57). This seems like an exaggeration. There is considerable talk along these lines to be found in the proceedings of the earlier Radio Conferences, the idea of some form of public good associated with radio was clearly articulated in the 1912 act, and as Douglas has shown, the amateurs had argued something to that effect more than a decade before.

73. Douglas, *Inventing American Broadcasting,* 67.

not thought of as in opposition to commercial organization. Rather, it was a criterion for use by knowledgeable experts to help make complicated decisions in the process of serving the larger business system.

The language of the "public interest" did in fact help solve some difficult problems associated with the policy of promoting commercial broadcasting, but those problems were as much political as they were technological. On the one hand, roughly half the usable spectrum remained in the hands of the military, which had not long before lobbied for total government control. Treatment of the spectrum as public property regulated in the public interest both satisfied the compelling interest in the use of radio for national security and maintained the compromise struck between the military and private interests that protected corporate users of the spectrum from attempts to expand military control.

On the other hand, the "public interest" was used to justify eliminating private interests from the spectrum if those interests happened to be nonprofits. There were numerous nongovernmental private interests associated with radio that had to be constrained in order to give maximum maneuverability to commercial operations. Even after the amateurs were eliminated from the broadcast band in 1921, roughly two hundred nonprofit educational and religious stations entered the field, comprising nearly 30 percent of all broadcasters in the early 1920s. Small nonprofit operations, when they were not forced off the air altogether, were typically required to share crowded and less desirable channels.[74] Throughout the decade, the federal government systematically used spectrum regulation to marginalize or eliminate these stations; by the end of the decade almost all had left the air.

Summary

Between 1920 and 1926 the corporate liberal establishment successfully met the challenge of the grassroots broadcasting phenomenon. This success involved classic corporate liberal procedures—a mixture of behind-the-scenes and public associational activities—and typical corporate liberal terms and concepts—broad liberal goals pursued in terms of technical necessity, expertise, and an overriding public good. Along the way, useful techniques were refined. The use of government power to allocate and assign radio channels was discovered to have a practical, material component: it could be used in such a way as to push the industry toward a center-periphery pattern of development, with corpora-

74. After 1927 many other small operations were forced to abandon the spectrum by the high costs of keeping up with the Federal Radio Commission's constantly escalating technical requirements (Sterling and Kittross, *Stay Tuned,* 110–11).

tions at the privileged center, surrounded by smaller businesses and non-profit organizations. An ideological component existed as well: channel allocation cohered nicely with the aura of technological mystery surrounding radio, which in turn helped obscure the political nature of the decisions of the time. A political and legal rhetoric was also refined to legitimate these activities, the rhetoric of broadcasting in the public interest, where the public interest was associated with an orderly, managed, corporate capitalism that worked hand in hand with a modest and cooperative government agency for the good of all.

By means of these procedures and techniques, a potential threat to existing institutional arrangements was corralled, and conditions were created for harnessing that threat to the corporate liberal establishment's own purposes. Details remained, of course: matters of content were still in the experimental stages, as were the interrelated questions of funding and relations among the emerging industry's component parts, that is, the manufacturers, networks, stations, advertisers, and entertainers, and so forth. But by 1926 the political questions of allocating and legitimating power relations were largely resolved: it was clear to those inside the system who would have the power to shape subsequent institutional evolution. The remaining questions hence could be treated as merely practical matters.

Finalization: The 1927 Radio Act

The Act as Legitimization of Existing Practice

One legal technicality remained. The Department of Commerce's authority to regulate radio rested on the limited foundations of the 1912 Radio Act, which had not anticipated broadcasting. When industry members began asking Hoover to limit the number of broadcasters in order to reduce congestion, it was not clear that Commerce had the authority to exclude technically compliant license applicants from the broadcast portion of the spectrum simply on the grounds that no space was left. In 1925 a legal test came: a broadcaster defied the authority of the Department of Commerce, leading to a court decision in the broadcaster's favor. With the collapse of Hoover's authority, a ten-month period of chaos ensued. New and old broadcasters jostled for position on the broadcast band, and interference grew to intolerable levels. The problem was resolved only when Congress passed the 1927 Radio Act on February 23.

The "chaos" of 1925 and 1926, it should be emphasized, was not the product of a simple lack of order. It was a by-product of the already-established *for-profit order:* the commercial framework generated incentives to try to drown out competitors. The order had been created in the preceding decades: the 1912 act, the experience of the war, the corporate alliances of the war's aftermath, and the private and public arrangements arrived at in the first half of the 1920s. Nothing in the 1927 act departed substantially from the established patterns. The act's immediate practical function was largely to iron out a kink in the existing system.

Furthermore, the collapse of Commerce's authority came as no surprise. The Commerce Department had recognized the uncertainty of its legal powers as early as 1921, which in turn helped inspire Hoover to organize the First Radio Conference and begin the drive for legislation.[75] The act had been in process for several years, and some version of it most certainly would eventually have been passed even without the court decision and the ensuing "chaos." Largely as a result of the conferences, twenty bills to regulate broadcasting were introduced between 1921 and 1923, thirteen between 1923 and 1925, and eighteen between 1925 and 1927—a total of fifty-one. Many of these were nearly identical to the act that was passed in 1927.[76] The 1927 Radio Act is better understood, therefore, as the intentional product of the corporate liberal establishment's activities that began in 1921 immediately on the heels of the broadcast boom.

The Content of the Act: Expertise and the Public Interest

In the long view, therefore, the 1927 act was important, not because it created anything new, but because it legitimated and solidified the established order. It gave legal sanction to Hoover's practice of classifying and licensing nongovernmental stations, enabling and giving tacit congressional blessing to the corporate broadcast order that Hoover's pattern of licensing was helping to create. The act thus elevated the corporate liberal legitimating framework for radio to the status of national law.

This required some discursive refinements. The act needed to ensure, both politically and constitutionally, the Federal Radio Commission's (FRC) power to deny technically capable applicants to the airwaves. Legitimacy for this arbitrary and nondemocratic power was generated in the intertwined corporate liberal ideas of apolitical expertise and a neutral, ascertainable public interest.

The consensus was that decision making should reflect expertise, not political points of view. Congress did debate over where the experts should be located, resulting in a compromise that created a temporary

75. Bensman, "Regulation of Broadcasting," 549.

76. Ibid., 544–45.

independent commission appointed by the president, the FRC, whose powers were supposed to revert to the Department of Commerce after a year. But the belief that the decision making should be by experts was shared by both sides. Democratic senator Charles C. Dill, the leading proponent of an independent commission, argued that the complexities of the situation required a strong regulatory body made up of "men of big ability and big vision."[77] Hoover, who had favored keeping the process within Commerce, nonetheless said with pride of the men he eventually chose to serve on the FRC, "They were all men of technical and legal experience in the art, and none of them were politicians."[78]

The other key legitimating technique inscribed in the 1927 act was the construct of an ascertainable public good. The FRC's experts were not going to transcend politics by way of formal legal categories and the adjudication of rights. They were going to transcend politics by using their expertise to determine the public interest. FRC decisions would show no political partisanship, reflect no special interest, and embody no subjective point of view because, in the words of the act, they would be made according to the criteria of the "public interest, convenience, and necessity." The vagueness of the phrase was quite deliberate: the FRC's experts would need broad latitude to deal with the shifting complexities of this new technology-based industry. But the vagueness also performed an important political function: it allowed different groups with different interests to read it as consistent with their own point of view, and thus obscured political differences at the same time that it helped generate broad political support.

The idea of the public interest also led to a very particular characterization of the relationship between broadcasters, the radio spectrum, and the government. Congress took pains to describe radio channels not as property but as something analogous to public waterways; the act's stated intention was "to maintain the control of the United States over all the channels of . . . radio transmission; and to provide for the use of such channels, but not the ownership thereof, by individuals, firms, or corporations . . . no such license shall be construed to create any right, beyond the terms, conditions, and periods of the license."[79] A license to broadcast, therefore, involved a fiduciary responsibility to serve the public interest more than a right to broadcast or a right to ownership of a channel.

The act contained some provisions reflecting anxieties of that era's

77. Lucas A. Powe, Jr., *American Broadcasting and the First Amendment* (Berkeley: University of California Press, 1987), 60, quoted in McChesney, *Telecommunications,* 18.

78. President Coolidge delegated the selection of the first members of the FRC to Hoover (Hoover, *Memoirs,* 145).

79. Preamble to Public Law 632, 69th Cong., 2d sess., February 23, 1927.

liberal culture. In section 13 of the act, Democrats suspicious of RCA's "radio trust" made sure the FRC was directed to include antitrust considerations in its definition of the public interest. In section 29, the commission was directed not to "interfere with the right of free speech by means of radio communication." And in section 18, broadcasters were prohibited from showing favoritism toward particular political candidates during electoral campaigns. Although, over the years, these provisions have served as focal points for struggles within the general framework, none of them challenged the basic institutional arrangements and structures of power that had been established in the preceding decades.

Aftermath: The Consolidation of 1927–1934

Consolidation of the Network-Advertising System

The next five or six years were a period of dramatic growth in the corporate structure of broadcasting. The now-confident industry settled on namebrand advertising of consumer products as the preferred source of revenue, and a tight, symbiotic relationship developed between the consumer product and broadcast industries, a relationship consummated and mediated by the newly burgeoning fields of marketing and advertising. In 1928 CBS joined RCA's NBC in the field of what was then called "chain broadcasting," and together the two organizations quickly established the basic outlines of the network system. The networks became the undisputed kingpins of broadcasting by establishing a pattern of oligopolistic dominance where local broadcast stations serve as outlets for vertically integrated and nationally centralized systems of program production and distribution. By 1933 broadcast entertainment delivered by networks and saturated with advertising was already a fixture of American culture.

After 1927 the FRC, armed with its congressional mandate, did much to foster a congenial environment for these extraordinary institutional developments. As the FRC set about clearing the broadcast band and deciding who would be eliminated, it continued the previously established policy of favoring well-funded corporate operations and marginalizing others, particularly nonprofits. Networks, already in an advantageous position because of the economies of scale inherent to networking, were further favored with all but two of the forty exclusive high-powered "clear channels."[80] The decline of nonprofits was acceler-

80. Federal Radio Commission, General Order No. 40, August 30, 1928 (reprinted in Federal Radio Commission, *Annual Report* [1928], 75).

ated by forcing them to share time on channels with others, often with the larger share of time going to commercial operations. Saddled also with steadily escalating technical requirements, such stations were thus forced into a cycle of shrinking airtime, shrinking audiences, and shrinking budgets. Hence, although nonprofits comprised nearly 30 percent of all broadcasters in the early 1920s, by the end of the decade almost all had left the air.[81]

Squaring Procorporate Actions with the Public Interest

These events are not best construed as the result of the simple brute power of capitalist self-interest forcing an administrative agency to do its bidding. In its first official attempt to specify the meaning of the public interest, the FRC said: "It is unfortunate that in the past the most vociferous public expression has been made by broadcasters or by persons speaking in their behalf and the real voice of the listening public has not sufficiently been heard. . . . The interest of the broadcast listener is of superior importance to that of the broadcaster. . . ."[82] The members of the commission were in all probability sincere when, in this gentle rebuke, they differentiated themselves from the commercial broadcast lobby. As they used their licensing power to systematically marginalize noncorporate broadcasters, then, the commissioners did not think of themselves as acting on behalf of private interests. Like their predecessor Hoover, they understood themselves to be serving a public interest that was distinct from the simple desire for profit. They were serving the system, not individual units within it.

A rhetoric of technical necessity and a corporate liberal understanding of the "public interest" remained crucial to the policy's success. The new FRC provided a congenial environment for convening men who, like Hoover, moved easily between the worlds of corporate management, government, and engineering, and who envisioned an easy confluence of interest among these worlds.[83] Historian Robert McChesney has revealed how, in a cozy atmosphere largely closed off from political and press scrutiny, the initial post-1927 act channel assignment plan was construed by its creators as largely an "engineering" problem. The high-powered clear channel assignments coveted by the networks were con-

81. Sterling and Kittross, *Stay Tuned,* 110-11.

82. Federal Radio Commission, "Statement Made by the Commission on August 23, 1928, Relative to the Public Interest," in Federal Radio Commission, *Annual Report* (1928), 2:166, reprinted in Kahn, *Documents of American Broadcasting,* 60.

83. The FRC's engineers were selected by AT&T's chief engineer and were all from government, radio manufacturers, or commercial broadcasters (McChesney, *Telecommunications,* 22).

strued as technological necessities, not as any kind of favoritism. "The reason for this is purely physical fact," said the chief engineer of the allocating committee.[84]

The logic that linked corporate favoritism with technical necessity, in this case, was based on the belief that expensive high-powered transmitters linked to well-funded program providers were most likely to deliver original programs to Americans in remote locations. Hoover had articulated the principle in 1925. "It is the quality of program, location, and efficiency of transmission that count. . . . A half dozen good stations in any community operating full time will give as much service in quantity and a much better service in quality than 18."[85] Three years later, the FRC elaborated the principle:

> [I]t is better that there should be a few less broadcasters than that the listening public should suffer from undue interference. . . . furthermore . . . the commission feels that a certain number [of channels] should be devoted to stations so equipped and financed as to permit the giving of a high order of service over as large a territory as possible. This is the only manner in which the distant listener in the rural and sparsely populated portions of the country will be reached.[86]

The argument here is not just technological or economic. On the one hand, the goal of a "high order of service" suggests a cultural value judgment: polished professionalism is presumed to be of more value than, say, diversity or local origination. Even more important, the argument is functionalist in its structure: channel assignments are justified in terms of the expected outcome for the entire system of broadcasting, not in terms of intrinsic values such as freedom or fairness.

The key abstraction in this argument, therefore, is the idea of a unitary broadcast "system" characteristic of popular functionalism. It was clear to all involved that it was David Sarnoff and NBC and William Paley and CBS who controlled the "stations so equipped and financed" as to provide "a high order of service." But in the logic of the day, the FRC was not using political power to favor certain groups and individuals at the expense of others; it was expertly and impartially working to keep a sociotechnological machine running smoothly.

A similar logic was used to justify the burgeoning advertising system. Hoover and the Radio Conferences, although wary of some kinds of advertising, nonetheless endorsed it on functional grounds. "The desire

84. Ibid., 23.

85. Senate Committee, *Radio Control,* 55.

86. Federal Radio Commission, "Statement of August 23, 1928," 60.

for publicity is the basic motive and the financial support for almost all the broadcasting in the country today," Hoover stated in 1925.[87] Advertising, in other words, served the needs of the system by paying the bills. At Hoover's recommendation, the Fourth Radio Conference unanimously adopted a resolution on advertising which began from the premise that "the excellence and public-service value of radio programs is increased by the support of those seeking appropriate publicity."[88] By 1929 things had progressed to the point where the FRC could present the "need" for advertising as irrefutable fact: "The commission must . . . recognize that, without advertising, broadcasting would not exist. . . . If a rule against advertising were enforced, the public would be deprived of millions of dollars worth of programs which are being given out entirely by concerns simply for the resultant good will which is believed to accrue to the broadcaster or the advertiser by the announcement of his name and business in connection with programs."[89] Advertising in American broadcasting was justified on the grounds that it served the needs of the system, and thus the public interest.

If the danger of functionalist theory is tautology, the danger of functionalist social policy is that the tautology turns into self-fulfilling prophecy. Imagined as an integrated corporate machine, American broadcasting was turned into one; the nonprofits and the alternatives they embodied were expelled as if they were parts that didn't fit. The images of neutral machinery and integrated systems obscured the deeply political character of the choices that were being made at the time, even to many of those who were making them.

Gaps and Contradictions: Was the 1927 Act "Obsolete When Passed"?

To say that the vision of a smoothly operating social machine shaped and legitimated the reality of commercial broadcasting is not to say that the vision simply *became* the reality. On the contrary, the forms that broadcasting took between 1927 and 1933 differed substantially from what the corporate liberals of the mid-1920s had in mind. Commercial broadcasting was hardly the blueprint come to life for which corporate liberal social engineers hoped.

The 1927 act, for example, has been described as "obsolete when passed" because it ignored the coming emergence of networks as the

87. Senate Committee, *Radio Control,* 54.

88. Ibid., 67.

89. Federal Radio Commission, "In the Matter of the Application of Great Lakes Broadcasting," Docket 4900, vol. 3, in Federal Radio Commission, *Annual Report* (1929), 32, reprinted in Kahn, *Documents of American Broadcasting,* 68.

dominant force in broadcasting.[90] By designating local station owners as licensees, and licensees as trustees of the public interest, the act was incorrectly assuming that the station owners would control what went out over the airwaves. In part this oversight was due to a simple lack of understanding on the part of a Congress unfamiliar with the workings of a still-evolving institution. But it also probably reflects a persistent liberal individualism, wherein owners are individuals and ownership grants absolute sovereignty, and the kind of power generated by a place in a structure is invisible. In any case, as local stations increasingly became mere outlets for network programming in the 1930s, the regulatory structure was faced with a major jurisdictional uncertainty: can networks be subject to regulation? It took a 1940s Supreme Court decision to create some clarity on the matter, and even then ambiguity continues to linger over the question.[91]

An even more striking gap in the policy vision of the mid-1920s concerns advertising. The corporate liberal establishment was distinctly uneasy about the prospect of direct advertising for consumer products over the airwaves. While acknowledging the value of broadcasting for publicity, the policy makers of the day seem to have had in mind something more along the lines of a "tastefully" corporate-sponsored system like today's PBS than the advertising-saturated system that their efforts created.

RCA head David Sarnoff, for example, suggested as an alternative to advertising indirect subsidies from radio equipment manufacturers.[92] Hoover repeatedly cautioned against what he called "intrusive" or "direct" advertising, and in 1924 suggested paying for programs with a 2 percent tax on radio-set sales.[93] At the Third Radio Conference, for example, he said: "I believe that the quickest way to kill broadcasting would be to use it for direct advertising. The reader of the newspaper has an option whether he will read an ad or not, but if a speech by the President is to be used as the meat in a sandwich of two patent medicine advertisements there will be no radio left."[94] The Fourth Radio Confer-

90. Barnouw, *Tower in Babel,* 199.

91. Thomas Streeter, "Policy Discourse and Broadcast Practice: The FCC, the U.S. Broadcast Networks, and the Discourse of the Marketplace," *Media, Culture, and Society* 5 (July–October 1983): 247–62.

92. David Sarnoff, "Address to Chicago Chamber of Commerce, April 1924," in *Mass Communications,* ed. Wilbur Schramm (Urbana: University of Illinois Press, 1960), 43. See also McChesney, *Telecommunications,* 15.

93. McChesney, *Telecommunications,* 16.

94. Department of Commerce, *Recommendations for Regulation of Radio Adopted by the Third National Radio Conference* (Washington, DC: Government Printing Office, 1924); see also Sterling and Kittross, *Stay Tuned,* 49.

ence passed a resolution that attempted to distinguish between acceptable "publicity . . . which limits itself to the building of good will for the sponsor of the program" and unacceptable "direct sales effort."[95] And as late as 1928 the FRC announced that "broadcasters are not given these great privileges by the United States Government for the primary benefit of advertisers. Such benefit as is derived by advertisers must be incidental and entirely secondary to the interests of the public."[96]

Furthermore, if the attendees of the Radio Conferences might have been made uneasy by the differences between their vision of commercial broadcasting and that which emerged, at least a few of the members of Congress who voted for the 1927 Radio Act were positively distraught. Some of those who voted for the act did so because they believed the "public-interest" mandate would *limit* commercialism and corporate dominance.[97] Once the pattern of regulation became clear, as a result, congressional supporters of nonprofit broadcasting voiced strong objections about the FRC's support of networks and acquiescence to advertising.[98] For them, the public-interest clause had worked as a Trojan horse for the corporate interests they sought to restrain.

These gaps, contradictions, and surprises in the early policy development of commercial broadcasting, however, need not be interpreted as simply the result of a behind-the-scenes conspiracy, as if the public-interest ideal had been simply perverted by corporate shills within the FRC. The 1927 act did not seal anything in stone. Congress could and did act to modify the legislation when inadequacies were perceived: when it was discovered that the problems of radio regulation were too complex to be solved in a year, for example, Congress extended the life of the FRC several times, eventually giving it permanent status. If the direction of the broadcast industry was still murky in 1927, furthermore, in 1934 it was perfectly clear, yet in that year the 1927 act was transcribed almost word-for-word into the broadcast portion of the 1934 Communications Act.

One cannot discount nonideological forces in accounting for this congressional rubber stamp. Bureaucratic inertia, the power of corporate-financed lobbying, and the lure of lucrative future corporate jobs to underpaid government employees all undoubtedly played a role in generating acquiescence to corporate dominance. But there was much in the political environment between 1927 and 1934 that could have counteracted these forces. Hoover, after all, was president for most

95. Senate Committee, *Radio Control,* 67.

96. Federal Radio Commission, "Statement of August 23, 1928," 61.

97. McChesney, *Telecommunications,* 33.

98. Ibid., 32.

of this period; if he had seriously cared about the emergence of "intrusive" advertising, he was in a position to do something about it. And the 1934 act was passed in the midst of Roosevelt's legislative revolution; legislation was being passed against corporate opposition with a frequency unmatched in American history.

It is probably better, therefore, to explain the blind spots of the 1927 act as products of tensions and contradictions *within* the broad corporate liberal belief system than as departures from its ideals. For all of Hoover's derisive comments about advertising, he also acknowledged its value as a means of income and—in the name of minimal government interference—insisted that the question of funding be left entirely up to the industry, with no government involvement, thereby setting the stage for advertising's triumph.[99] His worldview was one based, not on a belief in a public good as against commercial self-interest, but in the hope that the two could be comfortably reconciled; all the hairsplitting based on divisions between "publicity" and "direct advertising" reiterated throughout the mid-1920s reflect this hope.

Similarly, the contradiction between the principle of local broadcaster control and network dominance was less the product of a simple neglect of networking in the act than it was a product of the visions of the day. Hoover anticipated the contradiction at the Fourth Radio Conference. He advocated "leaving to each community a large voice in determining who are to occupy the wavelengths assigned to that community," but in almost the same breath spoke approvingly of the way that "interconnection" (networking) had developed rapidly without government involvement, and was becoming "more systematized and has gone far toward the creation of long-linked systems which will finally give us universal broadcasting of nationwide events . . . one of the most astonishing landmarks of radio broadcasting."[100] Enlightened policy making, in Hoover's mind, could reconcile local control and national control; it was a matter of system integration, not political choice. The 1927 act's insensitivity to the tension between networking and the public-trustee concept was an insensitivity based, not in simple ignorance, but in the shared corporate liberal vision of the time.

Conclusion

Roughly between 1900 and 1930, a remarkable set of social practices developed, practices that involved the coordination of nationwide insti-

99. Senate Committee, *Radio Control,* 53–54.
100. Ibid.

tutional structures, legitimating conceptual systems, new cultural forms, and new habits of everyday life. Together these practices constituted the American system of commercial broadcasting.

The process of development was groping and haphazard. Many errors were made, and many strategies once adopted were later abandoned. But consistent corporate liberal patterns of decision making during the period are clear: among managers, military men, and government officials, characteristic corporate liberal habits of thought and action can be seen in the style of organizational behavior, in the insights and blind spots, in the possibilities adopted and those abandoned, and in the hopes that these habits expressed.

Systems logic in particular underlay much of the activity of the period. It figured, for example, in both the pre-1920 resistance to broadcasting and its post-1920 incorporation into the corporate system. The exclusive focus on radio as a wireless telegraph was a by-product of the vision of modern business as a matter of bureaucratically integrated systems performing clearly established functions. When broadcasting was belatedly recognized as of potential benefit to the system, the same logic dominated its corporate implementation: broadcasting was pushed toward a for-profit, center-periphery model, and efforts were directed toward developing orderly, planned patterns of funding, programming, and program distribution.

The specifics of the system were up for discussion among the corporate liberal leadership: in the early 1920s Sarnoff imagined a receiver tax while Hoover preferred indirect sponsorship and AT&T tried toll broadcasting. But the larger goal was agreed upon: whatever was to happen, matters were not to be left to the uncertainties of either an open market or fully democratic politics: an orderly system of funding would have to be worked out within the circumscribed world of corporate liberal leadership, a system that assured stability and profits while integrating broadcasting with the larger corporate system. Hence, when all three of these funding methods were swept aside by the rush to product advertising, none of their powerful proponents raised a finger in protest. Broadcasting's development as a vehicle for product advertising was not the only choice of corporate liberals, but it was nonetheless a corporate liberal choice.

Throughout, these decisions were arrived at through associational activities that brought together corporate management and government. The degree of publicness varied: at one extreme were the fully public congressional debates leading to the 1927 and 1934 acts, at the other were the thoroughly secret patent-pool negotiations of the late teens and early twenties. But those activities that fell somewhere in between the extremes reveal more of the corporate liberal mind-set. In the discussions that produced the 1912 Radio Act, the extensive corporate-government cooperation of the war, the formation of RCA in the war's aftermath, and the Radio Conferences, one can see exactly the kind of behavior with which corporate liberals were most comfortable. The preferred forum was neither so democratic as to threaten the autonomy of corporate capital nor so self-interested as to be thoroughly indifferent to the demands and needs of the larger polity. It was one where people who understood themselves to be thoughtful and experienced leaders and experts could rationally develop pragmatic solutions, insulated from direct popular political demands while remaining, in their own minds, paternalistically sensitive to the needs of society as a whole.

Within this world of experts, issues that to others might have seemed fundamentally political appeared as merely technical matters. How much control was going to rest with the government, and how much with private parties? This was a merely pragmatic question, not one of principle, as long as matters were kept comfortably within the circumscribed sphere of administrators and corporate engineers accustomed to working for government one minute and corporations the next. If government made a few decisions that constrained private managerial decisions, perhaps using its licensing powers to shape transmission content, this was hardly cause for alarm; corporate leadership was also heavily involved in the development of programming and in any case was always close at hand, ready to provide corrective input should the government stray in an unhelpful direction. Formal boundaries between government and business were of little concern here; what mattered was the successful development of the system. Similarly, if corporations, outside of government purview, quietly carved up the foundations of the developing system of networking and manufacture among themselves in a secret agreement, no great principles were violated. Networking and manufacture were technical matters, and in any case the corporations had efficiency, progress, and the interest of the public—as understood by both corporate management and the Hoover-era government officials—at heart.

Several quintessentially corporate liberal elements of faith were necessary for the success of these efforts. Underlying the entire set of structural decisions ranging from the 1912 act through the formation of RCA to the 1934 act was a belief in a public good. The public good was understood less as a moral absolute than as a flexible standard that could shift in accordance with technological and social developments. Though flexible, the public good was not subjective. It was not to be mistaken for the popular will. It was an objectively ascertainable value, most obviously so in cases that involved the public safety or military necessity—the justifications that allowed for the initial assertion of government powers over radio in the teens.

But it was not limited to these. In less obvious circumstances, the public good could be detected by the right people following the right procedures: hence another element of corporate liberal faith, the faith in expertise and administrative process. The public good was ascertainable by experts and subject to proper implementation within well-organized administrative systems. Should those systems be relatively public, such as independent regulatory commissions, or relatively private, like the informal corporate-government network that negotiated the formation of RCA? Again, these were technical matters. What was important was that they were efficiently organized and staffed by the right people, by experts.

Metaphors of technology were woven throughout the fabric of these activities, and did much to make them seem sensible to both protagonists and onlookers. Technology, understood as something at once orderly, progressive, and, to the common person, mysterious and out of reach, justified the allocation of power to a narrow circle of experts. Who else could deal with this mysterious wonder except people like Marconi, Sarnoff, Hoover, and their trained engineering staffs? Technology also lent an aura of neutrality: corporate privilege seemed more like a side effect than the principal purpose of allocation decisions, which spoke of kilohertz, wattage, and abstract service types, not Westinghouse or RCA. And subtly but noticeably, the aura of technology surrounding radio lent itself, by way of association, to the belief that radio's use could be organized in the same way as its circuits: the social institution of broadcasting was treated as something that could be created, improved, and perfected through a mixture of expert knowledge, experimentation, and planning. In general, the aura of technology allowed highly political decisions such as the banning of amateurs, the fatal marginalization of nonprofits, and the heavy corporate favoritism of the late 1920s to pass by with little notice from the larger polity.

The decisions that emerged from this corporate liberal environment had profound and lasting impact. In the first stage between 1900 and 1920, a myriad of social possibilities occasioned by the technology of radio were discovered and explored. Some of the possibilities raised by the technology, such as radio communication without legal regulation, were eliminated with the assertion of government power over the spectrum, internationally in 1906 and nationally in 1910 and 1912. Other possibilities, such as the widespread use of two-way radio telephones by the general public, were eliminated by corporate maneuvering, particularly AT&T's patent purchases of the teens.

The remaining possibilities were prioritized. During the teens, in keeping with the barely solidified victory of the corporate system in the economy overall, entrepreneurial control of technological innovation and manufacture was pushed aside in favor of large corporations. Control of use was assigned to a delicately constructed coalition of the military and corporations, which, though divided on who among them should have dominant control, were agreed that radio should be cultivated as a tool for the strategic coordination of large, dispersed, bureaucratic enterprises. The extraordinary possibilities radio presented for use as a popular means of communication among common people and small social institutions were carefully assigned to a secondary and marginalized place.

Broadcasting itself emerged in those margins. Because it was largely unanticipated by corporate and military leadership, existing relations had to be elaborated to accommodate the new phenomenon and develop it according to corporate principles. Having recently discovered the value of organized publicity itself for enhancing corporate profits and legitimacy, the corporate community discovered the value of broadcasting as a publicity vehicle. Government control, meanwhile, was refined to allow for detailed and specific control over the content and control of transmitters. Those powers were in turn used to disassociate broadcasting from the amateur community in which it originated and to then nurture the development of a strictly commercial broadcasting system with corporations at its core, entrepreneurial efforts on its periphery, and nonprofits thoroughly marginalized. By 1934 the system had achieved spectacular economic, legal, and cultural success.

These events cannot be explained as the direct outcome of technological necessity or spectrum scarcity. Much of the effort in the formative period was directed against certain of radio technology's inherent characteristics, particularly its omnidirectionality and lack of secrecy (and many existing institutional structures continue to be dedicated to the struggle against those technological traits). Problems of interference, furthermore, sprang as much from organizational conflicts as from technology. The military, amateurs, and monopoly corporations all had their own nongovernmental means for maintaining order in the spectrum; government was necessary more to mediate the relations *between* these groups than to maintain order within their respective spheres. The competitive, for-profit use of radio did seem to require legal intervention in its own right, but only because its particular *mode* of organization, not its lack of organization, created an incentive for destructive competition.

The marketplace is equally inadequate as an explanation. Many of the essential developments occurred in contexts not dominated by profit incentives or market relations: in universities, in the navy, in government, and among amateurs. More important, most of the early technological and organizational contributions to radio and broadcasting occurred in premarket conditions, before there was anything to buy and sell. Even those efforts undertaken in the hopes of creating markets in radio were not themselves products, strictly speaking, of existing markets: most of the system was solidly in place before anyone, including the corporations, began to return a reliable profit in broadcasting. Nor can a more generalized notion of self-interest fully explain the specifics of the system that emerged. Self-interest there was in abundance, but much of the impetus it provided drove many, particularly the early inventor-entrepreneurs, to adopt unfruitful economic and organizational strategies.

Finally, the developments of the 1900–1934 period are not best explained as the outcome of a conspiracy, of successful behind-the-scenes maneuvering and planning of a corporate cabal. One of the striking things about the early history of radio is the frequency with which corporate liberals misjudged the situation. The central elite oversight was broadcasting itself, but there were numerous others: continuous-wave, voice, and shortwave transmission techniques were all first developed outside the corporations and only later, at considerable corporate expense, brought under corporate control. And in the ensuing years, the full importance of networking and direct product advertising was only belatedly acknowledged.

The lesson to be learned from broadcasting's early development in the United States, therefore, is that commercial broadcasting is the product, neither of impersonal natural forces nor of a narrow conspiracy alone, but also of corporate liberal habits of thought and social organization. Commercial broadcasting, in other words, is what it is, not because it had to be that way, but because a community of leaders acted according to a shared value system to create and organize the use of a new set of technologies and social possibilities. Commercial broadcasting embodies the vision of those shared values: it embodies a faith in the broad liberal framework of property rights, the market, and minimal government, coupled to and qualified by a faith in expertise, administrative procedure, and a reified, paternalistic notion of the public good. Most of all, it embodies the hope that these principles and values can be reconciled, and that taken together they can be part of a just, better, and satisfying life. And as the American system of broadcasting approaches three-quarters of a century, it can be looked at as a test of those principles, values, and hopes.

The Politics of Broadcast Policy in a Corporate Liberal State

Inside the Beltway as an Interpretive Community: The Politics of Policy

When an issue is raised in society, the first (and often most momentous) move is the one which defines it as "policy" or "politics," for once done, the rules of the game, including who can play, are set.

ROLF KJOLSETH

Introduction

Broadcasting Policy versus the Policy of Commercial Broadcasting

This chapter is about a puzzle. Since the consolidation of the system in the early 1930s, there has been a great deal of discussion surrounding something called "broadcast policy" in the United States. That discussion, however, is not about the American policy for broadcasting, about the American broadcast system.

Broadcast policy is an accepted part of the institution of commercial broadcasting. Station owners, network executives, and program producers all devote considerable amounts of time to following broadcast policy developments, supporting and advising the lobbying activities of their trade organizations, and, when necessary, directly participating in efforts to influence the FCC and Congress. To commercial broadcasters such activities are as inevitable as maintaining an inventory or paying taxes. Broadcast policy also thrives in a series of nonprofit organizations, think tanks, foundations, and university programs and disciplinary specialties. Every year, research grants are given, studies commissioned, conferences held, courses taught, and dissertations written under the rubric of broadcast (or sometimes "telecommunications") policy.[1] And

1. The history of the shifting variety of terms (e.g., broadcasting, communications, electronic media, telecommunications) and their shifting referents (ranging from AM radio to the telephone to military remote control systems) is complex enough to be worthy of a monograph. Because the focus here is on the institution of broadcasting, I will use "broadcast policy," even though "communications" and "telecommunications" are just as frequently used in practice, probably because the Communications Act and corporate America both tend to treat broadcasting and common carriers as subcategories of a whole.

at the center of these activities is the community of Washington-based lobbyists, lawyers, and career bureaucrats whose professional raison d'être is the broadcast industry, whose theater of operations is the FCC, and whose horizons are set by the terms and procedures of the 1934 Communications Act. Far from disappearing from the agenda, then, broadcast policy has become the basis for a thriving set of activities.

Broadcast policy nonetheless leaves the underlying legal and institutional framework of the system untouched. Granted, grand principles like free speech, the public interest, and the marketplace are frequently mentioned and debated in textbooks, professional conferences, congressional and administrative hearings, and government reports. And historically there have been challenges to one or another element of the system: attacks on the autonomy of station ownership during the 1940s, for example, and attacks on the public-interest principle during the 1980s. Nonetheless, with one or two possible exceptions, the desirability of the advertising-supported system of broadcasting has never been the subject of policy debate.[2] As we will see, the corporate liberal foundations have been left untouched throughout: the for-profit character of broadcasting, government licensing in the system's behalf, advertising support, and the other integral components of the system remain simply taken for granted, unchanging givens of the broadcast policy universe.

So what is broadcast policy about, if it is not about the American policy for broadcasting? This chapter offers an analysis of the broadcast policy-making process in the United States, focusing on the patterns of shared meanings (with associated political and social values) implicit in and enacted by conventional policy procedures, organizational structures, and professional roles common to the policy arena. It suggests that policy making may be usefully understood as taking place within a specific interpretive community, a community of individuals that interact with one another in such a way as to generate a shared, relatively stable set of interpretations in the face of potentially unresolvable ambiguities. What makes a ruling appear practical, a legal decision seem sound, or a procedure appear fair, is the contingent, shared vision of the interpretive community itself, not simply rational policy analysis, legal reason, formal rules of process and procedure, or interest group pressures. Looking at the FCC and broadcast policy making this way, this chapter argues, suggests explanations for both shifts and continuities in broad-

2. Two exceptions to the general rule of nondiscussion of corporate fundamentals in broadcast policy might be the failed 1934 effort to grant one-quarter of the AM band to nonprofits, and the discussion that led to the creation of PBS in the 1960s.

cast policy, and allows for an analysis of policy within the context of broad historical trends.

This chapter looks at broadcast policy, then, as a way of thought embedded in a social and institutional context. And it argues that the way of thought peculiar to broadcast policy, while certainly not eliminating all debate and political struggle, nonetheless aids in the creation of broad corporate liberal boundaries outside of which debate cannot go. As an institution, therefore, broadcast policy supports the principles of corporate liberal broadcasting by legitimating those principles without calling them into question.

Communities, Rules of Discourse, and the Power of Interpretation

Meaning is contextual. It is created, kept alive, and changed by people acting within institutional, social, and historical contexts. Some scholars describe the basic unit of support for patterns of meaning in social life as an "interpretive community," a community of individuals that interact with one another in such a way as to generate a shared, relatively stable set of interpretations. The formation of interpretive communities is an ordinary, perhaps fundamental, human process: over time, any group tends to create informal, commonly shared interpretations of the meaning of phrases, words, and activities important to the group, and builds institutional structures in which to maintain those shared meanings.[3] Those structures, in turn, embody discursive rules, rules about what can be said and done and what can't be said and done, and more important, how to say and do them.

Within interpretive communities, imponderables that otherwise might be open to an infinite variety of interpretation—moral values, canons of aesthetic taste, religious matters—are given relatively stable, agreed-upon meanings. The child may wonder how God created the world in seven days, the undergraduate may question the value of Shakespeare, but the designation of authorities (priests, professors) and the creation of institutions for inculcating the doubtful with appropriate

3. The phrase "interpretive community" has been made famous by Stanley Fish in *Is There a Text in This Class? The Authority of Interpretive Communities* (Cambridge: Harvard University Press, 1980). The general idea that meanings are collectively created and stabilized in symbol use by interacting communities, however, has a much wider currency in anthropology and sociology, going back, on this side of the Atlantic, to the works of C. S. Peirce, G. H. Mead, and the symbolic interactionist school and, on the other side, to Husserl, Alfred Schutz, and the tradition of phenomenological sociology.

interpretations helps assure that by the time the child is an adult, by the time the undergraduate graduates, they will have come to share the interpretations they once doubted.

The broadcast policy world is rife with imponderables, of which the public-interest standard is only the best known.[4] Like other human communities, therefore, the maintenance of stable meanings is a matter of building interpretive communities appropriate to the social and institutional context. These meanings are sometimes articulated explicitly, but more often implicitly represented or enacted in institutional structures, organized activities, and patterns of interaction. In this way, the policy process itself becomes meaningful; it takes on meanings that are in some ways more important than the explicit issues that are discussed within policy debate. Policy making, then, is not just a goal-directed activity. It is a way of thought.

This is not to say that broadcast policy making is free of dissent. Not everyone in the policy arena thinks the same way. Broadcast policy is characterized more by constant struggles and disagreement, not by some monolithic ideology. Policy "issues," as they are called, are always being hotly contested; knowing what those issues are, who is involved, and what is at stake is part of the job of the broadcast policy expert. Broadcast policy is hardly a straightforward matter of engineering. It is not a clean, neutral, predictable, mechanical, or routine process.

The approach here is based in an observation from interpretive sociology: whether one is dealing with parking-lot brawls or parliamentary debates, in human social contexts certain things can and cannot be said and done, and they must be said and done in certain ways. Disagreement, when it does occur, must occur within a broad framework of underlying assumptions, of what might be called discursive rules. Those rules are not so much rigid requirements as they are structures of expectations—including expectations about conflict—that embody the underlying operating assumptions of any social order.

The power of discursive rules lies in this: their influence on how one's statements and actions are interpreted by others. Discursive rules thus do not restrain action; they determine whether one's arguments make sense to others, and what sense others make of them. The heretic's artful violation of a rule or two, for example, might attract attention precisely because his or her actions would be labeled as violations from

4. For a discussion of indeterminate concepts in communications policy, see Thomas Streeter, "Beyond Freedom of Speech and the Public Interest: The Relevance of Critical Legal Studies to Communications Policy," *Journal of Communication* 40 (spring 1990): 43–63.

within the framework of discursive rules; the framework still operates in the heretic's arguments, even if negatively. But speaking from too far outside the framework renders one, not a heretic, but uninteresting, incomprehensible, or at best quaint—a far more effective way to silence opposition than condemnation. The world of broadcast policy is no exception: underlying all the (very real) disagreements and debates is a relatively constant structure of expectations that limit discussion, not by coercion, but by way of the subtle but profound power of interpretation.

Inside the Beltway as an Interpretive Community

The existence of this kind of interpretive community in American politics is informally acknowledged by the colloquialism known to all who are involved with the policy process: "inside the beltway." This phrase does not just refer to a place: many of the poor and working-class residents of Washington, DC, are geographically inside the freeway that circles the city, but they are not "inside the beltway," and some of those who *are* "inside" spend much of their time geographically elsewhere. Rather, the phrase stands for both an institutional context—the network of public and private organizations associated with the federal government—and a perspective—the point of view of Washington officials and bureaucrats, which is acknowledged to be peculiar and difficult to understand to those on the "outside."

The interpretive community of broadcast policy is basically a subset of the larger "inside the beltway" community in Washington. Like any interpretive community, it has its structures of authority, its masters and initiates, its roles and rituals. Taken together, these activities help generate stability of interpretation both on the level of contingent issues (e.g., Is it acceptable to discuss common-carrier regulation of cable TV this year? Is a rhetoric of "economic efficiency" necessary to being taken seriously today?) and on enduring patterns of thought (e.g., the belief that political problems can be resolved by expertise).

Of course, the interpretive community does not generate absolute unanimity on all issues. Particularly since its self-understanding includes the premise that its activities are on some level consistent with liberal democratic discussion, it frequently engages in heated debate and struggle over particular issues. But it organizes and circumscribes debate in very particular ways. One of the functions of any interpretive community is to designate which issues and which positions are properly subject to debate, and which issues are beyond the pale. A measure of the ideological strength of an interpretive community is the extent to which it can ignore its critics: if one resorts to denouncing those who speak from outside the community, the interpretive framework is troubled, but if one can afford to greet them with indifference, the power of the framework is secure.

The core institutions that maintain the particular interpretive community associated with broadcast policy are the FCC and similar government offices: that is, the Federal Trade Commission, the Office of Management and Budget, Congress's Office of Technology Assessment, and the National Telecommunications and Information Administration (NTIA) in the Department of Commerce. These institutions, in turn, share personnel and maintain ongoing relationships with a number of congressional committees and committee staffs.[5] Mastering the shifting labyrinth of organizations, congressional subcommittees, hearings, procedures, and terminologies in which broadcast policy is conducted has been the basis for many a distinguished and lucrative career.

Another premise shared by the broadcast policy community is that of the autonomy and neutrality of expertise. Although the FCC and related government institutions are to various degrees independent and neutral under law, they lack both the resources and the institutional distance from elected officials to successfully produce analyses that consistently appear autonomous and expert.[6] Other institutional homes for broadcast policy experts are needed.

Research universities, of course, are a natural institutional site for fostering the required neutral expertise. They regularly provide the society with a corps of individuals whose claim to authority and income rests on their degrees and professional training, and who thus are predisposed to careers as experts of one sort or another. Moreover, the tradition of the disciplinary specialty dovetails nicely with the sociopolitical need for expertise in policy matters; broadcast or telecommunications policy can and has become a subspecialty for academics in political science, communications, and law. Courses appear in the catalogs, articles appear in the journals, and academic conferences devote panels and subdivisions to matters of broadcast law and policy. In a few cases, universities have created institutes and programs specifically devoted to media policy.[7]

5. Key congressional subcommittees include the Subcommittee on Communications of the Senate Committee on Commerce, Science, and Transportation, and the Subcommittee on Telecommunications and Finance of the House Committee on Energy and Commerce.

6. There are exceptions. In the late 1980s the Office of Technology Assessment showed unusual autonomy, in some ways bucking the deregulatory trend with its report titled *Critical Connections: Communication for the Future,* OTA-CIT-407 (Washington, DC: Government Printing Office, 1990).

7. Well-known university programs include City University of New York's Center for

From the point of view of the broadcast policy world, however, universities can be *too* independent. Tenure and the principle of intellectual freedom allow for work that strays far outside the proper bounds of policy research. Psychologists, for example, have produced a steady stream of work that embarrasses television network executives with exhaustive studies of the negative effects of televised violence. Marxists and other malcontent tenured radicals rail against the for-profit foundations of the system in print and in front of their undergraduates. And even more frequently, academics produce work that is simply too technical and specialized to be of use to those inside Washington: the jargon, theorizing, and concern for obscure academic debates produces scholarship that does little more for policy participants than cause their eyes to glaze over.

The authors of this scholarship may believe that their work is ignored by the policy circles in Washington because of a cowardly resistance to hard truths or because of a conspiracy on the part of the powers that be. Those inside the policy world, on the other hand, are more likely to say that the problem is simply that this kind of academic work is too "impractical." There is a certain kind of truth to the latter explanation: to be practical in this context means that one somehow contribute to the larger project of using neutral expertise to integrate broad liberal principles within a corporate consumer economy. Being "practical" in this particular corporate liberal sense is a requirement of admission to the world of broadcast policy, and being "practical" is not exactly the same as being brilliant, wise, or insightful. Hence, if academic research doesn't successfully associate itself with this larger project, no amount of compelling evidence or elegant theory can gain the research a serious hearing in the policy world.

Because of the tendency toward "impracticality" in academia, a number of secondary institutions have evolved for circumscribing policy discussion: a few think tanks and foundations have made broadcast policy one of their specialties.[8] At least one well-funded annual conference (the

---

Public Policy and Telecommunications and Information Systems; Columbia Business School's Center for Telecommunications and Information Studies; Harvard's Program on Information Resources Policy; MIT's Research Program on Communications Policy; New York University Law School's Communications Media Center; Northwestern's Annenberg Washington Program in Communication Policy; and UCLA Law School's Communications Law Program. Graduate programs in media and communications with faculty interested in policy also are common in research universities throughout the country, particularly in land grant institutions.

8. The Rand Corporation, the Markle Foundation, the Gannett Foundation, and the

Telecommunications Policy Research Conference) devotes much of its energy to media regulation, and occasional blue-ribbon commissions (e.g., the Carnegie Commission on Educational Television)[9] all provide funding and outlets for policy expertise. The function of these organizations is to foster that special mix of practical yet expert activity that corporate liberal policy requires. They thus provide funding, outlets for research, and contexts that bring select academics and other experts together with Washington insiders around specific policy issues. The result is a steady supply of new research grants, conferences, research reports, and jobs for policy specialists, carefully screened and selected by the community of policy experts themselves. In this context, the shared meanings necessary for interpretive stability can be maintained.

It is tempting to understand the institutional context of broadcast policy instrumentally: after all, corporations pay corporate lawyers and lobbyists and create trade organizations in order to serve corporate interests in Washington.[10] Conflicts are generally between corporations, not between corporations and other "interests." Much of what goes on is thus fueled rather directly by corporate profit desires.

Yet the profit motive alone cannot account for all that goes on, at least not in a simple way. For, in a corporate liberal environment, administrative neutrality and expertise are political prerequisites of procorporate decisions. If there is going to be government intervention on the industry's behalf, it must be done in a way that at least suggests the presence of neutral principles and expert decision making, that is, some independence from corporate interests. As a result, even corporations have an interest—an ambivalent one—in fostering institutions that are not mechanically tied to corporate designs, institutions that demonstrate some autonomy.

Aspen Institute have all on occasion given substantial support to conferences and research programs in media policy areas.

9. See the Carnegie Commission for Educational Television, *Public TV: A Program for Action* (New York: Harper & Row, 1967). For a similar though lesser-known example of a blue-ribbon study, see the Sloan Foundation, *On the Cable: The Television of Abundance* (New York: McGraw-Hill, 1971).

10. The familiar "capture" theory of regulation can be interpreted this way, but the classic statement of this method of accounting for state action in capitalist societies is G. William Domhoff's Marxist instrumentalist account, *The Powers That Be: Processes of Ruling Class Domination in America* (New York: Vintage, 1979). For a direct and relatively compelling application of Domhoff's theory to communications policy, see Vincent Mosco, *Pushbutton Fantasies: Critical Perspectives on Videotex and Information Technology* (Norwood, NJ: Ablex, 1982), 24–37.

The Meaning of Broadcast Policy: Expertise Brings Order to Chaos on Behalf of Liberalism

Corporate Liberal History: The 1927 Act as the Rule of Law

One key to understanding any community is to look at the stories the community tells itself about itself. The community of broadcast policy is no exception. With remarkable frequency, textbooks, law journals, and legal decisions tell a particular version of the story of the 1927 Radio Act and the origin of broadcast law in the United States. In the early 1920s, the story goes, the fledgling broadcast industry lacked a proper institutional and legal order. As a consequence, broadcasters interfered with one another and chaos reigned. In response to the chaos, Congress stepped in and resolved the problem by passing the first legislation to govern broadcasting and by creating the FRC. The imposition of law and administrative structure thus brought order to chaos. This is the origin myth of American broadcast policy.[11]

Not surprisingly, these abbreviated historical accounts vary somewhat according to the agenda of their authors. The most common telling of the story describes the 1927 act and the resulting regulatory apparatus as the technologically necessary outcome of a period of preregulatory chaos in the 1920s.[12] Some conservative advocates of marketplace principles, on the other hand, have recently described the act as the ham-handed actions of marauding government bureaucrats restraining the efforts of plucky commercial entrepreneurs operating in a natural marketplace.[13] Significantly, however, none of these accounts discuss in any

11. A representative example of this version of the story can be found in Erwin G. Krasnow, Lawrence D. Longley, and Herbert A. Terry, *The Politics of Broadcast Regulation,* 3d ed. (New York: St. Martin's Press, 1982), 10-12.

12. The Supreme Court has tended to tell this version of the story when upholding broadcast regulations against charges of interference with the rights of broadcasters. See *National Broadcasting Co., Inc., et al. v. United States et al.,* 319 U.S. 190 (1943), reprinted in *Documents of American Broadcasting,* ed. Frank J. Kahn, 4th ed. (Englewood Cliffs, NJ: Prentice Hall, 1984), 138-41; the same argument is used in *Red Lion Broadcasting Co., Inc., et al. v. Federal Communications Commission et al.,* 395 U.S. 367 (1969), reprinted in Kahn, *Documents of American Broadcasting,* 275-93. Textbooks lean toward this version of the story as well. See, for example, Don R. Pember, *Mass Media Law,* 2d ed. (Dubuque, IA: William C. Brown, 1981), 424-25; and Marc A. Franklin, *The First Amendment and the Fourth Estate: Communications Law for Undergraduates* (Mineola, NY: Foundation Press, 1977), 461-64.

13. For example, Thomas Hazlett, "The Rationality of U.S. Regulation of the Broadcast Spectrum," *Journal of Law and Economics* 33 (April 1990): 133; and Matthew Spitzer, "The Constitutionality of Licensing Broadcasters," *New York University Law Review*

detail the pre-1920 history of broadcasting I discussed in chapter 3. All those events that set the stage for the both the regulatory patterns and the broadcast marketplaces of the 1920s—the assertion of legal control of the spectrum on behalf of a corporate-military alliance in 1912, the efforts of the amateurs during the teens, and so forth—are omitted from the accounts.

These omissions reveal a more general story that is being told in the context of the specific story of early broadcasting: the story that laws impose order on social relations, not the other way around. By starting the story in the early 1920s, one does not have to address the extralegal organizational patterns that presaged the 1927 act. Without the pre-1920 developments, the 1927 act is not a matter of asserting one *kind* of order at the expense of other possible forms of organization, nor a matter of using legislation to underwrite an already present but prelegal form of order. Rather, before there was simply chaos, and then in 1927 law brings order and justice.

Part of the implicit message here is a reassertion of the liberal belief in "the rule of law, not of men," that is, in the capacity of formal rules and procedures to transcend politics. The traditional story implies that the 1927 act and its 1934 successor represent an abstract, transcendent, impersonal order—the rule of law—not an assertion of the visions, designs, and interests of some specific groups and individuals at the expense of others—the rule of men. But it is also a particularly corporate liberal variant of that belief: it is less a story of lawyers and judges locating bright lines in the world of rights and responsibilities than a story of administrator-engineers making technical decisions on the basis of the public safety, kilohertz, signal-propagation characteristics, technological progress, and so forth.

Corporate Liberal Organization: The Federal Communications Commission

An obvious (though not the most cogent) sign of the corporate liberal principles underlying broadcast policy is the structure of the FCC that originated in the 1920s. By law, the FCC is an independent regulatory commission supposedly insulated from the winds of politics by formal institutional boundaries and rules. FCC decisions can be appealed to the

64 (November 1989): 990, 1046. Although Hazlett and Spitzer treat their revision of the history as highly original, its outlines are essentially the same as those used, with polemical clarity, in Ayn Rand, "The Property Status of the Airwaves," in *Capitalism: The Unknown Ideal* (New York: Signet, 1967), 122–29. These arguments are discussed in more detail in chapter 6.

federal courts, but only when the FCC can be claimed to have violated a legal or constitutional rule; the courts accept that, within its own sphere, the FCC's administrative expertise is to be respected. Commissioners must come from both political parties, and once appointed they are by law independent. The organization of the FCC thus embodies the corporate liberal faith that neutral expertise and social engineering can be brought into the service of liberal principles.

As any cultural anthropologist is quick to assert, underlying rules of behavior are rarely of a mechanical sort. Rather, their application is more a matter of art than science, and is typically rife with ambiguity and nuance. Participants must do a lot of work to creatively construct actions that uphold or celebrate the rules, often in contexts with which the rules seem to conflict. In the case of the FCC, for example, many decisions can be explained in terms of partisan politics; a newly elected administration in Washington typically appoints a majority of sympathetic commissioners to the FCC, who then make decisions reflecting the administration's views.[14] When the Reagan administration appointed Mark Fowler to be commission chair, for example, Fowler's many proindustry, "deregulatory" decisions at the FCC reflected the general political views of the Reagan administration.

Yet the rules of the game are such that the decisions cannot be officially justified on political grounds. Commissions are supposed to be staffed by experts, not politicians. A commissioner cannot defend a decision with the argument that "this is what the majority of the people want because they voted this way." FCC decisions must be justified within the framework of expertise: with references to expert testimony, statistical evidence, and a neutral public interest. More than one FCC decision has been overturned by the courts simply because these discursive rules were violated, because the political nature of its decision was not sufficiently couched in the trappings of neutral expertise.[15] In one particularly illustrative case, Fowler changed his vote on a broadcast policy issue after a visit to the White House. Eyebrows were raised throughout

14. James M. Graham and Victor H. Kramer, *Appointments to the Regulatory Agencies: The Federal Communications Commission and the Federal Trade Commission, 1949–1974,* printed for the use of the Senate Committee on Commerce, 94th Cong., 2d sess. (Washington, DC: Government Printing Office, April 1976), 385–86.

15. For example, Action for Children's Television was able to force the FCC to revisit its decision to allow program-length commercials, not because the courts thought the decision itself was a bad one, but because the courts thought that the FCC had not done a proper job of gathering evidence to support its decision; that is, it had not been properly expert (FCC, "Revision of Programming and Commercialization Policies, Ascertainment Requirements, and Program Log Requirements for Commercial Television Stations," 98 F.C.C. 2d 1012 [1984]).

Washington, and some suggested that his action constituted a violation of the law.[16] Fowler's mistake in the incident was not that he pursued policies shaped by the politics of the president that appointed him. No one would expect him to do otherwise. His mistake was to allow the politics of his decision to become blatant. He violated the discursive rules of policy.

Today, it must be pointed out, the belief in the administrative neutrality and expertise of the FCC has lost much of its cogency among participants in the policy world. This may in part be because of its obviousness: a particular ritual activity once cherished can over time become a stale cliché that no longer grips the imagination the way it once did. So today among seasoned policy experts it is a matter of insider wisdom that the FCC's autonomy is largely a chimera, that its activities are deeply political. But this does not mean that the policy insiders have abandoned the ideal of apolitical expertise. When symbols become clichés, communities are more likely to create new, more subtle symbols than they are to abandon the premises that the symbols embody.

The inhabitants of the contemporary broadcast policy world, therefore, follow more subtle, implicit versions of the discursive rules of expertise and apolitical objectivity. These rules exist on an implicit level throughout key sectors of twentieth-century American political culture, and by their implicitness are rendered all the more powerful. The rules extend to lobbying organizations, congressional committees, foundations, think tanks, the legal profession, and, at the policy world's outer perimeter, universities. This world is united by shared patterns of talk and action, by the set of expectations that come with the idea of expertise in a corporate liberal universe.

Corporate Liberal Semantics: Policy and Politics

Another interesting clue to the meaning of broadcast policy thus involves a pattern of talk, a habit of speech. Today, among those who inhabit the world of broadcast policy, it is often asserted as a matter of insider wisdom that broadcast policy is a highly *political* process. Curiously, however, one does not refer to the activities of this world as "broadcast politics." One does not hear of a "Telecommunications Politics Research Conference" or see courses on broadcast law and the FCC listed in university catalogs as "Broadcast Politics." None of the lawyers, lobbyists, bureaucrats, or academics whose careers focus on broadcast policy describe themselves as "broadcast politicians." In spite of the fact

16. Ann Cooper, "Fowler's FCC Learns Some Hard Lessons about What It Means to Be 'Independent,'" *National Journal,* April 6, 1985, 733.

that it is regularly described as political, in sum, the world of broadcast policy remains "policy."

To untangle this oddity, one needs to look at the two words "politics" and "policy." Over the centuries, the precise meanings of these two members of the *polis* family of words have shifted in complicated ways.[17] And in contemporary French and German, the distinction between "politics" and "policy" does not exist; each language has only a single word—"politique" and "Politik," respectively.[18] But in contemporary American discourse, policy is different from politics.

Basically, policy is spoken of as something quite distinct and opposed to the raucous clamor and maneuvering of open political struggle among self-interested parties. The most common ideal image of policy making is that of a neutral, calm, reasoned, carefully moderated process. Hence, there are frequent complaints about the "interference" of political concerns with policy making, and calls are frequently heard for replacing a chaotic "political" process with a rational "policy" process. As Harold Lasswell put it when he defined "the policy orientation," " 'policy' is free of many of the undesirable connotations clustered about the word *political*, which is often believed to imply 'partisanship' or 'corruption.' "[19]

Exactly what people have in mind when they discuss the policy process varies. For some it is envisioned as expert advice and guidance; for others, plans for management and coordination developed along scientific or rational principles; for others, rationalized structures for decision making such as administrative agencies and committees; and for many, it means some mixture of all of these. The yardstick of order and reason also fluctuates: there have been periods when the dominant model seems to have been legal reason, and others when it was social science, whereas today it is largely economics—though the model of electrical

17. According to the *Oxford English Dictionary*, the word "policy" itself apparently was once associated with the word "polish," and carried connotations of refinement, elegance, culture, and civilization. It has also been used to refer to formal documents that serve as evidence of money paid, as in "insurance policy." Throughout history, however, its most important meanings have come from its association with the complicated word "politics." At times the two words have been used more or less synonymously. The *OED* lists some of the older meanings of policy as "expedient . . . cunning, craftiness, dissimulation," and "a device, expedient, contrivance; a crafty stratagem, trick," senses today reserved for particular uses of "political."

18. Arnold J. Heidenheimer, " 'Politics,' 'Policy,' and 'Policey' as Concepts in English and Continental Languages: An Attempt to Explain Divergences," *Review of Politics* 48 (winter 1986): 3–30.

19. Harold D. Lasswell, "The Policy Orientation," in *The Policy Sciences* (Stanford: Stanford University Press, 1951), 5.

engineering seems to be a constant in the background of these shifting fashions. In any case, society, it is felt, sometimes needs something more orderly, more stable, more scientific, more rational than the simple struggle of self-interest or the uncertainty of political struggle. It needs, in other words, a process for making policy.[20]

In another context, Rolf Kjolseth has neatly captured the implications of the policy/politics distinction in contemporary discourse. A "political matter," he writes, "is one which has been socially defined as involving decisions where all those who are understood to be directly affected by the outcome are granted rights to influence the decision directly; by contrast, a 'policy matter' is one which has been defined as involving decisions in which only those certified as specially qualified (by training or office) are granted the right to have a direct influence upon the decision." Hence, these two terms are associated with different ways of organizing decision making:

> "politics" opens the door to participation by a wide range of persons and interest groups, . . . "policy" withdraws the matter to a narrow range of known and predictable experts . . . When an issue is raised in society, the first (and often most momentous) move is the one which defines it as "policy" or "politics," for once done, the rules of the game, including who can play, are set. Politics runs by popular democratic rules; policy follows elite, technocratic rules.[21]

The fact that it is broadcast *policy,* not broadcast *politics,* then, implies that what goes on in the FCC and related arenas is a neutral, technocratic activity. That the inhabitants of this world consistently refer to what they do as "policy" implies they are, by their own definition, specially qualified experts dealing with technical matters, not politicians dealing with matters of social value.

The larger point here is this: even on the level of language, today's broadcast policy experts are heirs (frequently unconsciously) to the faith, born in the utopian visions of Saint-Simon and Comte and imported to the United States by Charles Francis Adams, in economic management by government-appointed social engineers. And they

20. This ideal, associated with the Weberian ideal of administrative rationality, was embodied in Lasswell's seminal vision of "policy science," and was declared an actuality in the field of communications by Ithiel de Sola Pool in the early 1970s. See Ithiel de Sola Pool, "The Rise of Communications Policy Research," *Journal of Communications* 24 (spring 1974): 31–42.

21. Rolf Kjolseth, "Cultural Politics of Bilingualism," *Society* 22 (May–June 1983): 47.

implicitly follow the ideals of Senator Charles C. Dill and Herbert Hoover, who hoped for radio regulators that were not "politicians," but "men of technical and legal experience." For all their claims to the contrary, in their language and demeanor they cling to the supposition that the government's relation to the broadcast industry is a matter of policy, not politics.

Of course, the world of broadcast policy is bounded, not simply by the faith in neutral expertise, but by that faith in its corporate liberal form. Broadcast policy experts do not use their expertise to question the value or coherence of basic liberal values in broadcasting. They take for granted that broadcasting can and should operate on a commercial marketplace basis. The world of broadcast policy, therefore, is corporate liberal at its core: to be taken seriously as a broadcast policy expert, one's style, language, and behavior must uphold one or another version of the argument that classical liberal principles can be squared with government intervention by way of expertise.

What is meant, then, by the word "politics" inside this framework, when "policy" experts assert that "broadcast policy is political"? "Politics," it can be argued, has always carried two connotations: "high" politics and "low" politics. The two connotations are often hard to separate; since Machiavelli they have been inextricably intertwined. Roughly speaking, however, high politics is associated with the state, a body of citizens, or government affairs, particularly of a secular, democratic nature—it is what Kjolseth calls "involving decisions where all those who are understood to be directly affected by the outcome are granted rights to influence the decision directly." Low politics, in contrast, is politics in the sense of maneuvering and strategizing for the gain of oneself or one's group. In its most negative sense, low politics connotes scheming, craftiness, and the smoke-filled room. Hence, although "policy" was also once used in both the low and high senses of "politics," today it would seem incongruous to use "policy" in this way. Perhaps due to its secondary associations with civilized refinement and formal documents, "policy" has been moved into a position of contradistinction to low politics. Policy has become politics with the low sense removed. Hence, we can now speak of "politics interfering with policy," and call for a "rational policy" as opposed to "mere politics."

When broadcast policy experts talk about "politics interfering with policy," and speak of the "political" nature of policy-making processes, they mean politics in the low sense, politics in the sense of maneuvering for gain, in the sense of the smoke-filled room. They do *not* mean politics in the high sense, politics in the sense of democratic decision making by citizens. "Politics" in this world means a messy departure from the policy ideal—an inevitable one, perhaps, but nonetheless a departure.

Put this way, the character of American "broadcast policy" becomes clear: broadcast policy is a realm for experts, not for "politics" in the broad sense of governance in a democratic society. High political questions are not on the agenda; they are considered to be resolved, and thus to be taken for granted or at least best left to others. And when those inside the delimited broadcast policy world knowingly acknowledge that policy is political, they mean political in the sense of maneuvering for gain—low politics. The world of policy, they readily acknowledge, has become infected by the processes associated with self-interested strategizing and struggles. But they don't describe the subject matter of their conferences and research grants as "broadcast politics" because this might imply high politics: matters of value, structure, and legitimacy that they and their sponsors have little interest in opening for consideration.

Corporate Liberal Discourse: The Slide from Rights to Measurement

There have been those who, exasperated with the arid formalism of classical legal thought, argue that law schools should be turned into policy institutes, and lawyers turned into sociologists.[22] Occasional hubristic flights of social engineering aside, however, policy terms and procedures generally figure as subordinate and complementary to traditional liberal principles, not as alternatives to them. A useful illustration of the way that policy discourse relates to liberal principles can be seen in a favorite debating topic in broadcast policy for the last twenty years: the fairness doctrine. The doctrine was an FCC rule that between 1949 and 1987 required broadcast station owners to provide balanced news coverage of controversial issues. As of this writing it seems possible that it will be revived in some form by Congress.

All sides in the fairness doctrine debate conform to a basic legal liberal framework. All agree that the ultimate purpose of broadcast regulation is to protect the freedom of individual communication from constraint by others, and thus to maintain what the Supreme Court calls "an uninhibited marketplace of ideas in which truth will ultimately prevail."[23] (If one were to make an argument that fell completely outside the existing framework, for example suggesting that any form of the mar-

22. Harold D. Lasswell and Myres S. McDougal, "Legal Education and Public Policy: Professional Training in the Public Interest," *Yale Law Journal* 52 (March 1943): 203–95.

23. *Red Lion*, 287.

ketplace is inherently against the social good, one would be ignored by courts, the FCC, and, in all probability, funding agencies.)

Both opponents and proponents of the doctrine tend to begin by framing their arguments within something resembling classical liberal terms, such as the language of absolute free speech rights. Opponents might describe the doctrine as a case of state tyranny constraining the free speech rights of an individual broadcaster, or of marauding government bureaucrats and special interest groups limiting the self-expression of a licensee. Proponents might begin by depicting the situation as one in which, in the absence of the doctrine, individuals with alternative opinions are constrained from exercising their free speech rights by the notoriously centrist and politically timid television industry, or in which powerful broadcasters prevent minorities, activists, and people with unpopular opinions from expressing their views. In each case, the situation is framed in terms of active individuals with rights struggling against constraints to those rights.

Framed in these classical terms, however, the debate is threatened with some ambiguities. Everyone within the policy arena agrees that we should protect individuals' freedom to communicate by limiting constraints on that freedom by others, for the purpose, ultimately, of upholding the marketplace of ideas and thus the public interest. But who has rights that are being constrained and who is interfering with those rights? Is the selection of a schedule of broadcast programming, the decision to include some programs and exclude other programs, an exercise of freedom or an act of constraint? What about forcing a station to add one's own point of view to that schedule? It is hard not to conclude that both actions involve both freedom and constraint, and that the idea of inviolable rights cannot be applied in this situation.

The typical route out of this kind of dilemma is illustrated in the *Red Lion* case of 1969, in which the Supreme Court upheld the fairness doctrine. The Court responded to the dilemmas of absolute rights by turning to "policy" arguments. The Court made it clear that in the case of broadcasting the normal exercise of rights can restrict the rights of others; ownership of a television station, after all, gives one the power to prevent others from using it. Broadcasters, the Court reasoned, are necessarily much fewer in number than the audience, yet have a much greater opportunity to make their views heard. Arguing functionally, the Court suggested that treatment of broadcasters' free speech rights as absolute would clearly violate the purpose, if not the letter, of the First Amendment. So, the Court shifted the terms of discussion away from moral principles toward technical, "empirical" questions: Is there some technological necessity that can help decide the question, such as a scarcity of broadcast spectrum? Is there sociological evidence that provides an answer, such as a measurable chilling effect of the doctrine on broadcasters? The Court, in other words, sought to rescue the marketplace of ideas by recasting the problem in terms of empirical questions amenable to resolution by the logic of sociological expertise.

Broadcasters and audiences, the Court assumed, possess competing rights. If treated as absolutes, one set of rights proves fatal to the other. The Court's solution to this problem involves, in part, *balancing* broadcasters' rights against those of others; this is implicit in the Court's support of the fairness doctrine on the grounds that "[i]t is the right of the viewers and listeners, not the right of the broadcasters, which is paramount" (287). To make the problem into an empirically testable one instead of a purely moral matter of rights, the Court in a sense symbolically calibrated the space between absolute rights for broadcasters and absolute rights for the audience into a series of measured increments along a continuum, and then calculated and weighed the resulting relative values against one another. By taking upon themselves this task of balancing, the Court acted as if they and the FCC were a body of social scientists or judicial engineers. Hence, they were able to conclude that, if the fairness doctrine does interfere somewhat with broadcasters' free speech rights, such marginal interference is justified when empirically calculated against the rights of the audience.

Of course, the Court's *Red Lion* decision did not resolve the question. The fairness doctrine was thrown out by the Reagan-era FCC, has been passed in a legislative version twice by Congress only to be vetoed by Presidents Bush and Reagan, and, if in the likelihood it appears again under Clinton, very well might be found unconstitutional by the current Supreme Court. The debate goes on. The point is that, on both sides of the ongoing debate, matters begin with hallowed constitutional principles and then proceed into questions theoretically subject to resolution by experts: opponents muster data and experts to prove that spectrum scarcity no longer exists but chilling effects do, and proponents muster data and experts to prove otherwise.

This discursive pattern is repeated on a regular basis throughout the world of activity called broadcast policy. The ratio of traditional legal argument as opposed to policy argument differs according to context and political climate. Discussions about the allocation of frequencies for new technologies such as high-definition television, for example, tend to be overwhelmingly technical. They usually stay within the FCC and related policy institutions, and traditional liberal principles are left largely implicit (in the form of, say, an underlying assumption that corporate domination of the technology is in the society's best interest). Decisions involving free speech or minority ownership of broadcast stations, conversely, contain explicit discussions of rights and other constitutional matters, and are consequently more likely to work their way from the FCC into the federal court system (and to become topics for college debate teams).

Furthermore, policy arguments can be used against rights arguments and vice versa. They are not so much rigid requirements as they are available rhetorical tactics. In the early 1960s, during the civil rights movement, a successful effort to deny a broadcast license renewal to an overtly racist television station was construed by those on the left largely as a matter of protecting the rights of the station's large minority audience.[24] In the 1980s, conversely, the Supreme Court upheld licensing preferences for minorities more by reference to policy arguments, particularly to minority station ownership data compiled by social scientists.[25]

But participants in the policy arena quickly learn to operate within the general assumption that policy expertise ultimately serves traditional liberal values. Stray too far in the direction of either rights or expertise and you drop off the policy map. As an example of the former, in the early 1980s the cable television industry turned to conservative First Amendment and marketplace purists to lobby against FCC regulations that hobbled cable's growth. But when free market logic led to arguments against municipal regulations that protected cable monopolies on the local level in the mid-1980s, the cable industry quickly lost interest; the free market purists were no longer as useful.[26] An example of the dangers of "excessive" expertise at the expense of liberal values, conversely, can be found in the fate of an elegant and ingenious proposal, published in the 1970s. The proposal called for completely redesigning the American broadcast system in a way that would separate transmission from production, and thereby effectively remove much of both the networks' and local broadcasters' power. Nowhere in American law does it say that existing broadcasters have a right to government

24. *Office of Communication of the United Church of Christ v. Federal Communications Commission,* 359 F. 2d 994 (D.C. Cir. 1966).

25. *Metro Broadcasting, Inc., v. Federal Communications Commission,* 110 S. Ct. 2997 (1990).

26. Thomas Streeter, "The Cable Fable Revisited: Discourse, Policy, and the Making of Cable Television," *Critical Studies in Mass Communication* 4 (June 1987): 195; *City of Los Angeles et al. v. Preferred Communications, Inc.,* 476 U.S. 488.

protection of their economic interests in the existing system of spectrum allocations, yet the proposal is known today in the policy arena only as an archetype of madcap impracticality.[27]

How to Be a Policy Expert: Roles and Methods

Roles in the Policy World: Staging the Subjective/Objective Distinction

Generating interpretive stability is not simply a matter of declaring things to be true or of prohibiting the participation of those with alternative interpretations. Preferred interpretations need to be rendered compelling, and this involves ritual and drama. Meanings need to be acted and reenacted in the everyday activities of a community of people, in this case the community of broadcast policy experts. And regularized actions in human communities involve acting in a very real sense. They involve roles, systematic patterns of activity that are known to participants. Roles both enact underlying meanings in their form and structure, and serve as tickets to entry to the interpretive community. To take part in the broadcast policy world, to be taken seriously, one must compellingly act out a predefined role, and in playing that role, one gives expression to the underlying meanings that hold the community together.

There are several roles to play in the broadcast policy world, different positions to take up in order to participate. Although the roles are highly varied, the way that they operate can be illustrated by discussing four representative and archetypal roles in the policy community: commissioner, lawyer, lobbyist, and policy analyst. The same people often play different roles at different times, and people with different views and goals can play the same role. But entry into the broadcast policy world requires at least some knowledge of these different roles and the rules of behavior. And those rules of behavior, in turn, reveal and help reproduce the corporate liberal belief system underlying commercial broadcasting.

The general meaning of these roles was once suggested by a leading economist in the field of telecommunications policy. The role of policy research, he wrote, "is to tell decisionmakers how to maximize output with given resources, or how to realize given objectives at least cost, or at least to quantify the costs and benefits of decisions made on arational

27. John M. Kittross, "A Fair and Equitable Service; or, A Modest Proposal to Restructure American Television to Have All the Advantages of Cable and UHF without Using Either," *Federal Communications Bar Journal* 29, no. 1 (1976): 91–116.

grounds."[28] This was said in exasperation; it is revealing that the author went on to describe how far in his estimation the actual policy process had declined from this ideal. Yet it is just as revealing that he thought this to be a worthwhile ideal in the first place. Policy research, in this definition, views itself as working strictly within a framework established by "given resources," "given objectives," and established authorities or "decisionmakers." Decisions, objectives, and the authority of decision makers, in other words, are all "given"; they are not up for discussion. What policy research is about, then, is something objective and uninvolved, which is why it conceives of problems in terms of a linear view of causality divided into distinct means and ends, and relies heavily on the language and imagery of mathematics, science, and technology ("maximize," "quantify," "output").

Not everyone in the world of broadcast policy speaks exactly this way. But the quote does illustrate a pattern that is characteristic of much of what is said in the policy world: matters subject to decision, to choice, are presented as problems of a neutral and technical nature, whereas subjective matters of value, of high politics, are treated as outside the reach of the speaker. The subjective is thus treated as a given by those inside the policy world; it is decided elsewhere. And policy issues, those matters properly subject to choice within the policy world, though readily acknowledged as tricky, complex, and controversial, are nonetheless presented as basically objective in their nature.

Implicit in the logic here is a belief that decision making is categorizable along a continuum ranging from subjective to objective. At the subjective end are those who make basic value judgments, that is, those who engage in politics. Their archetype is the elected official: the House member who introduces legislation, the influential member of the FCC's oversight committee in the Senate. These people, properly speaking, are not part of the policy world; they provide decisions ("inputs" in the argot) to the policy world from the outside.

Commissioners and Judges

The policy world proper begins somewhere in the middle of the spectrum with the appointed members of the FCC, appointees to similar bodies like state utility commissions, and judges whose work touches upon broadcast policy (including the "administrative judges" or hearings examiners that make most of the routine decisions at the FCC). The job

28. Bruce M. Owen, "A View from the President's Office of Telecommunications Policy," in *The Role of Analysis in Regulatory Decisionmaking: The Case of Cable Television,* ed. Rolla Edward Park (Lexington, MA: Rand Corporation and Lexington Books, 1973), 3.

of commissioners is legally constituted as independent yet subordinate to the general guidelines given them by the political process and by the boundaries set by the courts. Commissioners are thus expected to use neutral, rational principles to flesh out broad mandates given to them from elsewhere, specifically Congress; this expectation puts them properly within the policy world, and makes it necessary for them to operate according to the discursive rules of expertise. They are generally thought of as somewhere in the middle of the subjective-objective spectrum because as political appointees it is obvious to all that they have political views. Yet they are restrained from presenting their actions as political, both by law and, more importantly, by the belief system of expertise.

When a commissioner does enter the scene with a clear agenda, it can be justified only if it is couched in terms of one or another legal, technological, or economic theory or principle. A commissioner does not act on behalf of a particular group of underprivileged people; she acts according to the "true," and thus politically neutral, legal interpretation of the public-interest principle. Similarly, a commissioner does not act on behalf of NBC, ABC, and CBS; he acts according to, say, the neutral principle of economic efficiency or free speech rights.

Commissioners with clear agendas are often prominent, but are probably a minority.[29] A more typical characterization of a commissioner, both among scholars and among commissioners themselves, is that of the neutral arbiter of political struggles among interest groups. Commissioners, the story goes, are buffeted from all sides by political pressures from lobbyists, politicians, and the like. They struggle mightily to maintain neutrality, and to somehow balance all these groups against one another, with varying degrees of success. Commission members often speak with exasperation of how hard it is to maintain neutrality in the face of all this pressure, and critics of the policy process often cluck their tongues sadly at the frequency with which commissioners seem to succumb to the force of the lobbyists. Yet, in spite of the nearly universal agreement about the impossibility of maintaining true neutrality, the discursive rules of the policy world prevent its participants from taking the logical next step: saying out loud that commissioners are politicians.

Hired Guns: Lawyers and Lobbyists

The role of communications lawyers and lobbyists is of course central to the policy process. To an outsider, they might seem to be the opposite of the commissioner or judge: they are paid by what are called "stakeholders" or "interests"—typically corporations involved in broadcasting

29. Krasnow, Longley, and Terry, *Politics of Broadcast Regulation,* 42–48.

—to cajole out of policy-making institutions decisions that favor their clients. And their actions are described, especially in the business press, in terms that tend to emphasize their role as embodiers of the profit desires of various business interests: they are soldiers in epic battles between industry factions.

But the subjective, "interested" character of these hired guns of the policy world is much like the neutral, objective character of commissioners and judges: it is more a dramatic role than a simple fact. Lobbyists, after all, do not simply transmit their clients' desires to policy institutions; if a television network simply wanted to express opposition to a regulation, they could send the FCC a memo. Rather, lobbyists do the work of translating those desires into the language and practices of the policy world. A lobbyist for broadcasters will tell the FCC that eliminating restrictions on syndication will serve the public interest by increasing competition, not that it will make more money for her clients. A lawyer representing cable operators will say to a judge that restrictions on cable operators limit, not their profits, but the operators' constitutional rights of free speech. The hired guns of the policy world thus first and foremost translate corporate goals into appropriately neutral and expert policy language. They do not represent corporate interests as much as they transform them; they *re-present* interests.

Given their function as translators of corporate designs into the language of corporate liberal expertise, it is thus not surprising that within the community of lawyers and lobbyists, a coolly professional attitude toward issues is valued. If the religious evangelist hides his or her polished rhetorical calculations behind a stylized emotional spontaneity, the lobbyist does the reverse: whatever passionate commitment exists is best hidden behind a coolly professional demeanor. Like others in the policy world, the hired guns are first and foremost "experts." A broadcast industry lobbyist, for example, might deliver a virulent attack on the cable industry in the hearings room, and immediately afterward compliment his cable industry counterpart for her performance—perhaps over drinks, as if they had just finished a friendly tennis match. One can sometimes see this sense of mutual professional respect revealed in the knowing smiles opposing lobbyists will direct toward each other while delivering their attacks during a conference panel or public debate.

Lawyers and lobbyists, in sum, are as much committed to the community of the policy world, and to the discursive rules that constitute it, as they are to their clients. They are all part of the same professional community, follow similar career paths, and frequently come from similar or even identical career backgrounds. It is well known that a stint at the FCC is frequently a stepping stone on the way to a more lucrative position with a communications law firm or lobbying organization. But this famous "revolving door" between private and public policy institutions is not best seen as evidence that vaunted neutral principles are being corrupted by some kind of cronyism. Rather, the behavior of lobbyists and lawyers dramatize a sustained commitment, not just to the interests of corporations, but to the hypothesis that professionals can reconcile corporate power and profit seeking with "neutral" principles like free speech and the public interest.

Policy Analysts

At the objective end of the scale in the world of policy are the academics and staff members of government agencies and subcommittees who conduct policy research or policy analysis. Like the other positions, the analyst is more a role than a particular person or group of people; it is good form for commissioners, lawyers, and lobbyists to engage in policy analysis from time to time.

The analyst does not vote on policy decisions, but does produce a steady stream of official reports, books, and journal articles that are used to provide support for decisions and on occasion launch new policymaking trends. Often enough, it is possible to identify the political position of policy analysts: a study that analyzes the efficiency of various means of establishing a marketplace in broadcast frequencies is likely to come from an economic conservative, whereas one that finds constitutional justifications for policies that favor minority ownership is likely to come from somewhere more to the left of the spectrum. Yet there are plenty of cases where policy analysts produce research whose political implications are unclear, where a scholar will pursue a theory or some evidence for its own sake, and come to conclusions that seem to conflict with the scholar's own camp.

The art of policy analysis principally involves finding a way to come across as both practical and objective at the same time. One's research must be able to be plausibly understood as expert advice proffered to those in positions of decision-making power, yet it must somehow project an image of expert objectivity and neutrality. If one's work is too obviously designed to advance the cause of a particular official's political agenda, one may temporarily gain a cozy appointment but in the long term will be written out of the profession as a political shill. Conversely, if one's work consistently leads to politically impractical or obscure conclusions, the conference invitations and grants begin to disappear, and one finds oneself marginalized and ignored.

There are several tricks helpful to the policy analyst's balancing act. One is knowing which policy issues at a given moment are up for debate, and which are not. This is not easy. Not only do issues shift from time to time, but the demands of both political and policy rhetoric work to hide the reality of shifting issues from outsiders: journal articles and official reports tend to present themselves as concerned with time-honored universal principles and objective needs of society and the legal system, not with policy fashion. Policy fashions are shaped behind the scenes or in quasi-public settings: in political cloakrooms, think tanks, funding centers, academic appointments to journal and conference boards, and the intricate community of professionals inside the Washington beltway. Keeping in touch with the shifting winds of the "practical" is thus a key to the policy analyst's art.

Of course, knowing which issues are considered "practical" at any given moment does not necessarily mean taking a stand on those issues. The need to present oneself as objective means that it's often helpful to produce work that, rhetorically anyway, is neutral on controversial issues. The point is to produce research that could plausibly provide evidence or argument for those who do take a stand. This not only increases the likelihood that one's research will be noticed by those in positions of power, but more importantly, transmits the image of appropriate practicality that defines one as a policy analyst in the first place.

Method as the Emblem of Expertise

An equally important trick in policy analysis is the use of a method that safely positions one's work as neutral and objective. A method is one of the principal defining characteristics of the expert. Methods are thus central to the maintenance of (and thus participation in) the interpretive community of broadcast policy. Ordinary people have mere opinions, but experts have knowledge and reason, which entitles them to more privileges, more participation in policy decision making, than ordinary people: this is the message that must be conveyed.

The archetype of expert knowledge in the modern era is physical science, which, it is generally assumed, produces irrefutable truths. Those operating in the policy world, of course, have generally not yet had the good fortune to produce irrefutable truths; for each policy theory, argument, or analysis, there exists an alternative that has at least some adherents. So, lacking the irrefutable truths of hard science, one turns to that which the hard sciences appear to have used to gain those truths: a specialized method. As Lasswell put it, "the closer the social scientist [comes] to the methods of physical science the more certain his methods could be of acceptance."[30] The way our culture signifies that

30. Lasswell, "Policy Orientation," 5.

experts possess knowledge and reason is that experts couch what they say in the trappings of one or another method.

It is more important for the policy analyst to have a Ph.D. or academic appointment than it is for others in the policy world, and more important that the academic trappings—jargon, footnotes, references to the literature, and so forth—are prominently featured in one's discourse. Some methods seem to have been respected throughout the history of the broadcast policy world: the "method" of legal reason, maintained by the legal profession with the support of legal academics and law schools, is a constant; and there will always be some respect for the electrical engineer, and a concomitant respect for the trappings of science and engineering: charts, graphs, data, mathematics are always helpful in this regard. But there are fashions in methods just as there are fashions in policy issues. (The two, as we shall see below, are in complex ways closely related.) In the 1960s and 1970s, for example, the methods of social science gained a certain amount of respectability, whereas in the 1980s they were almost thoroughly eclipsed by neoclassical economics.

The complex tension between practicality and objectivity is illustrated by the fact that, the more popular a method, the less necessary it is for one's research to be of immediate practical value. In the 1970s the popularity of the behavioral sciences became the occasion for large amounts of research on the effect of televised violence on children, even though the research was associated with very little concrete policy activity.[31] In the 1980s, in contrast, when neoclassical economics became a dominant mode in policy analysis, sociological or political economic methods appeared in government studies and conference panels only when they were closely tied to very specific policy issues, such as minority ownership policies, or the development of new telecommunications networks, and even then had difficulty being taken seriously.[32] Conversely, the use of economic method was often enough to qualify one's research as policy research, even if the specifics of the analysis were politically impractical.[33]

31. Willard D. Rowland, Jr., *The Politics of TV Violence: Policy Uses of Communication Research* (Beverly Hills: Sage, 1983).

32. *Metro Broadcasting v. Federal Communications Commission* (1990) upheld minority preferences policies in broadcast licenses as constitutional and used survey data on minority ownership of broadcast stations to make its point (see especially 3017 nn. 31–33, citing survey data and sociological analyses).

33. For example, Tim Brennan, "Discrimination in Theory, by Vertically Integrated Regulated Firms," paper presented to the Twentieth Annual Telecommunications Policy Research Conference, Solomons, MD, September 12–14, 1992.

Law and Legal Reason

The policy process generally operates as subordinate to law, formally recognized by that blurry zone of contemporary legal practice called administrative law. Law is thus the outer framework of broadcast policy, and the legal system's enigmatic modus of "legal reason" remains a central model of method.

Much of what counts as legal reason is expressed in the rhetorical practices found in any law review: argument from precedent (betokened by copious footnoting of legal cases), argument by legislative intent (quotes of legislative hearings), and axiomatic deduction from abstract legal principle (besides the conventional Latinisms, eloquent quotes from federal judges, particularly past Supreme Court justices, are helpful here).

Yet the peculiarly twentieth-century idea of administrative law, of a special legal realm dominated by contingent problems to be solved by experts according to the complexities of the moment, separates policy from more traditional areas of law. The role of legal reason is thus not as prominent on the surface as it is in, say, constitutional law. Rather, the more typical pattern of argument begins with references to hallowed constitutional principles, but then progresses toward "technical" questions such as matters of spectrum scarcity, minority ownership data, or chilling effects. The initial language and terminology of classical legal reason is gradually supplanted by the trappings of policy science such as allocations tables and survey data.

Social Science, Social Engineering

The power, prestige, and character of what may be broadly construed as sociological expertise and argument in broadcast policy has gone through numerous permutations and has waxed and waned over the years. Though often having very little relationship to the methods taught in contemporary sociology programs, a loosely sociological logic nonetheless remains a constant presence in the policy arena.

There always have been some who argue that the confusions of broadcast policy can be overcome by making the FCC more independent, more expert, more scientific, and generally more rational—that is, more what was envisioned by nineteenth-century social engineers like Charles Adams. In the 1940s, for example, some New Deal FCC staffers released what came to be called the Blue Book, a proposal for regulatory revision accompanied by elaborate and compelling surveys of broadcast content and economic analyses. The Blue Book argued that the commercialism that resulted from an exclusive reliance on marketplace forces to determine broadcast content clearly fell short of fulfilling the "public interest."[34] This inadequacy of the marketplace, therefore, ought to be remedied by a modest effort of social engineering: specifying detailed requirements for informational and public-service programs without advertising support ("sustained" programming), according to percentages fixed by economic and social scientific analysis. As the political climate shifted right in the post–World War II era, the Blue Book's proposals were ignored, but the hopes attached to social scientific method continued to grow.

The heyday of positivist social science in Washington came in the 1960s. In 1960 the social scientist Paul Lazarsfeld made the extraordinary claim to the FCC that, given enough time and "a corps of trained minds, it would be possible to set up workable standards of excellence in television."[35] Social science could finally objectify that most elusive subjective entity, cultural taste. A more elaborate and influential effort was the Rostow Report of the late 1960s, which concerned, among other things, cable television. The report called for the creation of a centralized government agency charged with planning and coordinating the U.S. telecommunications system, on the theory that something was needed to lift broadcast policy making up above the petty, feudal squabbles that bogged down the FCC. The report called for "a well-conceived public policy," which involved such measures as

> operational experiments . . . to explore the feasibility and flexibility of full-scale systems, [programs designed] to provide useful technical, operational, and economic data as a basis for more permanent policy decisions, [and c]areful preliminary training and testing [to help] reach firm conclusions about the possible contribution of full-scale applications of telecommunications technology to major development problems.[36]

The benefits expected to flow from these "operational experiments," "full-scale systems," and "data" were not modest. For example, the report suggested that appropriately engineered social policies for cable television might help solve problems of crime and unrest in the inner cities.[37] Although in retrospect such a claim seems almost poignant in

34. FCC, "Public Service Responsibility of Broadcast Licensees," March 7, 1946, reprinted in Kahn, *Documents of American Broadcasting,* 148–64.

35. *Current Biography Yearbook* (New York: H. W. Wilson Co. 1964), 252.

36. President's Task Force on Communications Policy, *Final Report* (Washington, DC: Government Printing Office, 1968), 12–15.

37. Ibid., 16. The report suggested that its social-engineering approach might point to "the constructive possibilities for the use of television to help overcome some of the

the extremity of its naïveté, at the time the report was read with great seriousness.

Each of these efforts expresses a desire to resolve policy ambiguities by turning to social engineering in order to make the regulatory system more autonomous and better insulated from the "subjective" (and thus "irrational") winds of politics. Each of these proposals expresses a version of the hope that the contradictions of policy can be overcome by more rigorously restraining the chaos of private desire with the constraint of a rationally conceived public interest—by moving, in other words, more decisively in the technocratic direction established but only sheepishly pursued by the Hoover-era framework of the Communications Act.

Social scientific discourse, however, is most effective when it is "practical" in the sense peculiar to the policy arena, that is, when it can be used to address dilemmas of liberalism. Hence, sociological method was put into service of regulatory practice in the 1960s, not in any full-scale implementation of the grand plans of a Lazarsfeld or Rostow, but in a procedure called "ascertainment." The FCC in the early 1960s was struggling with its mandate to, on the one hand, make sure broadcasters "serve the public interest" and, on the other, uphold the principle of free speech. Its way out of this dilemma was to require of broadcast licensees that they conduct survey research to "ascertain" their community's needs and interests.[38] Social scientific survey methods, the reasoning went, would thus ensure that broadcasters serve the public without the FCC having to act as a censor. Sociological technique would let the FCC off the hook and square the circle of rights and regulation.

Economics

Liberalism as a whole is permeated with economic ideas: ideas about the social value of private property, about the invisible hand of the market, about the compatibility or even equivalence of capitalism with social and material progress. But in this century the profession of economics has achieved for itself the status of the "hardest" of the soft sciences and thus, in its own mind, the closest thing to truly scientific expertise in the sociopolitical arena. As the grander plans of social science ran aground

---

problems of urban ghetto dwellers. Isolated rural people such as the inhabitants of Indian reservations could benefit from similar undertakings."

38. FCC, "Report and Statement of Policy re: Commission en Banc Programming Inquiry (the 1960 Programming Policy Statement)," 25 Fed. Reg. 7291, 44 F.C.C. 2303 (1960).

on the failures of the Great Society in the late 1960s and early 1970s, economic expertise rose to take its place.

The rise to prominence of neoclassical economics in Washington policy circles is a major phenomenon of the last two decades and has played a major role in the deregulatory movement. On one level, neoclassical economic theory has served largely to signify a reassertion of long-standing principles of American politics: the faith in laissez-faire economics and the free marketplace. But within the policy arena, economics has come to be the predominant discourse of expertise. Within policy circles, the heroes of deregulation such as Alfred Kahn and Richard Posner embody a new form of fascination with expertise and social science, this time centering on the concept of "efficiency" and the tools of "post-Coasian" law and Chicago school economics. Economists now get government grants and FCC staff positions, and noneconomists in the system often find it fashionable to couch their arguments in economic language.

Economist Donald McCloskey has suggested that his field is better understood as rhetoric than as a science.[39] It certainly is the case that economic arguments have been mustered in favor of many divergent positions over the last decade. A central tension in contemporary economic discourse seems to revolve around an ambiguity concerning the relation of markets to government action. On the one hand, the traditional economic assumption is that markets are more efficient regulators than government intervention. On the other, economists are frequently attracted to matters of broadcast policy precisely because the high level of government involvement in the industry provides opportunities to administratively enact economic principles, that is, to use government to tinker with industry affairs in order to enhance the efficiency of market relations. Markets, after all, don't ask economists for advice, but government agencies do. So, for example, economic arguments about the inefficiency of regulation figured prominently in the deregulation of cable television, but once that deregulation was accomplished, economic arguments about cable's monopoly profits under deregulation were used to support the current reregulation of cable.[40]

39. Donald N. McCloskey, *The Rhetoric of Economics* (Madison: University of Wisconsin Press, 1985).

40. The effect of deregulation on cable prices is discussed in A. B. Jaffe and D. M. Kanter, "Market Power of Local Cable Television Franchises: Evidence from the Effects of Deregulation," *Rand Journal of Economics* 21 (summer 1990): 226; and J. W. Mayo and Y. Otsuka, "Demand, Pricing, and Regulation: Evidence from the Cable TV Industry," *Rand Journal of Economics* 22 (fall 1991): 396.

Meritocracy, Insider Wisdom, and the Institutional Maintenance of Unquestioned Assumptions

Corporate liberal broadcast policy in many senses has been extraordinarily successful. While the FCC remains modest-sized as government agencies go, it is encircled by a thriving network of meritocratic organizations that in their very structure share and maintain the belief that expertise can solve problems of broadcast policy: law firms, conferences, consulting firms, institutes, and university research programs. Within this community, the ritual of expert decision making is maintained. Legislators and lobbyists present commissioners with problems to be solved, analysts undertake analyses, and along the way grants are funded, articles are published, hearings are held, and the business of policy goes on. Ritually speaking, everything is in its place.

The meritocratic premises by which this community operates fulfill two important functions. On the one hand, the legitimacy of the community's expertise is maintained by the entry criteria of higher degrees, accepted methodologies, and other traits typically esteemed on academic curriculum vitae. On the other hand, meritocracy provides a mechanism for policing entry into the community by insiders, ensuring that discussion stays within certain bounds: invitations to a conference, research funding, and staff positions are forthcoming only to those judged by community insiders as being both sufficiently expert and appropriately "practical." A career, a point of view, a research proposal that, say, questions the fundamentals of the system or uses terms and ideas in radically unconventional ways is judged not practical because it cannot be easily construed as relevant to policy problems, that is, as helping to reconcile dilemmas faced by politicians and regulators. It is thus quietly passed over or marginalized, and the unsuitable ideas are filtered out.

To varying degrees, there is some recognition within the policy community that "practicality" involves a simple deference to power. Sometimes this is relatively overt. The policy arena is after all bounded by the coercive power of law and of the state. If the Supreme Court declares the principles one believes in unconstitutional, what recourse does one have except to twist one's arguments to fit the requirements of the courts? Many proponents of media access as a legal principle, for example, probably agree with Jerome Barron that the fact of economic concentration should create a First Amendment right of access on the part of the public to *all* concentrated media.[41] But because the Supreme

41. Jerome A. Barron, "Access to the Press: A New First Amendment Right," *Harvard*

Court declared this general argument invalid, access advocates have fallen back on the tenuous but still acceptable argument that at least broadcast media are susceptible to access arguments on technological, if not economic, grounds. Broadcasting, the Court has held, can suffer from the technological condition of spectrum scarcity, and is thus susceptible to access arguments where print media are not.[42] Over the near term, at least, one is forced to either argue from within this framework or abdicate one's right to participation in the legal process altogether. Such is the coercive power of law.

Often, however, the deference to power is less overt, and extends to matters that are *not* explicitly stated in law. While rarely said in public forums, some policy experts will privately point out that if one is sincere about having influence with one's ideas, one must take into account the fact that for ideas to have influence they must be attended to by people in positions of power. There is a conservative version of this acknowledgment, which can afford to be relatively frank: it is based on the assumption that existing power relations are inherently legitimate, that some are destined to rule, others to follow. Yet, after a few years in Washington or state government, even former sixties activists will often express a desire "to do something successful for once," to no longer feel like one is howling in the wilderness; if compromise with the powers that be is necessary to get anything done, then compromise is the wise choice.

This limitation of debate, however, is not simply the result of a coercive power, such as the power of capital or of legal force. Rather, to a large degree, it can be seen as an unconscious product of the assumptions shared by the interpretive community of broadcast policy, as a product of the power of unquestioned beliefs. Certainly most if not all of the participants in the process are sincere in their actions, and do not go out of their way to exclude alternative points of view. They believe in what they are doing, and are not above occasionally inviting, say, a Marxist to their conferences.

But any community of people develops shared understandings, and the policy community is no exception. Those shared understandings become the insider wisdom, the insider's sense of what's practical, interesting, and original and what's foolish, trivial, and outdated. Policy issues, any insider will tell you, are complex; understanding them

---

Law Review 80 (June 1967): 1641–78; and Jerome A. Barron, *Freedom of the Press for Whom? The Right of Access to Mass Media* (Bloomington: Indiana University Press, 1973).

42. *Miami Herald Publishing Co. v. Tornillo,* 418 U.S. 241 (1974).

involves experience and judgment. For the insider, particularly after years of experience, certain things come to seem obvious. Insiders "know" that a given argument is either interesting or old hat, that the correct interpretation of a principle is this and not that. After all, everyone else inside the community agrees, and they are the experts. These shared understandings, then, are experienced by insiders as simply the wisdom of experience.

Of course, over time, the insider wisdom shifts dramatically. In the 1930s (as we will see in the next chapter) the law's insistence that broadcast licenses do not confer ownership was understood by many to mean that the exchange of money for a license as part of the sale of a station was illegal or at least problematic. Since the late 1940s, however, no one even thinks to question the practice; the inclusion of license value in station sales has been understood to be thoroughly normal, as if the practice were intended all along. Similarly, in the late 1930s the belief that broadcast networking constituted a natural monopoly was taken quite seriously by both regulators and industry executives and used as an argument against regulation, whereas today, economically similar levels of industry concentration are generally interpreted as "highly competitive," and natural monopoly arguments are considered ridiculous by all sides in debates.[43] In the mid-1980s, insiders "knew" that arguments couched in the language of free markets supported by quantitative economic data would fly while predictions that properly managed broadcast policy could solve problems of inner-city unrest would be scoffed at, just as in the late 1960s they "knew" something of the reverse to be true.

Yet for the most part, these distant historical differences are either ignored or attributed to early regulatory error or naïveté. At any given moment, the ideas that are taken for granted by insiders seem like the obvious, reasonable ones. Hence, matters that to others might look like irreconcilable contradictions or fundamental ambiguities are given stable meanings, are made to make sense.

The assumptions of the policy world are rendered all the more invisible by the fact that there are always matters that *are* contested, that are

43. In 1941 the fact that 61 percent of all radio stations were network affiliates was taken by the FCC to be a sign of outrageous monopoly, to which the networks replied, not that there was no monopoly, but simply that the monopoly was a natural and fair one. Since the early 1950s television station network affiliation has yet to drop that low, yet the situation has never been viewed as so monopolistic as it was generally agreed to be in the 1940s (Thomas Streeter, "Policy Discourse and Broadcast Practice: The FCC, the U.S. Broadcast Networks, and the Discourse of the Marketplace," *Media, Culture, and Society* 5 [July-October 1983]: 247-62).

debated and struggled over. If the Right continues to struggle for more business autonomy, the Left of the policy community generally couches its efforts in terms of "access." Access arguments can take various forms: the fairness doctrine is a famous example, but they also turn up in arguments for common-carrier-style regulation of the electronic media (an argument made all the more compelling by the coming of cable). During the Reagan era, the phenomenal success of the Right almost pushed common-carrier arguments off the map: one major policy figure announced that the idea that cable be treated as a common carrier "is about as likely to get a second hearing as the Articles of Confederation."[44] Yet the need to dismiss the common-carrier solution belies the claim of its demise, and as neoclassical economics has lost much of its sheen, the left edge of the policy community has made an effort to reintroduce the concept.

That these debates are bounded is seen when one searches for policy efforts that question the principle of for-profit organization in broadcasting. For the most part, they are nowhere to be found; one can successfully challenge the power of this or that industry segment in certain spheres, but not to the extent that one challenges or limits the profit imperative. Deference to some form of corporate hegemony in broadcasting, though rarely stated explicitly, is nonetheless a tacit assumption of any policy effort that is to be taken seriously.

This deference is evident in all successful policy initiatives that have come from those typically thought to be opposed to business interests. In 1941, for example, when an aggressive New Deal–era FCC staffed by Roosevelt appointees successfully broke NBC's near monopoly of network radio, its actions had to be justified with the declaration that the free market was "the essence of the American system of broadcasting."[45] In what was probably the most famous verbal assault on the commercial broadcast establishment, Newton Minow's "vast wasteland" speech, the Kennedy-era FCC chair felt compelled to add to his stinging attack on the low quality of broadcast television, "I believe in the free enterprise system. I believe that most of television's problems stem from lack of competition."[46] As this is being written, moderately left-wing policy activists, encouraged by the Clinton victory, are launching new access-based initiatives. A recent conference entitled "Breaking the Barriers to

44. Daniel L. Brenner, "Cable Television and the Freedom of Expression," *Duke Law Journal* (April–June 1988): 329.

45. FCC, *Report on Chain Broadcasting,* Docket 5060 (1941), 95.

46. Newton Minow, address to the National Association of Broadcasters (the "vast wasteland" speech), Washington, DC, May 9, 1961, reprinted in Kahn, *Documents of American Broadcasting,* 207–17.

Universal Telecommunications Access" speaks in terms of achieving "balance" between regulation that will "keep the quality of existing services high and prices low" and providing "incentives for the development of new services," that is, between equity-based regulation and the principle of profit.[47]

The strength of the corporate liberal way of thought is such that challenges to it are not even acknowledged as such; they are invisible. This is illustrated by the community's reaction to those rare cases that do stray in the direction of challenging the profit imperative in small ways: these efforts, while acknowledged to be controversial, are not even recognized as the small heresies that they are. The academic Marxist is a classic example. He or she may gain a seat on a panel at a policy conference, but is often treated by the other panelists as a naive, dewy-eyed idealist interested in helping the downtrodden but lacking a sense of the "hard realities" of the modern world; that the Marxist's paper is an almost overly grim analysis of exactly those "hard realities" is invisible from within the policy community's interpretive framework.

The systematic character of this obliviousness can be seen in the invisibility of the few historical cases when events within the policy community proper have strayed in heretical directions. When the Blue Book was published in the 1940s, for example, it couched its arguments in technocratic terms, but it also committed a heresy: it announced that the right to unlimited profits and the public-interest standard were in conflict. The Blue Book argued for relatively aggressive regulation of broadcast content in the form of, for example, requirements for sustained programming. The authors granted that broadcasters were entitled to a profit, but were frank that the regulations they suggested might reduce the levels of profit then current in the industry.[48]

The Blue Book was thoroughly rejected. It met with vociferous opposition from the industry, none of its proposed rules were adopted, and the station singled out as a bad example in the report's analysis won its license renewal shortly thereafter.[49] It fell outside the bounds of the corporate liberal parameters of broadcast policy.

Yet this is not how the Blue Book is described in much of the policy literature. As one textbook puts it, the Blue Book's conclusions were "neither regulations nor proposals for new rules but rather . . . codifica-

47. The quotes are from a promotional pamphlet for a conference presented by the Alliance for Public Technology on February 25–27, 1993, titled "Technologies of Freedom: Breaking the Barriers to Universal Telecommunications Access: A Conference on Achieving Telecommunications Equity in the 21st Century."

48. FCC, "Public Service Responsibility of Broadcast Licensees."

49. *FCC Annual Report,* 15 F.C.C. 1149 (1951).

tion of FCC thinking to help licensees and regulators alike. . . . the report had some solid results over time. . . . the FCC showed that it had the backbone for once to speak out if not act in a controversial area, and the 'Blue Book' still provides the commission with a useful precedent and the industry with a rallying point."[50] Perhaps this Pollyannaish characterization of the Blue Book episode is a simple matter of historical interpretation. Yet it is just as plausibly seen as a projection of the expectation that policy experts are reasonable, and reasonable policy experts cannot disagree over fundamentals, only over details of implementation. The history of policy by its own definition is a steady accumulation of rational, expert solutions to problems. Underlying givens of the system, such as for-profit principles and the public interest, do not conflict. Broadcast law is not beset by insoluble contradictions or fundamentally political discontinuities. Policy problems can be solved—hence, evidence of deeper fissures in the system of broadcast policy gets ignored.

The fundamental assumptions of the policy community, in sum, are like water to fish: so much a part of the environment as to be invisible. They are simply common sense, beyond questioning. It is more by the maintenance of unquestioned corporate liberal assumptions than by deliberate exclusion that the discussion of fundamentals is kept off the agenda of broadcast policy.

Angst at the Edges: Theorizing Regulatory Disappointment

For all its success as an institution, broadcast policy has been the subject of an enormous amount of academic criticism. Since the beginning in 1927, broadcast policy has been the object of a steady stream of complaints, a bibliography of which would be a book in itself. Books on broadcast regulation that describe the situation as a "crisis" or describe the agency as a "reluctant regulator" are met with very little refuta-

50. Christopher H. Sterling and John M. Kittross, *Stay Tuned: A Concise History of American Broadcasting,* 2d ed. (Belmont, CA: Wadsworth, 1990), 304-5. The idea that the Blue Book "provides the commission with a useful precedent" seems to have originated in two articles by Richard J. Meyer: "The 'Blue Book,'" *Journal of Broadcasting* 6 (summer 1962): 197-207; and "Reaction to the 'Blue Book,'" *Journal of Broadcasting* 6 (fall 1962): 295-312. The suggestion that the Blue Book involved "neither regulations nor proposals for new rules" seems to be based on a relatively trivial distinction between rules that are automatically required of all licensees (which the Blue Book did not suggest) and rules that are applied in deciding license renewals (which the Blue Book did suggest, and which over time would influence all licensees anyway).

tion.[51] It is now almost axiomatic that the nominally independent FCC is anything but independent, that the phrase "public interest" is extraordinarily vague, that existing legislation was already "obsolete when passed" and is even more obsolete today, and that FCC policy is systematically biased toward the industry it is supposed to regulate. It is practically impossible to find an article in the literature that does not criticize one or another aspect of the system. So common is a negative evaluation of FCC performance that one illuminating survey of the literature classifies the bulk of the writing about the "public-interest" principle, not as public-interest theory, but as *perverted* public-interest theory; most observers seem to believe that regulation, at least in the case they are discussing, has failed to live up to the public-interest ideal.[52]

A 1960 report by James Landis, a major figure in the history of U.S. regulation, is as good an example as any of the general tone of the commentary: "The Federal Communications Commission presents a somewhat extraordinary spectacle. Despite considerable technical skill on the part of its staff, the Commission has drifted, vacillated, and stalled in almost every major area. It seems incapable of policy planning, of disposing within a reasonable period of time the business before it, of fashioning procedures that are effective to deal with its problems."[53] This opprobrium is very pronounced in broadcast regulation, but can be heard in other areas of regulation as well. As one writer put it, "the prevailing burden of judgment holds overwhelmingly that regulation in America has been a failure."[54]

Why the pervasive negative tone? After all, broadcasting under American regulation has become a massively successful and powerful institution. There is no comparison between broadcast regulation and, say, the crises in the welfare system or efforts at international develop-

51. Don R. Le Duc, *Cable Television and the FCC: A Crisis in Media Control* (Philadelphia: Temple University Press, 1973); Barry G. Cole and Mal Oettinger, *Reluctant Regulators: The FCC and the Broadcast Audience* (Reading, MA: Addison-Wesley, 1978).

52. Robert Britt Horwitz, *The Irony of Regulatory Reform: The Deregulation of American Telecommunications* (New York: Oxford University Press, 1989), 27.

53. James Landis, *Report on Regulatory Agencies to the President-Elect,* printed for use of the Senate Committee on the Judiciary, 86th Cong., 2d sess., 1960 (Washington, DC: Government Printing Office). For a brief survey of some major studies of the FCC over the years illustrating the typicality of Landis's evaluation, see Le Duc, *Cable Television and the FCC,* 29–30. Le Duc quotes from reports with conclusions similar to Landis's released in 1941, 1951, 1958, 1962, 1964, and 1969.

54. Thomas K. McCraw, *Prophets of Regulation: Charles Francis Adams, Louis D. Brandeis, James M. Landis, Alfred E. Kahn* (Cambridge: Harvard University Press, 1984), viii.

ment, where large-scale human suffering is at stake. And it is not the case that critics of regulation are motivated by disdain for the cultural product of broadcasting, at least not overtly.

One source of the negative tone may be the expectations associated with the role of the policy analyst. Authors of scholarly work on broadcast policy for the most part are, or are aspiring to be, policy experts. They want to be heard by the system. And the job of experts, particularly analysts, is to bring order to chaos. There's little point in publishing an article that says, "Everything's fine, leave it as it is." To fill the role to which they aspire, authors thus need some problem to which they can apply their legal, sociological, or economic expertise. They need some chaos to bring order to. So it is to a degree a matter of rhetorical form that the typical policy study locates a flaw or confusion somewhere in the regulatory process for which the author can then propose a solution.

Many of the most successful campaigns to change federal policy in broadcasting have in fact come from "Young Turk" policy activists who claim to offer a cure, an idea or approach that can transcend the irrational quality of business as usual in Washington. In the 1980s the best known of these were generally the deregulatory economists, offering transaction cost analysis, new technologies, and market forces as the way out of what they often astutely analyzed as the irrationalities of existing policies. But many have preceded them. In the late 1930s, for example, left-wing Roosevelt-era trustbusters were the Young Turks, and in the late 1960s and early 1970s, "wired nation" fans of cable television's utopian promises filled that role. In each case, regulators with ideas and principles marched into the policy arena in hopes of overcoming the resistance of entrenched power bases with the force of better ideas.

Each of these groups had considerable impact on the course of events, arguably more impact than can be attributed to any industry member's self-interest. But the fate of the motivating ideas that each wave of regulatory reformers brought with them is less clear. The Roosevelt-era trustbusters successfully overcame vociferous industry opposition and, among other things, forced NBC to divest itself of its second radio network, which became ABC. Yet the ideas that motivated their actions—a populist desire to overthrow the entrenched monopoly powers of the corporations in the name of an open society and marketplace competition—were not realized: network dominance of both the industry and the airwaves only increased in the ensuing years.[55]

Three decades later, "wired nation" advocates of cable television successfully spearheaded the drive to remove FCC constraints on cable

55. Streeter, "Policy Discourse and Broadcast Practice," 247–62.

expansion against the objections of the networks and over-the-air broadcasters. The visions that informed their efforts, however—a utopian hope that cable's limitless channels would bring interactive democracy and openness to the centralized, one-way world of television—have met with lukewarm success at best.[56] In the 1980s, proponents of marketplace solutions to regulatory problems undoubtedly transformed the character of regulation and industry relations in numerous ways, from broadcast station prices to program content. Whether or not the resulting changes amount to dramatically more efficient market relations or merely a rearrangement of relations within generally oligopoly conditions, however, is arguable. Many of the initial deregulatory efforts are now under attack for having had unintended effects: the deregulation of cable that culminated in the Cable Act of 1984, for example, is now accused of having created local monopolies instead of markets. And as we will see, most of the intellectually driven efforts to institute "pure" marketplace relations such as spectrum auctions have been used at most experimentally, and as of this writing are falling out of favor.

Although professional self-concept and ambition provides some impetus to criticize the existing state of affairs, therefore, it cannot account for all of the negative tone in policy work. While the continuities between today's regulators and those of the first decades of this century are striking, one can also detect a marked shift in tone between then and now. It is still common to call on expertise to solve political problems, but it is rarely done with the same optimism or confidence that was expressed by Charles Adams, Woodrow Wilson, Herbert Hoover, or Charles C. Dill. On the one hand, expertise has become an assumption; it is part of the background, not a rallying cry. At the same time, a kind of skepticism has crept into the process, wherein once-vaunted ideals are treated as simple practical necessities. Bringing a Ph.D. to a hearing to testify on your behalf, commissioning a research project, or couching your goals in the language of the public interest—these are simply means to end, they are necessary to get one's way, but they no longer stir the same enthusiasm they once did.

The Secret Fate of Regulatory Dropouts

Throughout the century numerous individuals have optimistically marched into the policy arena to do battle on behalf of the public good against entrenched interests and irrational thinking, only to emerge a few years later, not so much defeated as disappointed. The list is long: just a few of the most prominent include the proponents of educational

56. Streeter, "Cable Fable Revisited," 174–200.

broadcasting in the early 1930s, trust-busting FCC commissioner James Fly of the late 1930s, the Blue Book advocates of the 1940s, law professor Bernard Schwartz in the 1950s, Newton Minow in the early 1960s, Commissioner Nicholas Johnson in the late 1960s.[57] In different ways, each of these crusaders began by raising questions about the fundamental structures of the system, and in the end at best were able to accomplish only modest reforms within that system.

The policy insider might describe these people as representing extreme points of view—moderate and conservative ones might add, extreme points of view from somewhere on the left. They failed to bring fundamental change, the insider might say, because they fell outside the dominant center of the American political system, at least the center as it lay at the time of their efforts: that's how the American political system works.

Yet this political-center-of-gravity view of what's going on doesn't exactly capture the character of this pattern when looked at historically. Looked at individually, it may make sense to interpret each case as a simple struggle between one view of the public good and another interpretation, as democratic debate, as the gradual struggle of truth to assert itself over time. Yet taken together, it's harder to view matters as simple struggles between conservatives and liberals, or between heroic visionaries and the status quo. Things look less like a story of heroic struggle than a case of patterned contortion, less like *Pilgrim's Progress* and more like Kafka's *The Trial.*

One reason the Kafkaesque character of policy is not often noted within the policy arena is the self-policing function of the community's self-definition as "practical." A policy expert by definition should be working toward finding solutions to policy problems, and books and articles that suggest that the problems can't be solved aren't of much help in achieving that goal. The experiences of those who drop out of direct participation in the policy process in frustration are thereby automatically filtered out of the system because what they have to say after they

57. For the educational broadcast proponents, see Robert McChesney, *Telecommunications, Mass Media, and Democracy: The Battle for the Control of U.S. Broadcasting, 1928–1935* (New York: Oxford University Press, 1993). For the story of James Fly, see Streeter, "Policy Discourse and Broadcast Practice," and Joon-Mann Kang, "Franklin Roosevelt and James L. Fly: The Politics of Broadcast Regulation," *Journal of American Culture* 10 (summer 1987): 23–33. Another revealing discussion of the period is found in Frank C. Waldrop and Joseph Borkin, *Television: A Struggle for Power* (1938; reprint, New York: Arno Press, 1971). For the Blue Book advocates, see Meyer, "Blue Book," and the discussion above. For the story of a dismayed professor who tried and failed to introduce some integrity into the policy-making process, see Bernard Schwartz, *The Professor and the Commissions* (New York: Knopf, 1959).

have dropped out is not "practical," that is, not helping anyone engaged in policy activity to solve problems. So while the works of such authors are cited often enough in the literature, their criticisms of the system as a whole are more often ignored than refuted in the policy literature.

Media scholar Vincent Mosco, for example, began his career with a book that took the policy reformer's approach of identifying a policy problem and offering a solution.[58] Since then, Mosco has concluded that, as he puts it, turning to the FCC to solve the problems of broadcasting is like expecting the Wizard of Oz to get you back to Kansas. When Toto finally pulls back the curtain, you realize that nothing's there but a rather smooth-talking old man.[59] Mosco has gone on to become a thoughtful critic of the media policy-formation system, well read in academic circles.[60] Similarly, political scientist Murray Edelman's first work was a look at the FCC and broadcast policy making that, although critical, was written as though his criticisms might be heard and acted upon.[61] He subsequently took a much more skeptical approach to policy formation, and made a career of studying what he calls the "symbolic uses of politics," with an emphasis on the ways that American politics often serves purposes other than those officially stated.[62] Former FCC economist and Blue Book contributor Dallas Smythe, after his initial stint "in the system," became a prolific and unpredictably innovative neo-Marxist critic of media structure.[63] Each of these individuals has a wide following within the academy, but their skeptical analyses of the system as a whole are for the most part ignored or trivialized by the policy community out of which they emerged.[64]

58. Vincent Mosco, *Broadcasting in the United States: Innovative Challenge and Organizational Control* (Norwood, NJ: Ablex, 1979).

59. Vincent Mosco, paper presented to the Mass Communications Division of the International Communications Association Convention, Honolulu, May 1985.

60. Mosco's critical works include *Pushbutton Fantasies,* and *The Pay-per Society: Computers and Communication in the Information Age: Essays in Critical Theory and Public Policy* (Norwood, NJ: Ablex, 1989).

61. Murray Edelman, *The Licensing of Radio Services in the United States, 1927–1947: A Study in Administrative Policy Formation* (Urbana: University of Illinois Press, 1950), reprinted in *Administration of American Telecommunications Policy,* vol. 1, ed. John M. Kittross (New York: Arno Press, 1980).

62. Murray Edelman, *The Symbolic Uses of Politics* (Urbana: University of Illinois Press, 1964).

63. Dallas Smythe's best-known works include *Dependency Road: Communication, Capitalism, Consciousness, and Canada* (Norwood, NJ: Ablex, 1981), and *Counterclockwise: Perspectives on Communication,* ed. Thomas Guback (Boulder: Westview Press, 1994).

64. Krasnow, Longley, and Terry, in *Politics of Broadcast Regulation,* for example, cite Mosco's and Edelman's early works in their annotated bibliography (289) but do not

Expert Explanations of the Failure of Expertise: Interest Group Theories of Regulatory Behavior

The phenomenon of government regulation in capitalist economies in the twentieth century raises some of the most crucial questions of contemporary life: questions about the character of bureaucracy and its relation to democracy, questions about the relation of government to business and to citizens, and so forth. At its best, the literature on regulatory behavior contributes to the discussion of these larger questions. The discussion taking place in that literature has informed many of the schools of thought important to this book, including revisionist historiography and legal realism. This is not the place for a full review of the literature on regulatory behavior, however.[65]

Instead, I will focus on one important way that these theories have filtered into public life and become part of the institution that they are trying to describe. While the skeptical critics have generally withdrawn into academic subcultures, another midlevel sort of criticism has developed in the form of theories of regulatory behavior offered as a means to better master, and thus improve and participate in, the policy arena. Theories of regulation, both formal and informal, have come to serve as social-scientific systems that can help legitimate policy arguments and certify an expert's authority. Academic theories that provide explanations for why regulatory systems do what they do thus can, with varying degrees of explicitness, function as a way to refurbish the official explanation of regulatory behavior, that is, the belief that regulatory agencies like the FCC are staffed by neutral experts who make apolitical, rational decisions. To the extent that these theories are used as tools for experts within the policy arena, they thus can have an effect opposite that of the

---

refer to any of the later works (a particularly glaring omission in the case of Mosco's *Pushbutton Fantasies*). A related pattern can be found in surveys of the literature: a recent survey of locations and outlets for communications policy making, while listing "communications policy journals," failed to include *Media, Culture, and Society,* a progressive journal that regularly publishes in the area, and the interpretive *Critical Studies in Mass Communications,* while including journals as far afield as the *Computer Lawyer* and the *Rand Journal on Economics.* The measure that determined inclusion seems to have been neither scholarly respect nor amount of content directly relevant to communications policy, but the seriousness with which journals were taken within the policy community, that is, "practicality." See Mark S. Nadel, "U.S. Communications Policymaking: Who and Where," *Comm/Ent* 13 (winter 1991): 273–323.

65. For the best overview of these theories to date, see Horwitz, *Irony of Regulatory Reform,* 22–43, especially his discussion of "perverted public interest," "conspiracy," and "capture" theories of regulation.

skeptics and policy dropouts. Even though these theories raise profound questions about the official view of the policy process, in other words, in the end they can serve to keep the policy faith alive.

A classic example of this pattern is found in Krasnow, Longley, and Terry's *Politics of Broadcast Regulation,* a very useful textbook regularly used in classes on broadcast policy in universities across the country. The book, as its title suggests, is quite willing to admit that broadcast policy is fraught with politics, at least in the low political sense of that word. It opens with the following passage:

> [T]he regulation of American broadcasting is often portrayed as if it takes place within a cozy vacuum of administrative "independence." In reality, the making of broadcast policy by the FCC, an ostensibly independent agency, is an intensely political process. . . . Too frequently, the participants [in the regulatory process] are viewed in a way that suggests an impersonal mechanical operation. Witness the description of their activities by the term "government regulation." Realistically, there is no such thing as "government regulation"; there is only regulation by government officials. . . . Thus a major problem for regulatory agencies like the FCC is not just to conform to the letter of the law but, beyond that, to find ways to attune their behavior to the requirements imposed by its political environment. (9–10)

Right from the start, the authors express skepticism about the idea of administrative neutrality and independent expertise, and assert the insider wisdom about the "political" nature of policy making. They go on to acknowledge the well-documented fact of what they call the "complex web" of industry-commission relationships, and that relations "between some Washington lawyers and officials of the regulatory agencies can be so intimate they embarrass an onlooker" (50).

Thus the FCC, they warn the reader, is heavily tainted by the subjective winds of politics. Yet they do not then go in the direction of, say, Theodore Lowi's *End of Liberalism,* which argues that the administrative discretion that enables "regulation by government officials" has come to undermine the rule of law and the democratic process.[66] For them, liberalism has not ended. Rather, the "letter of the law" has not been abandoned but, more benignly, has been supplemented by a process of "attuning behavior to political environments."

66. Theodore J. Lowi, *The End of Liberalism: The Second Republic of the United States,* 2d ed. (New York: Norton, 1979).

The FCC, in the authors' view, is basically an embattled arbiter caught in the midst of struggles between a variety of interest groups—the "political environment" to which the FCC "attunes" itself. For although "interest groups" pull the FCC this way and that, the agency itself is not inherently biased toward one or another view. They dismiss the criticism that the agency has been "captured" by industry interests with the observation that most policy conflicts are *between* members of the industry (49). They conclude that, happily, in the "pluralist" process of policy making, "nobody dominates the process consistently" (139). Lobbyists, Congress, the president, even community groups have a chance to jump into the fray and, if not always get their way, at least influence the process.

If this state of affairs is safe for democracy, it also has a place for experts. The book goes on to provide a "systems analysis" of the FCC accompanied by a diagram bristling with boxes connected by dotted lines and arrows by which one can trace the "inputs" and "outputs" of the policy process and a series of case studies that exemplify the workings of the "policy system" (136). Policy experts are no longer imagined as producing grand industrial blueprints for society to gratefully effectuate. Yet they have a role to play as advisers to bewildered "actors" in the policy arena, charting a course through the complexities of the process, perhaps suggesting procedures to "better attune the FCC's behavior to its political environment." The authors, in other words, suggest that the admittedly hurly-burly "politics" of broadcasting can be approached as a "management problem," a situation that can be rationally analyzed and managed, if not completely controlled.

The vision presented here thus responds to the disappointments of the original corporate liberal framework by reproducing it in new terms. The citizens of a democratic polity are replaced by "interest groups," and the "men of big ability and big vision" that Senator Dill hoped would staff the FCC are replaced by modest policy scientists offering their expert knowledge of the complexities of the policy "system" to those very interest groups. Charles Francis Adams's heroic social engineers have been superseded by liberal Machiavellians armed with theories of regulatory behavior.

The continuities between contemporary interest group theory and earlier corporate liberal visions run deeper than their differences, however. Interest group theory shares much with Herbert Hoover's original vision of society as functionally interrelated units of capital, labor, and the public. Although the public-spirited optimism of Hoover's day has since dampened considerably, today's interest group approach is similarly functionalist, and similarly sets up policy making as a matter of achieving homeostasis through a mixture of careful arbitration, balancing of interests, and right reason. And the vision is similarly tautological: the different, competing "interests" (also known as "stakeholders") are treated as self-explanatory givens of the social universe, not as socially constructed, certainly not as changeable.

The most central given of applied interest group theory, of course, is the belief that a corporate capitalist system of broadcasting is the best choice or at least inevitable. Perhaps it is not surprising that an American book such as *The Politics of Broadcast Regulation* does not once seriously address the for-profit character of the broadcast system; the book takes it to be so obvious as to be not worth mentioning, that is, not having any political implications. That assumption, furthermore, is necessary to the authors' sanguine assertion that the industry has not been captured by industry interests on the grounds that there are struggles between industry factions. That the general interests of industry as a whole should be the focus of FCC policy is treated as a given, and "capture" is interpreted to mean capture by one industry faction over the interests of another (49).

But the structure of the book's argument also takes for granted some of the principles of liberal metaphysics, particularly the theoretical separability of politics from law and policy. The authors maintain a dichotomy between "the letter of the law" and political pressure, between the objective character of formal rules and the subjective character of politics. Also present is the terminological distinction between politics and policy: the authors ask how politics "imposes" itself on policy, but nonetheless maintain the assumption that these two things are distinct. They do not ask, in other words, about the politics of policy itself.

The Politics of Broadcast Regulation is a textbook, generally thought of as useful but not as a contribution to the most sophisticated theories of regulatory process. The book's arguments are nonetheless telling because, although the role of this kind of interest group theory in scholarship is complex and partial, its assumptions have become a part of standard operating procedures in the policy world. Informal interest group theory's construction of the role of expertise in the policy-making process, I would submit, is symptomatic of patterns of thought that underlie much of the practical wisdom about contemporary broadcast policy.

Ever since the late 1940s, FCC commissioners have increasingly come to understand their role in the world as one of mediating disputes between subgroups of regulated industries. Both observers and participants in policy making regularly describe rule making as the product of a compromise between particular factions. Cable must-carry rules which pit broadcasters against cable operators, financial syndication regulations which pit Hollywood television producers against the networks, and video carriage policies which pit cable operators against telephone companies: all of these are generally treated as self-interested industry struggles carried out, not in the marketplace, but in the arena of federal policy. The FCC and other policy-making bodies, in turn, understand their principal role to be mediating such disputes. On more than one occasion, the FCC has made its arbiters' self-image official, announcing proceedings designed to find a compromise between interested parties. With unconscious irony, a theory that began as a criticism of the policy process has become a tool of that process. If not in academe, then in the trenches of the Washington bureaucracy, interest group theory becomes a working reality; functionalist analysis becomes a self-fulfilling prophecy.[67]

Living with the Legitimation Crisis: Policy Practice as Theater

At a recent conference on communications policy in the Washington area, a plenary session was devoted to the role of policy research in policy making.[68] A panel of staff members and commissioners of various regulatory bodies described what they wanted from policy research. The panel members spoke in familiar corporate liberal terms: one of them, for example, asserted the apolitical character of his work as a state utility commissioner by claiming that he was not a "policy maker," but merely a "regulator," an implementer of policies established elsewhere. While the panel members had their differences, they all seemed to agree that policy research tended to be too arcane to be understood by those who weren't academic specialists, not directed at solving problems of immediate relevance to regulators, and lacking in solid, irrefutable data and conclusions that could be used to back up policy decisions.

Several of the panelists were quite explicit about the difficulty of get-

67. Another example of the extent to which pluralist habits of thought have become second nature in policy circles can be found in Nadel ("U.S. Communications Policymaking," 296), who writes, "Although the pluralist theory of policymaking often neglects the early stages of policy formation (e.g., journal articles . . .) these are the forums where it is easiest for stakeholders to participate in the policymaking process and where proposals are most susceptible to modifications." He thinks he is criticizing pluralist theory, yet he still understands the process as one where self-evident "stakeholders" simply "participate" in the policy-making process.

68. Twentieth Annual Telecommunications Policy Research Conference, Solomons, MD, September 12-14, 1992. The panel was titled "Telecommunications Policy Research: Policymakers' Perspectives," and featured representatives from the FCC, the Canadian Radio-Television and Telecommunications Commission, and the Maine Public Utilities Commission.

ting things done when any given decision inevitably invoked the ire of one or another powerful industry or political "interest." If research was going to be useful in this context, they said, it should be accessible and unambiguous enough to silence potential opposition; work that was obscurely presented or equivocal in its policy implications just provided more fodder for debate. In order to be useful, therefore, policy research must be accessible, directed toward resolving problems in the lives of regulators, and unequivocal in its conclusions.

The audience reaction was varied, but there was a lot of grumbling, particularly among the many academics present—each of whom, to some degree, had staked their careers on being policy experts. Perhaps part of the tension was simply the product of conflicting operating principles inherent in the different policy roles of commissioner and analyst. One gets tenured for producing sophisticated work that deals with leading-edge issues in a scholarly field, not for being accessible to non-academic politicians and bureaucrats. Yet the tension seemed to go a little deeper than that. An economist in the audience pointed out that, for all the research conducted, most past major policy decisions were not based on research. Most of the major policy actions of the 1980s, for example—the elimination of the prohibition against program-length television ads, the elimination of the station-trafficking rule, changes in station ownership rules, and the extension of license terms—were loosely justified by promarket economic logic but were not based on any detailed analysis. The panelists, it seemed, were not asking the researchers for advice about what to do; they were asking the researchers to help smooth the way for decisions already made, which was not exactly comforting to academic egos.

What was surfacing at that moment was a contradiction within the policy process. The implicit model of policy making suggests that, given broad guidelines, policy analysis produces solutions to policy problems based on expert, rational analysis. More often than not, however, research serves the largely rhetorical purpose of helping to justify decisions made politically. NBC's network duopoly was broken up in the 1940s because of the political savvy and moral fervor of FCC chair Fly combined with the residual New Deal political climate; the facts collected in the case were persuasive but not irrefutable, and by today's standards would not be understood to demonstrate monopoly.[69] Broadcast-industry resistance to limits on cable growth were overcome, not because of the irrefutability of the often grandiose "wired nation" policy analyses of the time but because the rhetoric of the "cable fable" helped introduce

69. See footnote 43.

new ways of thinking and enabled a political realignment in the policy arena.[70]

The data show, in sum, that the data often don't matter. As often as not, what matters in policy analyses are not the careful, well-reasoned empirical analyses but the catch phrases, the sweeping introductory and concluding paragraphs, and the broad patterns of thought to which the analyses lend authority. What has a real effect in policy analyses, in other words, is the rhetoric, the window dressing, whereas the information and reasoning merely lend expert authority to that rhetoric; the content is in the window dressing, and the rhetoric in the details.

If the policy process sometimes fails as a rational example of social engineering, however, it generally succeeds as theater, as a symbolic enactment of the procedures of social engineering. In this case Murray Edelman's analysis of the symbolic uses of politics is accurate. The ambiguities of policy allow for any number of policy questions: Is this or that interpretation of the public interest the constitutional, efficient, or reasonable—that is, the "correct"—one? Should this or that criteria be used to select broadcast licensees? Is the spectrum scarce or plentiful, and does scarcity in any case justify restricting the rights of one or another "interest" in broadcasting? Should this or that industry faction be favored with supportive regulation, or punished by favors granted its competitors? These questions, furthermore, can be made to overlap with numerous interesting academic ones: What is economic efficiency, and how best might it be measured? How can constitutional principles be applied to new electronic media? How does one measure the relationship between media ownership and media content? What is media diversity and how can it be measured?

As long as professionals are employed in regulatory agencies to raise these questions, studies are conducted in an effort to answer them, and academics continue to design policy courses, write policy analyses, and attend policy conferences around them, policy activity goes on. Every few years a new policy trend surfaces that promises to correct all the errors of the old ways, and amid struggle and debate we are treated once again to the drama of heroic public servants doing battle on behalf of truth and the public interest against entrenched interests. As a result, a general image of a successful policy activity is presented that, though perhaps a little ragged around the edges, is sufficient to comfort those inside the apparatus and generate acquiescence on the part of those outside it, at least to the degree necessary to keep the process going.

70. Streeter, "Cable Fable Revisited."

Conclusion: Policy and the Deferral of Politics

Policy insiders are not unaware of many of the criticisms that have been advanced here, and would undoubtedly provide compelling and thoughtful counterarguments. True, broadcast policy has its problems, they might argue, but a less-than-flawless record need not be understood as indicative of total intellectual bankruptcy. Let's not throw the baby out with the bathwater, it might be said. There are cases where good policy ideas helped improve things. There is still room for reasonable and qualified people to present thoughtful advice to elected and appointed officials with an eye toward helping make things work a little better.

Without wishing to discount the occasional strategic importance of policy actions in the short term, it is nonetheless difficult to remain sanguine about existing patterns in American broadcast policy if one looks at things over the long term. In its seventy-five-year history, the hope that reconciling corporate liberal logic and procedures with broad liberal goals in broadcast policy has received very little support, even in its own terms. The definition of the public interest, even within corporate liberal parameters, has proven to be highly unstable and subject to constant wrangling; many have pronounced the phrase dead, though it continues to live on in the official laws of the land. As we will see in subsequent chapters, the search for some kind of marketplace competition that satisfies all the traditional liberal criteria—easy access, large numbers of competitors, lack of government interference, lack of privileged players—has fueled any number of policy initiatives but has yet to produce anything that is universally accepted as an industrywide open marketplace. And the hope that expertise and administrative procedure would insulate broadcast regulators from both political winds and corporate self-interest has had to take refuge in a series of policy fads: from the trustbusters of the thirties to the sociologists of the sixties to the deregulators of the eighties, each of which entered the scene with a promise to achieve the technical clarity and neutrality that its predecessor failed to deliver and each of which, under the force of experience, eventually succumbed to the same forces that it proposed to transcend.

The point here is not that corporate liberal broadcast policy should be abandoned in some kind of revolutionary purge. The point is rather that there is plenty of justification for inquiry into the fundamental assumptions that are generally left untouched by existing discussions. Broadcast policy *should* be about the American policy for broadcasting.

Corporate liberal policy may not need to be abandoned, but it should be allowed to be open to question.

The process of inquiry into underlying premises, it should be emphasized, need not be seen as a perhaps interesting but relatively "impractical" theoretical exercise. If being practical means seeking understanding that might be brought to bear on the improvement of collective life, then in the current context exploration of premises is altogether practical. It is impractical only if one narrowly defines "practicality" as contributing to the corporate liberal project of enacting liberal principles with recourse to neutral expertise. After seventy-five years of at best mixed success in trying to work within those corporate liberal premises, opening those premises themselves to questioning might be as practical as anything else.

The most obvious of the never-discussed questions is the social value of commercial organization itself. Yet inquiry into commercial organization is not a simple matter of debating the values of untrammeled greed versus elevated public principles (or of free markets versus government regulation). That these seem to be the fundamental questions is in turn a product of other fundamental liberal premises: assumptions about the nature of social organization, markets, government, rights, bureaucracy, communication, and property. So the following chapters investigate the practical character of those assumptions, particularly the last, as they have operated in broadcasting.

It should be remembered, however, that this investigation also calls into question another characteristic liberal principle, the principle underlying formalist understandings of the rule of law and scientistic understandings of expertise: the wish to transcend politics. The point in exploring the role of property and other key liberal categories in commercial broadcasting, therefore, is not merely to neutrally dissect them, or to discover the correct, most efficient, or coherent understanding of them. The point is to open them up, to help make them available, in all their complexity and fluidity, for a broader discussion—to politicize them in the "high" sense of that word.

FIVE

Postmodern Property: Toward a New Political Economy of Broadcasting

As the professions of the middleman lose their economic basis, the private lives of countless people are becoming those of agents and go-betweens; indeed the entire private domain is being engulfed by a mysterious activity that bears all the features of commercial life without there being actually any business to transact.

THEODOR ADORNO

Introduction

In a society of private property, much of our experience involves bounded objects. We walk down a residential street and, by way of walls, driveways, and picket fences, are made instantly aware of the boundaries that separate homes into finite, ownable things. We walk through a grocery store, and are surrounded by thousands of packaged objects, each distinct, each with a price.

What broadcasting illustrates particularly clearly, however, is the degree to which ownership boundaries can be anything but obvious. In commercial broadcasting the fundamental questions of private property —who has control over what, who owes whom what for which individual item—are becoming an increasingly indeterminate, blurry matter. To be sure, most of us have at times, in a late-night stupor, watched the closing credits scroll by all the way to that final moment when the copyright notice—the textual equivalent of a picket fence—levitates onto the screen. But to most viewers, this is just a bit of legal flotsam, the debris of complex machinations that have taken place elsewhere. The broadcast media provide us less with discrete objects than with an unending rush of images, sounds, and messages; as Raymond Williams has pointed out, television is distinguished by "flow," by the degree to which both its production and reception are characterized not so much by a series of distinct programs as by a complex stream of juxtaposed texts.[1]

1. Raymond Williams was one of the first to call attention to the centrality of juxtaposition to television aesthetics with his concept of "flow" (*Television: Technology and Cultural Form* [New York: Schocken, 1977]). For further discussion of this phenomenon,

Of course, viewers buy boxes of cereal and sell houses, but don't buy and sell programs or stations; perhaps they can't be expected to understand systems of ownership in which they are not directly involved. What *is* peculiar about the broadcast media, however, is that ownership boundaries are similarly obscure and fluid from the other side of the camera, from the point of view of those who *do* buy and sell programs and stations. Even to viewers who care, the copyright notice that punctuates a program's closing credits hardly begins to chart the labyrinthine and often systematically blurred "boundaries" that delineate the exchange relations embodied in that broadcast.

This chapter argues for the importance of the politics of property creation to broadcast policy and the political economics of electronic media. On the one hand, the argument is fairly straightforward: the most important form of political intervention in the electronic media is property creation. Commercial broadcasting, if it is to be commercial, involves taking a set of activities—sending signals through the air to unseen audiences in a highly organized way—and somehow constituting those activities as things that can be bought, owned, and sold: as property. Commercial broadcasting is not just the product of an absence of political or social control; it is not the result of some elemental state where you simply take the lid off and let the market run its course. It involves an ongoing, collective effort of commodification, of turning social activities into property. And the ways in which this is done define the basic ground rules of marketplace activity in broadcasting; the laws and policies that create property designate who has control over what in what circumstances, and thus allocate power in ways more profound than the much-debated matters of free speech and the public interest. On the other hand, property creation in broadcasting is not a simple

---

see Jane Caputi, "Charting the Flow: The Construction of Meaning through Juxtaposition in Media Texts," *Journal of Communication Inquiry* 15 (summer 1991): 32–47. It is not just television texts that differ from the traditional model of a linear, coherent book. For reasons linked but not reducible to the bureaucratic structures of the television industry, television audiences also use and experience the medium in a thoroughly nonbooklike way. As every network executive knows only too well, the bulk of the audience turns on the set to watch television itself, not programs; their channel choice is simply a matter of finding the least objectionable of what's available at the time. People seldom turn to television to watch a particular program, and even less often seek out the "work" of the television equivalent of an "auteur," such as a television producer or writer. The evidence suggests, furthermore, that many, perhaps most, of the audience use television as an accompaniment to other activities, and thus give the medium a highly selective, idiosyncratic, deliberately divided attention. The classic description of this pattern is found in Paul Klein, "Why You Watch What You Watch When You Watch," in *Television Today: A Close-up View*, ed. Barry G. Cole (Oxford: Oxford University Press, 1981), 214–17.

mechanical process that, once accomplished, simply rolls along of its own accord. On the contrary, reconciling the activities necessary to broadcasting with the idea of property is no mean task. In the first place, broadcasting is almost entirely ephemeral. Unlike beans or ball bearings, for the most part broadcasting cannot be held in one's hand; it was hard to resist the temptation to title this chapter "How to Sell Nothing and Get Rich." But a focus on the sheer ephemerality of broadcasting might obscure a more profound and more general point: broadcasting is also fundamentally social. Technologically, the difference between broadcasting and other uses of radio is relatively trivial; the engineer who knows enough to build a ham radio or remote control detonation device, for example, could apply that knowledge, and many of the same or similar components, without much difficulty to building broadcast transmitters or receivers. What distinguishes broadcasting from other uses of radio is less material or technological than it is a matter of collective organization, of the social conditions and structures that make broadcasting possible. And it is within and out of those social conditions and structures that something like property must be constituted if commercial broadcasting is to work.

To justify making so much out of property, the second part of this chapter discusses the substantive limitations of traditional broadcast policy discourse, which typically focuses on markets, competition, and the public interest. As competition is the main theme of regulatory discourse in this century, this part shows how, historically, the search for open markets seems always to lead to closed ones. This is not, I will argue, the product only of something peculiar to broadcasting such as cartel behaviors, technological constraints, or economies of scale (though these factors may exacerbate the problem). Rather, it is a byproduct of traditional economic discourse, with its assumption that marketplace competition and nonmarket social behaviors, such as government intervention and business cooperation, are in opposition to one another. Many progressive efforts to change commercial broadcasting have similarly foundered on the assumption that government's function is or should be to serve as a check or restraint on "naturally" occurring economic relations.

What's missing from traditional inside-the-beltway policy discussions of both the Left and the Right is a sense of how cooperation and competition, government "interference" and private initiative, are ultimately inseparable. The history of regulatory zigzagging between "competitive" and "public interest" policies may be less a product of political pendulum swings than of the fact that a policy language that presents government and markets as incommensurable cannot help but flutter back and forth over a terrain of social activities in which the two are intertwined.

After discussing the limits of traditional approaches, then, the chapter goes on to lay out an alternative framework for understanding the relations among competition, government, and public and private action in the electronic media. Borrowing from the field of economic sociology, I argue that what traditional economists call "externalities" or "exogenous factors" are better understood as internal, and often prerequisite, to marketplace conditions, competitive or otherwise. Broadcast policy, therefore, is best understood, not as government regulation, but as the mix of private and public social arrangements that undergird market relations. And those private and public social arrangements are usefully approached through the question of property creation. The category of property is the point where private and public most clearly and forcefully intersect. Property is a kind of nexus of culture, economics, ideology, and state power.

While property is central, the argument of this chapter is not that property is a blunt, obvious reality, or that it is *the* determining force, the underlying secret that unlocks the meaning of everything else. Property is fluid, discursive, and bound up with broad problems of legitimacy, driven as much by the desire to make sense as by the desire to make money. It is a way into the complexity of institutions, not a way to eliminate that complexity. As a way to emphasize this, then, the chapter concludes by selectively borrowing a concept from postmodernism. Much of the high drama of commercial broadcasting, I suggest—the ratings race, the rise and fall of series, rocketing and plummeting careers, corporate struggles over regulation—might be understood not as straightforward market behavior, but as a set of bureaucratic rituals that represent market behaviors using the codes of administrative logic, as bureaucratic simulations of markets.

The Limits of Traditional Approaches to Broadcast Policy

Policies of Competition

The history of mainstream U.S. broadcast regulation can be seen for the most part as a nearly century-long search for competition in the electronic mass media. Over the years, both supporters and critics of the broadcast industry have tended to view its behavior through the lens of a dichotomy of competition and monopoly: a lack of competition is the principal source of problems in broadcasting, increased competition is the solution to problems, and the test of government regulation is thus whether or not it enhances competition. When it is not engaged in the routine bureaucratic tasks associated with spectrum management, therefore, the FCC spends most of its broadcast-related regulatory energy trying to enhance competition in the broadcast industry in various ways. FCC cases and reports concerned with competition and broadcasting outnumber those concerned with free speech and broadcasting by more than ten to one.[2] Although the precise terms of these efforts vary depending on time, place, and goals, the general pattern is consistent.

This is consistent with American political discourse as a whole, which tends to view matters through the lens of liberalism's version of the public/private dichotomy, a dichotomization of social constraint and individual actions. *Either* we have government regulation *or* we have a private marketplace; *either* we have a competitive industry with freedom for individual actions, *or* we have a monopoly that socially constrains individual actions. Critical legal studies is just one of the modernist currents of thought that have pointed out the "indeterminacy" of the public/private distinction, that is, the extent to which the public and the private are more like two sides of the same coin than they are distinct types of social relations.[3] It is in the nature of the public/private distinction, the critical legal studies argument goes, that there will always be a reasonable point of view from which a supposedly public issue will appear to be really a private one, and vice versa.

To a large degree, the goal of designing a government policy for business that encourages marketplace competition became a centerpiece of American politics during the late nineteenth and early twentieth centuries, and thus was already on the broad political agenda in the 1920s, broadcasting's early years. At the time, the "trust question" had only recently subsided, and the ways in which it was settled were reflected in broadcast regulation. The ferocious late-nineteenth-century legal and political struggles over trusts had been resolved by the political consensus arrived at based on the Supreme Court's "rule of reason" deci-

2. A NEXIS/LEXIS search conducted on March 28, 1994, of the file "FCC" (which includes both FCC cases and reports) turned up 2,568 entries for the search terms "competition and broadcasting" and 208 for "free speech and broadcasting" for the years 1957–94.

3. For a discussion of indeterminacy, see Thomas Streeter, "Beyond Freedom of Speech and the Public Interest: The Relevance of Critical Legal Studies to Communications Policy," *Journal of Communication* 40 (spring 1990): 43–63. See also Mark. G. Kelman, "Trashing," *Stanford Law Review* 36 (January 1984): 293.

sions in the teens.[4] In the emerging consensus, corporations were not to be allowed entirely free rein, but by the same token, big was not bad: what was actionable was "unreasonable restraints of trade," where these were understood to be more a matter of deliberate actions like price-fixing than a matter of structural advantage. Although a clear definition of "unreasonable" has proven elusive over the years, the consensus of the time set a broad agenda for the politics of big business that has persisted (with variations) to this day.

The general consensus has been that large corporations and marketplace competition can coexist (with a few exceptions, perhaps, such as natural monopolies in utilities). The problem, according to the consensus, is not that the simple existence of giant oligopoly and monopoly corporations challenges the liberal vision of an entrepreneurial competitive marketplace. It is not the case, the theory goes, that we must choose between a liberal competitive economy and a corporate one. Rather, with the proper precautions, we can have an economy that is both competitive and populated by giant vertically integrated corporations. If there is a problem with large corporations, it can be solved by taking steps to enhance competition: prohibiting practices that restrain it, policing industry boundaries in ways that enhance it, and on rare occasions going so far as to break up monopoly or near-monopoly companies. The central expression of this regulatory philosophy is the Sherman Antitrust Act as interpreted since the teens, but it has been repeated in numerous other areas of law and regulation as well.

One of the places that this philosophy was reiterated was in the law of communications. As the language that eventually became the Communications Act took shape in the 1920s, congressional grumbling about the RCA "radio trust" led to a passage in the Radio Act directing the FRC to deny licenses to antitrust violators. The 1934 Communications Act similarly contained antitrust principles in sections 313 and 314. And enhancing competition has been the principal nonroutine activity of the FCC ever since.[5]

4. Martin J. Sklar, *The Corporate Reconstruction of American Capitalism, 1890–1916: The Market, the Law, and Politics* (Cambridge: Cambridge University Press, 1988), 173.

5. The following discussion of the history of competitive regulation (through the FCC report, *New Television Networks: Entry, Jurisdiction, Ownership, and Regulation: Final Report*, by the Network Inquiry Special Staff [Washington, DC: Government Printing Office, October 1980]) is derived from Thomas Streeter, "Policy Discourse and Broadcast Practice: The FCC, the U.S. Broadcast Networks, and the Discourse of the Marketplace," *Media, Culture, and Society* 5 (July–October 1983): 247–62.

The Chain Broadcasting Investigations

Even among American regulatory agencies, the FCC is known as particularly acquiescent, more the industry lapdog than watchdog. The single major exception to this rule occurred in the late 1930s and early 1940s, when, against strenuous industry objection, a temporarily aggressive commission introduced a series of rules designed to limit network dominance of the industry, including one that forced RCA's NBC to divest itself of its second "blue" network. The episode is interesting both for what did happen—in the name of fostering economic competition, the FCC took an aggressively oppositional stance toward the industry—and for what didn't happen—in spite of the regulations, broad industry patterns of behavior continued largely unchanged.

As the network system consolidated and grew throughout the 1930s, congressional and public criticism of the existing state of affairs began to focus on questions of network dominance. Traditional progressive antimonopolistic sentiments were aroused, and were joined by complaints from the major networks' weak but principal competitors, the Mutual Broadcast Network and the Transcontinental Broadcasting System. In this context Congress passed legislation directing the FCC to investigate AT&T's monopoly of the telephone, and hinted it might do the same for network radio. And in 1939, Roosevelt appointed New Deal-enthusiast James Lawrence Fly to be chair of the commission.

In 1938 the FCC launched an investigation of what was then called "chain broadcasting," which under Fly's aggressive leadership took an unusually adversarial stance toward the networks. A long series of hearings ensued, accompanied by extensive research, which eventually led to the release of the *Report on Chain Broadcasting* in 1941. The report contained dramatic condemnations of network behavior and recommended several strong network regulations, including the divestiture of all but one network-owned station per market and of NBC's second "blue" network. Citing evidence that the networks had introduced restrictive affiliate contracts and other strategies as a means to thwart competition, the report recommended that such practices, as well as the ownership of more than one network, should be abolished. Fly grandiosely described the report as a "Magna Carta for American broadcasting."[6]

Significantly, the report made three basic assumptions about competition in broadcasting: that competition was possible, that the networks deliberately thwarted competition, and that the best way to

6. "The Monopoly Report: Five Men against the Public" (editorial), *Broadcasting*, May 12, 1941, 18.

enhance the public interest would be to increase competition in the industry. The logic, in other words, was classic antitrust: the free market, the report declared, was "the essence of the American system of broadcasting" and the regulations should be introduced restricting "practices or agreements in restraint of trade or furtherance of monopoly."[7]

The industry reaction was vociferous. CBS claimed that the regulations would "cripple, if . . . not paralyze, broadcasting as a national service."[8] NBC agreed. *Broadcasting* magazine darkly concluded that "[r]adio by the American plan, as we know it today, will go into the discard. What is today the best and the freest radio system in the world will begin rotting away—the prelude to a government-operated system for which the public then will clamor."[9] The networks took the FCC to court, and a legal battle ensued, resulting in a Supreme Court ruling in favor of the commission.[10]

The regulations inspired by the report had less effect than either side had predicted. While they did force the divestment of a number of stations in duopoly markets and of NBC's second network, in the end the networks were neither "crippled" nor challenged by new levels of fresh competition. The percentage of stations that were network affiliated, at 61 percent in the year of the *Report,* continued to climb after the regulations, reaching a highpoint of 97 percent in 1947.[11] The only change was where previously two organizations, NBC and CBS, reigned over the continuing trend toward concentration in broadcasting, now, with the addition of ABC operating the divested NBC network, there were three. The story of the 1941 *Report on Chain Broadcasting* and the resulting rulings, therefore, clearly illustrates a pattern that has been repeated several times since: with much fanfare, legal steps were taken to increase competition, yet the oligopoly, center-periphery structure of the system remained basically unchanged.

The 1950s

The cycle was repeated, with less drama, when the network system moved to television in the 1950s. The capital-intensive character of television, combined with high production costs and some regulatory blunders on the part of the FCC,[12] resulted in a rigid, nationally centralized

7. FCC, *Report on Chain Broadcasting,* Docket 5060 (1941), 95.

8. Christopher H. Sterling and John M. Kittross, *Stay Tuned: A Concise History of American Broadcasting,* 2d ed. (Belmont, CA: Wadsworth, 1990), 190.

9. "Monopoly Report," 18.

10. *National Broadcasting Co., Inc., et al. v. United States et al.,* 319 U.S. 190 (1943).

11. Sterling and Kittross, *Stay Tuned,* 634.

12. Because of technical interference problems, the FCC halted processing of television licenses in 1948, at which time all of the few stations on the air were network owned

production and distribution system. Consequently, American television was not simply dominated by, but virtually under the complete control of, NBC, CBS, and ABC. The political reaction to network dominance was similar to the reaction of the thirties: again, a combination of complaints were reduced in policy discussions to complaints about a lack of competition in the network-dominated new medium. Again a series of hearings, rule makings, and inquiries took place. And again, the situation remained basically unchanged.

The central stage for the first television version of this policy cycle was the Barrow investigation, an FCC study of the networks begun in 1955, published in 1958, and named after its chief investigator. The study appeared as a response to a combination of congressional concerns and complaints from the major networks' struggling competitors, especially the DuMont Television Network, whose significant attempt at forming a fourth network had collapsed in the same year. The resulting *Barrow Report,* while several times longer, more detailed, and more timid than its 1941 predecessor, still shared with the *Report on Chain Broadcasting* the presupposition of the possibility of competition in nationwide broadcasting, and the goal of determining the extent to which industry practices intentionally inhibited that competition.[13] However, while the first investigation was willing to seriously address the effect of heavy industry concentration to the point of forcing the divestiture of NBC's blue network and the sale of a number of network stations, the Barrow investigation accepted as given the even heavier concentration that characterized television in the 1950s. The structural factors that placed the networks in a privileged, noncompetitive position—principally the economies of scale of television production and distribution—while clearly the principal causes of undue network power, were simply ignored by the *Barrow Report.* Committed to its presupposition of a competitive broadcast industry, the investigation ignored the evidence, which clearly challenged that presupposition.

The 1960s

Throughout the 1960s and into the early 1970s, the issue of network dominance continued to be pursued by the FCC, largely under the auspices of

---

or affiliated. Free from competition, those few stations on the air were able to consolidate their market positions during the following four-year "freeze." When the freeze was finally lifted, moreover, most of the new licenses available were in the technically inferior UHF-band (ibid., 295–96).

13. FCC, *Barrow Report (Network Broadcasting: Report of the Network Study Staff to the Network Study Committee),* presented to the House Committee on Interstate and Foreign Commerce, 85th Cong., 2d sess., 1958.

the in-house Office of Network Study, an offshoot of the *Barrow Report.* A complex series of hearings, rules, and revisions eventually led to the adoption in the early 1970s of several rules that are still being debated today: the prime-time access, financial interest, and syndication rules.

The prime-time access rule in particular provides a clear illustration of mechanisms by which public complaints are subsumed under the question of competition, and by which the social aspects of the "public interest" are supplanted by the interests of various members of the broadcast industry. In the 1960s the role played earlier by first Mutual Broadcasting and then DuMont was played by Westinghouse Broadcasting (Group W), at the time the largest broadcast corporation after the three networks. Group W not only owned five affiliated television stations, but had become active in program production; it knew that if the network stranglehold on prime-time broadcasting were loosened, Group W could expect to develop a much larger syndication market for its wares. In submitting the original draft of the prime-time access rule to the FCC, however, Westinghouse predictably chose not to emphasize its own financial interests. Instead, the request for a rule barring the major networks from a portion of prime time took on the rhetoric of localism, emphasizing the inability of "local" stations to gain access to prime time, which in turn eroded the "community" nature of television. Although the prime-time access slot has since then hardly fostered a marked increase in local "community"-based programming, it has become an outlet for off-network, nationally syndicated materials, such as game shows, and not coincidentally, some of Group W's material: its prime-time program *PM Magazine* became very lucrative well into the 1980s.

Deregulation

In 1978 the FCC embarked on yet another study of the networks, again under prompting from Westinghouse, which submitted a petition complaining about the networks' dominance of broadcasting. The petition argued that the networks maintained unfair economic dominance over their affiliates, with the result that "[e]ach year local affiliated stations have less involvement in and responsibility for the totality of the programming carried on *their facilities* to the public in *their communities.*" Moreover, the petition argued, the excess of crime, sex, and violence in network programming was partly the result of the lack of affiliate input into programming decisions; if affiliates were given plenty of time to clear network programs and a chance to provide some "grass roots reaction" to network decisions, the problem would be reduced.[14] What surprised

14. "Top of the Week: Group W Asks FCC to Cut TV Networks Down to Size," *Broadcasting,* September 6, 1976, 25, original emphasis.

many about the resulting 1980 *New Television Networks: Final Report* was that it was very critical of past regulatory efforts and suggested that in many cases problems would be solved, not by more regulation, but by elimination of current rules.

As is often the case with deregulatory tracts, the report's criticisms of past efforts were in many ways incisive. Existing rules, the report concluded, "do nothing to promote competition."[15] This is because the primary determinants of network relations to affiliates and other industry members are not restrictive contracts or practices, but the economic efficiencies of networking. Since the cost of program reproduction and distribution are insignificant when compared with the high cost of production, and because sales of advertising time are greatly facilitated by the ability to simultaneously transmit through a nationwide network, a network-dominated system of broadcasting is inevitable. In other words, the enormous bargaining power of the networks over their affiliates, the primary source of concern, is a product of what the networks are, not of what they do. This fact, the report argued, had been largely ignored by previous regulatory thinking. Hence, the long-standing assumption that restrictive network practices force affiliates to accept network programming ignored that affiliates tend to accept network programs simply because they are more profitable, regardless of whether or not the affiliates are contractually obligated to accept the programs. The prime-time access rule, for example, "ignores the fact that the programming incentives of the three affiliates are in general identical to those of the three networks" (IV-82). Similarly, "the minimal impact of the rules on affiliate clearances is not surprising in light of the incentives both the networks and their affiliates have to maximize the joint profits from network exhibition and in light of the generally more profitable nature of network programs, attributable to the efficiencies of networking" (IV-47). "The great economic advantages of networking," observes the 1980 report, "are simply too great to expect economic concentration to be reduced through restrictions on network conduct" (I-3). From this perspective regulations such as the ban on option time and the prime-time access rule are therefore basically pointless.

The report thus acknowledged, in a way that the previous fifty years of broadcast regulation had not, the importance of economies of scale, vertical integration, and associated centralizing, oligopolistic pressures that militate against full-fledged open competition in the broadcast industry. Yet the report's proposed solutions were not to rethink the premise of naturally occurring competition. Rather, in a case of classic

15. FCC, *New Television Networks,* IV-47.

deregulatory logic, it argued that since regulatory efforts were either ineffective or counterproductive in countering the centralizing economic forces in broadcasting, regulation in general should be eliminated in favor of "the systematic disciplining and eroding forces of competition" (I-29). In other words, since regulation does not work, deregulation will. As syllogisms go, this is less than airtight—unless one takes as given the belief in the natural character of the marketplace. The report's deregulatory argument made sense only if one assumed that the marketplace is not a legal and political accomplishment, but the product of a state of nature, something that inevitably flourishes in the absence of political intervention.

Cable Television

The story of cable television is frequently told by supporters of deregulatory economics. Cable's life as a trope in political economic debate began in the 1970s, when the fact of broadcaster-supported FCC restrictions on cable growth were held up as a classic case of government regulation being used to hobble innovation and competition. And since then cable's dramatic growth and erosion of network dominance has become a favorite example of what can be accomplished when marketplace forces are allowed to replace government regulation.

At first glance, the case of cable television does seem convincing evidence for the value of unregulated markets—more convincing, for example, than the problematic case of airline deregulation. Not only do many homes have many more channels today, but it is undeniably the case that, for better or worse, a new aesthetic and political openness has come to the medium of television in the wake of the spread of cable in the 1980s. Channel surfers can now easily hop between the right-wing social conservatism of the Family Channel and the sexual liberalism of a Dr. Ruth Westheimer—perhaps not the best that has been thought and said in either camp, but at least a range of values much broader than was ever common on the politically timid networks in their first three decades. And culturally, new things are happening as well. For all the sexism and clichéd teenage exploitation on MTV, for example, it is also the case that the rock video has become a new cultural form, making popular and elaborating on an entire universe of visual techniques once known only among the art school avant-garde. And the networks have been forced to respond: they have had to grope for new ways to retain audiences, many times taking the low road, but sometimes also experimenting with more interesting routes, allowing for a little more cultural exploration and innovation in programming. Even if it's true that the bulk of what's available in the new video environment consists of tired reruns, low-budget advertising vehicles like fishing shows, and lowest-common-denominator sensationalism, one cannot deny that there is also, amid all the clutter, more genuine diversity and innovation than there used to be in the days of complete network dominance.

So what could be wrong with the idea that cable is a classic example of the benefits of deregulation? A closer look at the case of cable television suggests that cable's success is both less a product of deregulation and less structurally unique than is typically imagined. Deregulation is only one among many factors that led to cable's success, and even then it is an oversimplification to describe what happened as a simple shift from government regulation to laissez-faire. Cable was reregulated more than it was deregulated. Cable was able to grow, not so much because regulations were simply eliminated, but because, beginning in the early 1970s, cable's status among the policy community was changed from industry threat to industry component; haltingly, sometimes awkwardly, but nonetheless systematically, those with influence surrounding the FCC came to bring cable into the fold and to consider cable's survival and health part of the legitimate goals of industrial system management. The result was not a radical change in industry structure toward entrepreneurialism but rather a series of incremental adjustments within the existing oligopolistic, center-periphery, advertising-supported system of electronic media. Cable has not revolutionized the basic corporate liberal structure of television; it has been integrated within it.

The industry we now call cable began life around 1950 as Community Antenna Television (CATV), a service providing improved television-signal reception in remote areas. In the early years, CATV helped fill in the gaps in the ragged periphery of the television system dominated at the center by the three television networks, which distributed their signals nationwide via coaxial cable and microwave relay to broadcast transmitters in local communities. When the tiny but growing CATV industry set off a squabble in the broadcast system's periphery by threatening the profits of small local broadcasters, the broadcasting lobby persuaded the FCC to generate a set of regulations that effectively halted CATV's growth. By the mid-1960s CATV was thus locked out of television's economic mother lode, the top hundred television markets. In both law and informal insider wisdom, cable was exterior to the "system" whose healthy functioning was the appropriate goal of regulation.

The process by which cable was imaginatively and practically brought into the system was gradual, but the watershed moment occurred in 1972, long before anyone at the FCC uttered the word "deregulation." By the late 1960s a number of gradually building economic and political trends combined to generate pressures for something new. First, the library of commercial film and videotaped programs, including old movies and reruns, had grown dramatically, and with the increase in supply came a predictable decrease in price: filling a schedule with material was becoming a much less expensive proposition than it had been in the early days of television. Second, technological evolution in several areas was making the distribution of video programs by nontraditional means gradually easier, and potentially less expensive; improvements in satellite distribution, videotape technology, and coaxial-cable carrying capacity all made the prospect of going around the network system of distribution a less expensive and thus less risky venture.[16] Third, network programming had lost whatever novelty and glitter it once had to much of the public; if in the 1950s simply having a television was enthralling enough in itself, by the late 1960s much of the public was wondering if there might be alternatives to the network formulas that had become so familiar. In this environment a coalition of interests ranging from the cable industry itself to left-wing enthusiasts of cable's "new" technological potentials for diversity and interactivity brought pressure to bear on the policy community's consensus against cable.[17]

After several years of cable's rhetorical repositioning as a "new technology," in March 1971 the FCC held a series of public hearings on cable for the first time, and in March 1972 the FCC adopted the "Third Report and Order on Docket 18397," which in theory eliminated the principal restrictions on cable's expansion into the top hundred markets.[18] The "Third Report and Order" of 1972 marked a reversal in policy: for the first time the FCC took concrete action based on the idea that cable ought to be encouraged to expand and develop.

It took nearly a decade, however, for this policy reversal to bear fruit; numerous adjustments had to be made. Arguably, the regulatory logic over the next decade was not so much one of deregulation, of simply removing regulations, but of taking actions to support cable's growth as an element of the media system. Throughout the 1970s the policy community did gradually lift a series of cable regulations, but

16. The maximum channel-carrying capacity of broadband cable technology grew from three to eight in the 1950s, from eight to twelve in the first half of the 1960s, from twelve to more than twenty in the early 1970s, and by the early 1980s was reaching one hundred and more.

17. Patrick R. Parsons, "Defining Cable Television: Structuration and Public Policy," *Journal of Communication* 39 (spring 1989): 10–26; Thomas Streeter, "The Cable Fable Revisited: Discourse, Policy, and the Making of Cable Television," *Critical Studies in Mass Communication* 4 (June 1987): 174–200.

18. Don R. Le Duc, *Cable Television and the FCC: A Crisis in Media Control* (Philadelphia: Temple University Press, 1973), 193.

more on behalf of cable-industry executives than on behalf of the marketplace: the intention in most cases was to create an appropriate environment for the cultivation of cable. For example, the maximum channel requirements required by the 1972 rules proved difficult for many operators to achieve, and thus the rules were suspended in 1975, and in 1976 the date for compliance was postponed to 1986.[19] Access channel requirements were also substantially relaxed in 1976, and the local origination requirements originally included in the 1972 regulations were later dropped for similar reasons.

The most important developments, however, involved the creation of satellite program networks. When it became clear that major market access was not enough, cable operators began to look for ways to attain nationally distributed programming as the route to success. In November 1975 the FCC adapted its regulations to this goal by abolishing its "leapfrogging" rules against distant signal importation,[20] thereby eliminating a major legal barrier to cable's ability to import inexpensive programming, and making possible the phenomenon of superstations. In December 1976 the FCC decided to license satellite earth stations as small as 4.5 meters, which were dramatically cheaper than the previously minimal 9 meter stations. This reduction, it was estimated, made it possible for systems with as few as fifteen hundred subscribers to afford the earth stations; between eight and nine hundred more systems, therefore, were instantly brought within reach of the satellite program distributors such as Home Box Office (HBO) and Ted Turner.[21] What brought cable to the point of takeoff, in sum, was not cable itself, but the possibility of cheap networking via satellites.

19. George H. Shapiro, "Federal Regulation of Cable TV: History and Outlook," in *The Cable/Broadband Communications Book, 1977–1978,* ed. Mary Louise Hollowell (Washington, DC: Communications Press, 1977), 12.

20. The FCC's original ban on "leapfrogging" required cable operators to import all broadcast signals within the radius of the most distant signal they imported; importing a signal from a distant location in another state or another region of the country was thus effectively prohibited.

21. Another force in the changes in cable policy was the courts, which were acting more out of deregulatory principle than out of a logic of industrial management. The central case here was *Home Box Office, Inc., v. Federal Communications Commission,* wherein an appeals court ruled that the FCC's "ancillary principle" for regulating cable—the principle that cable was subject to regulation on the grounds that it impinged upon the FCC's primary responsibility of local broadcasters—was ruled invalid. The Supreme Court refused to review the case, thereby effectively removing most of the FCC's direct regulations of cable (Shapiro, "Federal Regulation of Cable TV," 11). Significantly, however, the HBO case was in 1977, *after* most of the FCC's major changes in attitude toward cable; the legal climate of deregulation may have accelerated, but did not create, the policy community's shift in attitude toward cable television.

The culmination of the process of cable reregulation (and the event that most sharply belies the vision of cable as a product of unfettered free markets) was the Cable Communications Policy Act of 1984, an amendment to the Communications Act intended to clarify cable's regulatory status.[22] The 1984 act, passed largely in response to cable-industry lobbying, fits nicely within the corporate liberal mode of industry regulation. It is yet another case of legislative power used to stabilize, organize, and protect an industry.

As the FCC gradually removed itself from detailed regulation of cable television during the 1970s, cable operators' legal environment came to be defined largely by city governments. Because of the municipal franchise that cable operators must obtain from cities to string up wires, cities had become accustomed to extracting favors such as public-access channels and franchise fees from cable operators in exchange for the franchise. The 1984 act gave legislative blessing to this informally evolved practice, while carefully circumscribing the amount of legal intervention available to cities in negotiating franchise terms. Cable operators thereby gained legal stability and protection for their monopoly status on the local level (the act effectively prohibited overbuilds), while the political trade-offs for that protection were minimized: cities were prohibited from regulating content and subscription fees, and franchise fees were capped at 5 percent. Shortly after passage of the act, the value of cable stocks leaped upwards, largely because all understood that the unregulated subscription fees charged consumers would soon do the same.[23]

The corporate liberal premise that government should act on behalf of corporate growth is fairly explicit in the 1984 act: among its stated goals is that of establishing "franchise procedures and standards which encourage the growth and development of cable systems" (sec. 601). Also in typical corporate liberal fashion, the act gestures toward noncorporate interests and values, as long as those values are implemented, of course, "in a manner consistent with growth and development of cable systems" (sec. 612). For example, largely due to the efforts of the National League of Cities and the U.S. Conference of Mayors, the act is supposed to "assure that cable systems are responsive to the needs and interests of the local community," primarily by granting cities the power to require public-access channels (sec. 611).

Although the 1984 cable act itself is hardly remarkable as an example of the tradition of twentieth-century corporate liberal legislation,

22. Title VI of the Amended Communications Act of 1934, 47 U.S.C. 601 (1984).

23. Alex Ben Block, "Fat, Wired Cats," *Forbes,* February 25, 1985, 84.

what is remarkable is that at the time it was heralded as an act of *de*regulation, and was passed when the popularity of deregulatory policy was at its zenith. To some extent this interesting construction of industry-protectionism-as-deregulation was just another case of government intervention qualified by the language of antitrust: the act was intended to "promote competition in cable communications" (sec. 601). But the act also managed to differentiate itself from previous regulation by replacing the traditional language of the public interest with language about channel diversity: in several places the act called for providing "the widest possible diversity of information sources to the public" (e.g., secs. 601, 612a). Congress, according to this logic, was neither doling out favors to a protected industry nor interfering with private business affairs. In the popular wisdom it was merely clarifying some legal uncertainties while letting loose an explosion in media diversity, an explosion driven by the "natural" forces of the market and technology.

The most obvious irony in the notion that the 1984 cable act was an act of deregulation is that it created monopolies at the local level, allowing cable franchises to be exclusive. It was entirely predictable that, in the absence of either regulation or competition, subscription fees would climb dramatically in the years following the act. (Congress would eventually pass the 1992 Cable Act, which motivated the FCC to reregulate cable rates precisely on the grounds that cable operators are insulated from competition.) But if local cable monopolies are the most obvious irony, the more important irony is that the economic conditions driving cable and generating pressures for passage of the 1984 act are the same ones that have motivated most regulation of the corporate economy in general: economies of scale that generate pressures toward vertical integration, oligopolistic behavior, and the practice of using regulation to generate industry stability and political legitimacy.

In its first two decades, cable was a relatively entrepreneurial industry, comprising large numbers of relatively small, often locally held companies delivering signals to small communities. Cable is a classically capital-intensive business, however, on both a local and national level. It was inevitable that, as it grew, it would interact with the corporate worlds of Hollywood and network television. When cable was conceptually brought into the fold in 1972, therefore, the corporate world gradually began to involve itself with the cable industry, and brought with it traditional corporate vision: industry planning began to look for forms of integration on a national level.[24] Today, most of the pre-1972 players in

24. One of the first corporate players to enter the field was Time-Life, Inc., which embarked on what was to become Home Box Office in the early 1970s. HBO began as a

the cable industry are gone or absorbed (e.g., Teleprompter), and the key players in recent years bear names familiar from other contexts (Time, Hearst, CBS, Paramount, Warner, Westinghouse). The few new names that did emerge have gradually shed their entrepreneurial roots and have become increasingly corporate in their approach. For example, in the mid-1980s Ted Turner, who had been heavily mythologized in the press for his swashbuckling, entrepreneurial approach, sought to vertically integrate his operations by buying the MGM/United Artists library of films; in turn, the high cost of the purchase forced him to sell a large portion of his company's stock to a coalition of fourteen of the nation's largest cable operators, further integrating the industry as a whole while limiting his individual control.[25] Concurrently, dominance of the industry by a shrinking number of large corporations has steadily increased for the last twenty years.[26]

This is not best understood, however, as simply a problem of monopoly or increasing concentration. The complex series of interindustry mergers, spin-offs, cross-ownerships, cooperative deals, and institutional experiments that characterize the last twenty years have so far produced less economic concentration than in the days of exclusive three-network dominance, and probably less concentration than, say, the auto or mainframe computer industries. But the issue is no more a matter of simple monopoly dominance than it is a matter of the unfettered marketplace. The history of cable television is a solid example of corporate liberal industrial logic at work: over the last fifteen years, cable has been gradually, if occasionally awkwardly, integrated into the American corporate system of electronic media and communications technologies.

---

distributor of movies to cable subscribers in the Northeast via microwave relays, but it soon set its sights on becoming the first nationwide pay television system, inaugurating its satellite-to-cable network in 1975. See *CATV*, November 24, 1975, 18.

25. Al Delugach, "Turner to Keep Control of Firm with $550-Million Bailout Deal," *Los Angeles Times*, January 23, 1987, business sec., part 4, p. 1.

26. The six largest multiple-system operators (Tele-Communications Inc., Liberty Media, Time Warner, Viacom, Cablevision Systems, and Comcast) together serve almost half of all cable subscribers and are heavily involved in programming. TCI, for example, has a stake in Turner and Discovery. Turner, in turn, controls Turner Network, CNN, Headline News, and superstation TBS. Viacom has substantial interests in MTV, Nickelodeon, VH-1, Showtime, The Movie Channel, and Lifetime. Many of these relations are sealed with corporate interlocks: John C. Malone, for example, doubles as TCI president and Liberty Media chair, and six of the fifteen directors of Turner Broadcasting System represent part owners Time Warner and TCI. And predictably, cable has become increasingly intertwined with media interests in general: Capital Cities/ABC Inc. has dominant interests in ESPN and shares Lifetime with Viacom and Hearst Corporation (Kathryn Harris, "Reordering the Cable Universe," *Los Angeles Times*, July 25, 1993, business sec., part D, p. 1.

On one level, all this regulatory activity was functional. The release of certain regulatory restraints in the 1970s did loosen the regulatory framework at strategic moments, allowing cable to be gradually ratcheted into its place between the usually calcified, tightly joined elements of the corporate industrial system of electronic media. Similarly, the 1984 cable act did introduce stability and predictability into the industry that allowed it to become a key piece of some of the fastest growing and most profitable industrial combinations of the late 1980s and early 1990s. But this latest promenade in the elaborate, century-long industry/government dance to the tune of enhancing competition within oligopoly structures suggests a systematic gap between the rhetoric and the phenomena it is being used to describe.

The 1990s and the Search for Level Playing Fields

In the first half of the 1990s, the language of the public interest has yet to make much of a comeback,[27] though the deregulatory enthusiasm for unfettered marketplace competition has waned. The media industries, faced with competition from abroad and technological confusion at home, have rather predictably returned to Congress and the FCC in search of protective and coordinating legal regulation. In the new context of an industry eager for protective government action, a new phrase has appeared: "level playing field." For example, a cable operator faced with threats to his franchise as well as potential competition from a "wireless cable" (microwave) system recently said, "I'm confident that we have the best service as long as we're competing on a level playing field."[28] Similarly, the head of Viacom recently said of a government report recommending telephone-company access into cable's turf, "[T]hat's going to create a lot of aggravation, and it's going to create an unfair playing field. . . . They really haven't thought through the consistency of their recommendations in terms of a level playing field."[29]

In a way, talk of level playing fields is simply another way to advocate legal and policy decisions that favor me and not the other person. The term is most often heard from representatives of industry factions

27. The lone advocate of public-interest principles on the FCC in the late 1980s and early 1990s was Commissioner Ervin Duggan. See Ervin S. Duggan, "The Once and Future Public Interest Standard," paper presented to the International Communication Association, Washington, DC, May 28, 1993.

28. TCI general manager Scott Brown of Westchester County, NY, quoted in R. Thomas Umstead, "Overbuild Threat, Complaints Don't Faze New TCI Manager," *Multichannel News*, November 29, 1993, 110.

29. President and chief executive officer of Viacom, John Goddard, quoted in "NTIA Opens Pandora's Box for Change in Cable: Beginning with Telco Entry," *Broadcasting*, June 20, 1988, 38.

(broadcasters, cable companies, etc.) seeking government regulation that will protect or enhance their competitive positioning. The cable operator is arguing that his franchise be renewed on terms that are favorable to its continued profitable existence. Viacom wants government regulation to continue to restrain telephone companies' efforts to compete with it.

But the context and connotations of the phrase are worth elaborating. The term was hardly heard in the policy community before 1986, but its use has grown almost exponentially ever since, to the point where as of this writing it has become a regular incantation.[30] The term became popular, then, in the second half of the 1980s, in the immediate aftermath of the deregulatory changes of the early 1980s, particularly the breakup of AT&T and the successful reregulation of television to make room for cable television. In the mid-1980s, having achieved major political successes with the language of free markets and deregulatory economics, industry members were faced with the problem of finding a language to articulate something whose very existence they had been denying: their dependence on government regulation.

The language of level playing fields has proven effective in this regard. On the one hand, this metaphor from team sports has a ring of healthy competition about it; in an environment still suspicious of the "public interest" and still redolent with the 1980s enthusiasm for marketplace competition, the rhetoric of level playing fields provides a way to call for protective government regulation while suggesting a willingness to compete. On the other hand, this informal metaphor nicely makes clear what all the volumes of neoclassical economic analyses obscured: first, markets occur not in a state of nature but when underlying conditions and rules of behavior—that is, the rules of the game—are socially constructed; and second, the ways those underlying conditions are constructed can "tilt the field," that is, privilege some players over others. The rhetoric of level playing fields thus informally acknowledges the constructedness and political character of market relations to a much larger degree than was the case in formal deregulatory discourse, and thus makes calls for government regulation more palatable.

The mythical component of the rhetoric of level playing fields, however, consists of the idea of "levelness," the implication that there exists a neutral, universal yardstick, a legal equivalent of the carpenter's

30. A search of the COMPUB file of communications trade journals in the NEXIS/LEXIS online database showed only 6 articles in the trade press containing the phrase "level playing field" before 1984, 81 articles in the two years after 1984, 154 in the two years after 1986, 358 after 1988, 719 after 1990, and 1,244 in the two years since 1992 (through February 18, 1994).

level, against which the tilt of the economic playing field can be measured. The problem is that, in the socially constructed world of economic relations, "level" will always be a relative term. The definition of a level playing field always depends on one's point of view. Granted, in specific circumstances it is sometimes possible to demonstrate that one or another regulatory decision will generate more or less competition in a specific area. Allowing telephone companies to compete directly with cable operators will indeed generate ferocious competition for cable operators, quite possibly so much competition that some or all cable operators will be forced to choose between merging with the phone companies or being driven out of business. But these are matters of degree, and involve taking for granted certain existing conditions. In the larger scale of things, what distinguishes "tilt" from level in the marketplace is a political balance of power, not gravity. No one speaks of the unlevel playing fields between media corporations and ordinary citizens, for example, or between the United States and developing nations, or between those with inherited wealth and those without.

Ultimately, then, there will always be a reasonable point of view from which any act of regulatory "leveling" appears as a granting of privilege, as a government favor on behalf of a private industry or industry segment. Maintaining regulatory restrictions on phone-company access to video delivery will be a favor granted to cable companies, and removing those restrictions will be a favor granted to telephone companies. And the regulatory compromises that are the most probable outcomes of current struggles over this issue will amount to a favor granted to both sides: profitable territory will be carved up, and a handful of established corporations will be granted the privilege of cultivating the territory for profit. In sum, the current rhetoric of level playing fields undoubtedly marks a pendulum swing away from deregulation, but it is a pendulum swing that remains quite within the general bounds of corporate liberal practice overall.

Summary

In 1925, as the first broadcast legislation was still being drafted, Herbert Hoover confidently predicted that, if the industry were left to fend for itself, "intrusive" product advertising would never become the norm in broadcasting because of the nature of open competition. Since then, the history of broadcast regulation has been filled with countless examples of regulators acting on behalf of what they describe as the marketplace, only to be pilloried by the next generation of regulators for undermining the free market that they had imagined they were upholding. Hoover's generation acted on behalf of free enterprise. The results of their efforts were then attacked by New Dealers for having created monopolies at the expense of market competition. In the late 1970s and 1980s, deregulators in turn attacked precisely the regulatory patterns established by the New Dealers (and their timid descendants of the 1950s and 1960s). And today, amid calls for "level playing fields" and complaints about unfair cable monopolies, new regulations are being gradually introduced, many in the name of competition. And if history is any judge, it is a safe bet both that many of those regulations will work to the benefit of industry overall, and that in a few years many of those same regulations will come under attack for hindering marketplace competition.

Policies of Reform

Populist Antimonopoly Efforts

If competition has been the dominant goal of broadcast policy, at most times in the history of commercial broadcasting there also have been important efforts on behalf of reform that to various degrees have something more in mind than fine-tuning marketplace competition. Perhaps the most common strategy for those seeking change in the electronic media has been to appeal to the dominant discourse, the discourse of competition and monopoly, for nondominant ends. The current system, the argument goes, is or is headed toward monopoly; this is *both* undemocratic and economically inefficient, and so restraints on corporate action or other forms of change are called for because they will make broadcasting available to the grass roots *and* enhance competition. This was a central concern of reform-minded legislators in the 1920s who inserted the antitrust clause into the Radio Act, it was central to the network regulations of the late 1930s, and it has remained a central refrain of left-of-center reformers ever since: Ben Bagdikian's compelling *Media Monopoly* remains a favorite of both media activists and leftist college professors.

The obvious appeal of this discourse on behalf of reform is that it resonates with core American values. In American politics it is much more palatable to argue for competition and against monopoly than, say, for the poor and against the rich, or for nonprofits and against private enterprise. Inside the beltway no one wants to be construed as restricting competition. Intervention in the name of competition, furthermore, has broad legal support in the Sherman Act and its many legislative echoes, such as the Communications Act.

The limitations to the discourse of competition are clear, however. Suggesting that the key problem with the media industries is monopoly or concentration does have the effect of calling into question the industry's claim that it is fully democratic and open. If the discussion is left at that level, however, in the end the commercial principles of the system are upheld; the discourse of competition as it is normally presented takes for granted the legitimacy of private, for-profit control of the system itself. The American polity, as a result, seems inevitably to respond to accusations of monopoly, not with alternatives to commercialism, but with efforts to regulate the existing system so as to make it more "competitive."

The dichotomy of competition and monopoly, furthermore, cannot adequately grasp the center-periphery, oligopolistic structures characteristic of corporate industries. On the one hand, the discourse tends to suggest that the dangers of monopoly are largely a problem of ownership, which lends itself easily to the common belief that more and different owners will substantially change the character of the system (as in FCC regulations designed to encourage minority ownership of stations). While ownership is important, an exclusive focus on it obscures the pervasive, collective powers of corporate managerial culture, which regularly extends across ownership boundaries, exerting its influence in both "competitive" and "concentrated" industry segments alike.

On the other hand, given the imprecision of the discourse of competition, it will in most cases be plausible to claim both that any given situation is competitive and that it is the reverse. In corporate industries there will always be elements of economic struggle and competition, both between industry factions and, in the peripheries, between large numbers of enterprises. By the same token, conditions will never fully conform to Adam Smith's vision of wide-open markets as long as economies of scale and corporate organization are preferred by the legal and economic system. Neoclassical economic theories notwithstanding, in the rough and tumble of political discussions in which decisions are made, the discourse of competition is indeterminate and can lead to an infinite regress of argument and counterargument without really touching on the issues that matter most.

Without a clear sense of this pattern, broadcasting will remain "competitive" only in the periphery, and oligopolistic at its center. The chain broadcasting investigation of the late 1930s had little effect on industry structure and behavior, not because the resulting regulations had no teeth; they were the most aggressive thing the FCC has ever done. Rather, they were based on an assumption of naturally occurring competition, and thus attacked intentional competition-restricting actions rather than economic structures. Most subsequent competition-oriented efforts have similarly failed to touch underlying structures, and thus have amounted to little more than inside-the-beltway regulatory tinkering, whose principal effect has been to shift pie sharing among elements of the system.

The Discourse of the Public Interest

Alongside accusations of monopoly, progressive critics have also relied very heavily on appeals to various constructions of "the public interest." Like competition, notions of "the public interest" have the advantage of centrality to existing legal language, and a widespread popular appeal. And unlike competition, on its face the term "public interest" suggests something noneconomic, something outside the bounds of private property and market exchange. It is absolutely crucial, however, to acknowledge both the liberal roots of the term as it has been used and the specific corporate liberal uses of it that have historically played a key role in the construction of commercial broadcasting.

It is common these days among certain centrist and left-of-center policy activists to wax nostalgic about the good old days before Reagan-era deregulation when the public-interest principle was taken seriously, and calls have been heard for the term's revival.[31] Before such a strategy is pursued, however, it should be remembered just what it meant to take the term seriously: even under the reign of the more antiadvertising regulators such as Herbert Hoover, upholding the public interest in practice meant regulation clearly directed at favoring a corporate-dominated broadcast system. At the same time that Hoover was complaining to the Radio Conferences about "advertising chatter" in the airwaves, he was also distributing license privileges to the corporations who controlled "our great industrial laboratories" and ensuring them the power to make the fundamental decisions about content and funding in the new industry of broadcasting. To his mind, and to the minds of many in positions of authority today, this was thoroughly consistent. In this century Herbert Hoover's seminal use of the phrase still predominates: the dominant legal uses of the term suggest a functionalist, systemic vision of social relations, and are easily subsumed into a technocratic interpretation, as a general term for the extramarket social engineering imagined to be necessary to the smooth integration of the corporate system, which stands in a paternalistic relation to a consuming public.

True, there have been a number of struggles in the regulatory arena to push the public-interest principle beyond the corporate liberal boundaries of Hoover's vision, and alongside some failures there have been some important, if partial, successes. For example, the Blue Book,

31. See, for example, Newton N. Minow and Craig L. LaMay, "Licensing Filth: Radio and the FCC," *Chicago Tribune,* February 10, 1994, "Perspective" sec., 31; see also Duggan, "Once and Future Public Interest Standard."

seriously if unsuccessfully, attempted to introduce regulatory practices under the banner of the public interest that, if they had been successful, would have amounted to small incursions into for-profit control. Perhaps the most important successful uses of the public-interest principle to stretch the traditional bounds of corporate liberal broadcasting have been the noncommercial channel reservations in FM radio and television. Although an effort to amend the 1934 act so as to reserve one-quarter of the AM band to nonprofits failed, there have been less ambitious efforts since then, which met with success.[32] First with FM in 1941, and then with television beginning in 1952, the FCC reserved one or two channels in most markets for noncommercial broadcasters.[33] This was a foresighted action that created a platform upon which the public broadcast system could be built. When these bands were first opened up, nonprofit broadcasters did not have the resources to obtain and maintain station licenses. Later, when resources became available, the cost of purchasing by-then-privately-held stations would have been prohibitive. Without the channel reservations, American community and public television and radio, in all likelihood, would not exist, at least not to the degree they do today.

The noncommercial channel reservations nicely illustrate the fact that the FCC's most important power is creative, not regulative; the FCC's ability to create and define channels is more significant than its ability to govern the behavior of existing licensees. The channel reservations used that power on behalf of noncommercial broadcasting. The Blue Book was an after-the-fact regulatory strategy: the Blue Book sought to use the public-interest clause to force existing commercial stations to

32. The Wagner-Hatfield amendment to the Communications Act of 1934, had it passed, would have reserved one-quarter of the existing broadcasting frequencies for nonprofits. As McChesney has documented, when the FRC excluded educational and other nonprofit broadcasters from the spectrum in the late 1920s, a lively reform movement developed that culminated in the drive to pass the Wagner-Hatfield amendment. The strategy of the amendment is noteworthy for its awareness of the character of the power relations in the industry: rather than trying to engage in after-the-fact regulation of the behavior of existing commercial broadcasters, it addressed the basic structural issue of for-profit organization, and took action based on the government's fundamental power to create and define the character of broadcast channels. It did not, moreover, prevail on the FCC in the name of the public-interest clause, but instead went straight to Congress (Robert McChesney, *Telecommunications, Mass Media, and Democracy: The Battle for the Control of U.S. Broadcasting, 1928–1935* [New York: Oxford University Press, 1993]).

33. The first noncommercial channel reservations involved the FCC's first and ill-fated attempt to enact FM standards in 1941. When FM was reallocated in 1945, the lowest twenty of the allocated one hundred channels were reserved for educational use. Noncommercial reservations in television first appeared in the FCC's *Sixth Report and Order*, 17 Fed. Reg. 3905 (1952), 3908, 41 F.C.C. 148, 158.

engage in unprofitable activities. The channel reservations strategy, in contrast, gets in on the ground floor, so to speak.

Of course, if the channel reservations stretched the boundaries of corporate liberalism, they did not go outside them. It is perhaps no coincidence that public broadcasting today rather strikingly resembles the kind of broadcasting that Herbert Hoover seems to have had in mind in the 1920s: generally noncontroversial, patrician programming supported to a large degree by corporate donations in exchange for low-key, image-enhancing corporate "publicity." The reservations created platforms, but did not address the problem of funding public stations. Ever since, the biggest problem facing public broadcasting in the United States has been a lack of funds for programming. But it is also significant that the channel reservations were legitimated largely through the argument that they should be alternatives to, not competitors with, existing broadcasters. It was in the public interest to support noncommercial stations, the FCC argued, because such stations would provide "programming of an entirely different character from that available on most commercial stations," programming that would "provide a valuable complement to commercial programming."[34] The licenses were thus defined as "noncommercial" instead of nonprofit, and prohibited from running advertising.[35] This understanding of public broadcasting as alternative has seriously constrained public broadcasting's behavior. In particular, it has prevented the exploration of a large range of organizational possibilities, such as product advertising and head-to-head competition with commercial broadcasters for viewers.

For all their limitations, however, it must be remembered that the public broadcasting institutions enabled by the noncommercial channel reservations have done far more to change the character of the broadcasting available to the American public than any of the content-oriented public-interest regulations of commercial broadcasters that have captured so much of the attention of the FCC and the law journals. The FCC's theoretical power under the public-interest clause to engage in

34. FCC, "Third Notice of Proposed Rulemaking: Appendix A," 16 Fed. Reg. 3072 (1951), 3079, reprinted in *Documents of American Broadcasting*, ed. Frank J. Kahn, 4th ed. (Englewood Cliffs, NJ: Prentice Hall, 1984), 181; FCC, *Sixth Report and Order*, 158.

35. In response to concerns about public broadcast funding in the early 1980s, Congress established a Temporary Commission on Alternative Financing for Public Telecommunications, which eventually led the FCC to allow for "enhanced underwriting," that is, between-program promotions that showed both company logos and simple descriptions of "representative" products and services (Sydney W. Head and Christopher H. Sterling, *Broadcasting in America: A Survey of Electronic Media*, 6th ed. [Boston: Houghton Mifflin, 1990], 264–65).

after-the-fact regulation of existing stations has produced much storm and fury and a few Supreme Court cases, but has had very little impact on the overall character of American broadcasting. Most of what has occurred has been "regulation by raised eyebrow," that is, unfulfilled regulatory threats that cajole industry members into slight modifications such as a few more documentaries or minor modifications in programming directed at children.[36] Arguably, public broadcasting has done much more for diversity of public dialogue in the United States than the fairness doctrine and section 315 ever have, and most would agree that public television's children's programming has been much more important and valuable than any of the "prosocial" children's programming efforts cajoled out of the commercial broadcasters by the threat of regulation.

There are, of course, a handful of cases in which the FCC did go beyond the raised eyebrow and invoked its power under the public-interest clause to revoke licenses on other than procedural or technical grounds. In the Brinkley case of the 1930s, for example, an unscrupulous snake-oil salesman was forced off the air for endangering the public health, and in the 1960s a radically right-wing religious radio broadcaster and a crudely racist and segregationist southern television station lost their licenses.[37] It is significant not only that these cases have been extremely rare, but that the targets have been politically and culturally marginal entrepreneurs, not stations engaging in typical corporate behaviors. Each of the offending broadcasters was violating, not only the FCC's sense of the public interest, but the corporate liberal principle of paternalistically addressing the public strictly as apolitical consumers. In each of these cases, if a network-owned station had tried to broadcast similar material, a New York–based network executive probably would have censored it without giving it a second thought; when some network executives and other major authorities in the industry objected to the FCC's actions, they were objecting to the fact that a government agency had taken the actions, but not to the character of the actions

36. Perhaps the most famous incidents of raised-eyebrow regulation include the minor network "documentary boom" prompted in part by the quiz-show scandals and Newton Minow's "vast wasteland" speech in the early 1960s (Michael Curtin, *Redeeming the Wasteland: Television Documentary and Cold War Politics* [New Brunswick, NJ: Rutgers University Press, 1995]).

37. For the Brinkley case, see *KFKB Broadcasting Association, Inc., v. Federal Radio Commission,* 47 F. 2d 670 (D.C. Cir. 1931); for the right-wing religious broadcaster, see *Red Lion Broadcasting Co., Inc., et al. v. Federal Communications Commission et al.,* 395 U.S. 367 (1969); and for the racist station (which was WLBT of Jackson, MS), see *Office of Communication of the United Church of Christ v. Federal Communications Commission,* 425 F. 2d 543 (D.C. Cir. 1969).

themselves. Even if the license revocations made some industry members nervous, in sum, they were consistent with the patterns of corporate liberal broadcasting as a whole.

Given the paternalistic connotations of the public-interest clause, it is probably no accident that one of the most compelling uses of the term involves real (rather than metaphorical) children: since the mid-1960s there has been a sustained effort to use regulation to change programming watched by children and to reduce the violent content of broadcast programs largely because of the effect that content is believed to have on children. The history of these efforts is complex and ongoing, and has been discussed in detail elsewhere.[38] What is important for this discussion is the way the debate is framed. One could frame this question as a moral and political one: How do we as a democracy want to use the possibilities of broadcasting for our children? Of course, what prevents matters from being addressed in these terms is the construction of commercial broadcasting as "private" and thus in its basic outlines insulated from collective control or public responsibility. If it could be shown to be the case that most people did not want to use children's television primarily as an advertising vehicle for toys, legislation or regulations that simply enacted that popular preference would be viewed as illegitimate interference in private business affairs (or, as is becoming more common, as violations of free speech).

In a corporate liberal political environment, qualifications of basic liberal principles are typically justified when they can be constructed as "scientific," and thus as neutral and impersonal instead of based in subjective preferences. Rowland has shown how, in the case of the famous surgeon general's inquiry into televised violence in the early 1970s, a principal motivation for the turn to scientific research was a sense that it provided the only way around the prohibition on violating the barrier between government and private entities. As the principal political player in the episode, U.S. senator John Pastore, put it, "when it comes to the format of a program you get into the area of censorship, and it is hard to legislate, if at all possible. . . . We need expert authority. You just can't legislate here, and you can't debate here in an area of darkness. We have to find out first scientifically what effect it does have so that we will have proof positive when we begin to move."[39] It is precisely this logic that

38. Willard D. Rowland, Jr., *The Politics of TV Violence: Policy Uses of Communication Research* (Beverly Hills: Sage, 1983).

39. Senator John O. Pastore to Secretary Finch, p. 31, in Senate Commerce Committee, Communications Subcommittee, *Federal Communications Commission Policy Matters and Television Programming,* Hearings, parts 1-2, March 1969, 337-38, quoted in Rowland, *Politics of TV Violence,* 146.

has driven the entire question of children's programming exclusively toward questions of scientifically provable negative effects that would justify after-the-fact restrictions on the behavior of existing commercial broadcasters. Within this construction of the problem, matters of structure are marginalized. An enormous volume of suggestive social scientific and psychological research has been conducted on the effects of television on children, and legislative hearings on the matter have become a regular ritual in Washington. To date, these efforts have had at best only marginal influence on the character of programming for children produced in the commercial context. Meanwhile, popular opinions on the matter remain largely unknown, and debate is largely confined to universities, inside-the-beltway meetings, and other "expert" forums.

The Discourse of Free Speech

These days, "free speech" has been almost entirely captured by economic conservatives, and it is treated as an opposing term to "the public interest." Free speech most often stands for industry independence from regulation (it has come to be almost a synonym for the marketplace), and the public-interest principle stands for increased government regulation.

This was not always the case. The drafters of the 1927 Radio Act spoke approvingly of both free speech and of the public interest, and did not seem to view the two terms as in tension with one another. It was probably more common at the time to see tensions, not between free speech and the public interest, but between free speech and commercialism. During the hearings leading up to the passage of the 1927 Radio Act, for example, Morris Ernst of the American Civil Liberties Union (ACLU) testified that "[w]e are deeply concerned in the bill in so far as it relates to the question of censorship and freedom of speech. Even the term 'free speech' is more or less of a misnomer when you have to pay $400 an hour in one of the good New York stations and are lucky if you can get on at all. . . . the whole bill is predicated on money."[40] Ernst gave a list of political activists who had been denied access to the air by private broadcasters, and pointed out that "Secretary Hoover's signature in New York City sells for $150,000 to $200,000," thus limiting access to the air on the part of labor unions and other underrepresented groups (125–27). Ernst proposed that, among other things, stations should be transferred by the FRC, not by private exchange; nonprofits should be given preferential treatment in the granting of licenses; no entity should be allowed to own more than one station; and the licensing process

40. Senate Committee on Interstate Commerce, *Radio Control: Hearings before the Committee on Interstate Commerce*, 69th Cong., 1st sess., 1926, at 42 (hearings on S. 1 and S. 1754), January 8 and 9, 125–29.

should be open to the public (128–29). In the name of free speech, the ACLU's Ernst was calling for a change in regulatory policy that not only would have led to the creation of a radically different system of broadcasting, but also embodied a substantive, political approach to free speech, with clear reference to political economic structure.

As an analysis, Ernst's approach suggested regulation with attention to fundamental structure so as to encourage public dialogue. He was not arguing on behalf of formal barriers between government and private entities. And as a tactic, Ernst's approach was political, not legalistic. His approach reflected the view of one of the ACLU's founders, Roger Baldwin. Of the ACLU's early approach, Baldwin has said:

> If we had been a legal aid society helping people get their constitutional rights, as such agencies do their personal rights, we would have behaved quite differently. We would have stuck to constitutional lawyers [and] arguments in courts. We would not have surrounded [the ACLU] with popular persons. But we did the opposite thing. We attached ourselves to the [labor and other] movements we defended. We identified ourselves with their demands.[41]

Baldwin himself came to be concerned with the restraints on political dialogue exerted by the commercial broadcast system, and in 1933 helped to establish the ACLU's Radio Committee. The committee lent its weight to the broadcast reform movement in the period leading up to the 1934 Communications Act, and continued to push for structural reforms for several years afterward.[42]

Yet by 1938, after meeting with failure, the ACLU changed its tactics. The Radio Committee sought to bring industry members into its ranks and gradually shifted in the direction of a formalist, legalistic position that lent itself to protecting commercial broadcasters from government harassment.[43] And since then, the legal establishment has elevated a formalist interpretation of the First Amendment to quasi-religious status in American consciousness, free speech has come to mean freedom for private communicating entities (in practice, usually media businesses) against any sort of unwanted political interference, and questions of media structure have become almost entirely disassociated from matters of free speech in mainstream discourse.

41. Roger Baldwin, interview by David Kairys, "Freedom of Speech," in *The Politics of Law: A Progressive Critique*, ed. David Kairys, rev. ed. (New York: Pantheon, 1990), 255.

42. McChesney, *Telecommunications*, 80–86.

43. Ibid., 236–39.

Since the 1920s, then, the political affiliation of free speech has been slowly moving from left to right, at the same time that the legal interpretation of free speech has become increasingly formalized and broadened to encompass an ever wider variety of corporate activities. Today, corporate donations to political campaigns, telephone-company entry into the video delivery business, and cable operators' elimination of public-access channels are being described as protected "speech" by the courts. In the popular imagination free speech still stands for dissent, for alternatives, for debate. But, practically speaking, free speech functions to structure industry relations and insulate them from political accountability.[44]

Free speech has come to function in much the same way that property and contract functioned in the nineteenth century: on a practical level, as a justification for prohibitions of any kind of unwanted political intervention with the affairs of business leaders, and on a theoretical level, as the archetype of bright-line legal neutrality. Questions of structure in association with free speech have all but disappeared. If a more substantive notion of free speech is to be revived, its alignment within contemporary legal and political logic will have to be fundamentally altered.

The Discourse of Access

If free speech no longer works for advocates of change, media access seems to have taken its place. Measures designed to allow outsiders some form of representation in radio and television are probably the most structurally oriented discursive strategies available in the current environment. The very notion of access calls attention to the fact that the power to communicate over radio and television is not evenly distributed. And the notion of access has led to some notable, if small, legal and institutional successes: section 315 of the Communications Act requires that broadcasters allow all political candidates equal opportunity to buy time during campaigns, and the principle of "cable access" has produced a tradition of public, educational, and government (PEG) access channels on most cable systems and an accompanying tradition of grassroots cable television programming.[45] And the principle of access may very well serve as a compelling tool for carving out open spaces in the ongoing transition to more computer-based forms of electronic media.

44. Monroe E. Price, "Speech, Structure, and Technology," *Cardozo Studies in Law and Literature* 2 (spring 1990): 113.

45. William Boddy, "Alternative Television in the United States," *Screen* 31 (spring 1990): 91–101; Patricia Aufderheide, "Cable Television and the Public Interest," *Journal of Communication* 42 (winter 1992): 52.

The principle of access, furthermore, like the public interest, has in the past lent itself persuasively to fairly dramatic forms of political intervention. In the late 1960s and early 1970s, for example, activist groups sought to use the legal system to force commercial broadcasters to grant access to outside groups and alternative points of view. For a while, these efforts seemed to hold up the potential for a serious opening up of the broadcast system. In the WLBT case the courts forced the FCC to grant legal "standing," not just to commercial broadcasters, but to groups representing audience members in license-renewal cases, thus opening the door to a series of license challenges from community groups that, when they did not lead to license revocations, at least encouraged broadcasters to take the concerns of the community groups more seriously. During the same period, in the wake of the *Red Lion* case that upheld the fairness doctrine, activists sought to extend the principle of fairness to advertising, to the commercial industry's lifeblood. Not just newscasts, but advertisements as well, the argument went, often present arguments about controversial issues of public importance, and thus should be subject to fairness doctrine requirements of balanced coverage. Hence, the *Banzhaf* case forced broadcasters for a few years to air antismoking public-service announcements free of cost as "replies" to the enormous amount of cigarette advertising that then saturated the airwaves.[46]

After some initial successes, however, these efforts were rebuffed. Attempts to extend the logic of *Banzhaf* further (e.g., to car ads) failed, and eventually even the question of cigarette advertising was rendered moot by legislation banning cigarette ads from broadcasting altogether.[47] Citizen access to license procedures was rendered impractical by an elaborate bulwark of FCC administrative requirements.[48] And contemporary formalist legal thinking increasingly threatens to label access requirements, not as tools for the enhancement of free speech, but as forms of government interference with the free speech rights of businesses.

Part of the problem with access arguments is purely tactical. Because the Supreme Court limited access arguments to broadcasting in the *Tornillo* case, access advocates have had to fall back on the weak notion of spectrum scarcity to justify access provisions, rather than broader and more plausible justifications based on concentrated power

46. *Banzhaf v. Federal Communications Commission,* 405 F. 2d 1082 (D.C. Cir. 1968), 14 Radio Regulation 2d 2061.

47. Public Law 91-222, 15 U.S.C. 1335 (1971).

48. Willard D. Rowland, Jr., "The Illusion of Fulfillment: The Broadcast Reform Movement," *Journalism Monographs* 79 (1982): 17.

of any sort.[49] The general retrenchment against many progressive political trends of the late 1960s, furthermore, has made both courts and legislatures more resistant to such efforts.

But, such tactical problems aside, the discourse of access remains assimilable within a general corporate liberal horizon. First, access arguments involve asking for access to something that someone else already owns or controls; the legitimacy of that ownership and control is thus presupposed. Second, access arguments tend to be part of a general discourse of administratively achieved balance and neutrality: the fairness doctrine, for example, was a derivative of the philosophy of broadcaster neutrality that was used to justify excluding labor and other groups from the airwaves in the first place. The granting of access is thus easily interpreted as one of the technocratic corporate liberal adjustments useful for maintaining smooth relations between corporations and the consuming public.

Summary

In his 1970 book, *How to Talk Back to Your Television Set,* maverick FCC commissioner and reformer Nicholas Johnson proclaimed: "[B]y understanding and using the right strategy the meekest among us can roll back the ocean. . . . You *can* fight city hall, the 'little man' [*sic*] *can* do effective battle with massive corporate and governmental institutions, the government *can* be made to be responsive to an individual citizen's desires."[50] Johnson was writing at a moment of optimism for those seeking to introduce democratic change into American broadcasting. There have been other such moments, such as the broadcast reform movement of the early 1930s, recently detailed by McChesney.

But the fact remains that today's system of electronic media and its relations to corporate consumer industries is remarkably similar in its outlines to the system of 1930. With a few important exceptions, furthermore, whatever change has come has tended to be driven by the corporate world itself. A decade after Johnson's confident call to arms, two former reformers reflected: "[W]e should no longer accept an abstract hope that an evolution from license challenge to rule making to lobbying in Congress will in and of itself produce multiperspectival mass media. That route has become a dance of delay, limits, cooptation, and quiescence."[51]

49. *Miami Herald Publishing Co. v. Tornillo,* 418 U.S. 241 (1974), 254.

50. Nicholas Johnson, *How to Talk Back to Your Television Set* (New York: Bantam, 1970), 188–91.

51. Tim Haight and Laurie Weinstein, "Changing Ideology on Television by Changing Telecommunications Policy: Notes on a Contradictory Situation," in *Communication and*

Antimonopoly, the public interest, free speech, and access cannot be dismissed. They speak to important American democratic traditions and ideals, and can be used to advantage in the political arena. But one should not expect that, because these words seem to connote exalted ideals, their presence in a piece of legislation or a policy directive will ensure that those ideals will be upheld in practice. Nor should one too readily grant the argument that, because these terms have force in existing law and politics, the practical approach is to rely on them. These terms must be approached with a sense of their context, their history and their role in institutional practices. They gain their meaning, their force, not from what one wants them to mean but from the frameworks used by others to interpret them, and those frameworks are in turn shaped by the way the terms have been effectively used in the dominant social formation. And for most of this century, these words have been used comfortably alongside, and in many cases on behalf of, corporate-dominated broadcasting. Antimonopoly, the public interest, free speech, and access are, in practice, corporate liberal terms.

As a result, progressive efforts to use this language on behalf of democratic change are systematically deflected. In the contemporary context, railing against the unregulated marketplace on behalf of the public interest, vilifying media monopolies, or promoting the free speech rights of the disenfranchised tend to be of limited consequence: it is easy for commercial broadcasters to claim that they *do* follow the public interest, they do compete, they do support free speech—in the corporate liberal senses of those terms.

The problem need not be seen in terms of a simple functionalist determinism, that is, that capitalism needs supportive regulation, and what capitalism needs it gets. Part of the difficulty in opening the politics of broadcasting in the United States is the mixed character of the dominant discourse. Sometimes arguments tend toward the classic liberal: the rather obvious structural constraints and limits of corporate organization and advertising support are hidden behind the tautologous assertion that, because our system is market driven, whatever happens is a product of the marketplace and is therefore by definition free and open. Other times arguments lean more in the corporate, administrative direction: functional drives toward "efficiency," "stability," and similar "natural" forces have produced a system that may have its faults but that seems economically productive, is preferable to alternatives, and is in any case inevitable. And supporters of the status quo have the option of hopping

---

Social Structure: Critical Studies in Mass Media Research, ed. Emile McAnany, Jorge Schnitman, and Noveene Janus (New York: Praeger, 1981), 141.

between the two as it suits their purposes. In response to logical extensions of administrative discourse, supporters fall back on classical logics: the Blue Book and the fairness doctrine violate private rights of free speech and the principle of business autonomy. Yet, in response to extensions of classical discourse, supporters turn to administrative arguments: granting audience members or nonprofit and minority groups "rights" to the broadcast spectrum, it is argued, would be inefficient and impractical.

As a result, with very few exceptions, efforts to enact positive changes in the system by way of these terms have been turned into after-the-fact tinkering with the existing system, into efforts to use government powers of regulation to force private business interests to modify their behavior in small ways. Businesses have been told to try to make their news coverage more "fair," that they can buy no more than twelve stations, that they can broadcast only so many commercials per hour during children's programming, and so on.

All of these after-the-fact efforts take for granted the legitimacy and character of the private entities created by government intervention in the first place, and for that reason have limited effect. But it is precisely because these entities are taken for granted and assumed to be part of the natural order of things that political action of any kind is popularly assumed to be "government interference," while the "interference" that created those entities in the first place is obscured. Progressive efforts, therefore, would do well to seek ways to open up this closed door in political discourse, ways to "defamiliarize" the obvious, taken-for-granted character of the institutional forces that constitute and drive commercial broadcasting.

How Competitors Cooperate: The Search for "Orderly Progress"

In this context, CBS chair William Paley's 1936 statement to the FCC is worth recalling: "[S]udden revolutionary twists and turns in our planning for the future must be avoided. Capital can adjust itself to orderly progress. It always does. But it retreats in the face of chaos." At the time, inventor Edwin Armstrong and his supporters were promoting the new technology of FM radio—a startling improvement over AM—and were requesting that the FCC enact technical standards and spectrum allocations to allow its commercial development. Paley was probably right to be worried about this threat to his current fiefdom based in the AM band: if the FCC were to create new spectrum territories for the better mousetrap of FM radio, CBS's relatively stable oligopoly with NBC could be seriously threatened. The battle over FM was a battle over politically created territories, not so much a struggle in the marketplace as a struggle about the *creation* of a marketplace. And it is telling that Paley, with the help of his economic competitor and political ally RCA, largely got his way: delays and vacillations in the implementation of FM standards in the 1930s and 1940s successfully prevented its commercial development until after the networks had transferred their empires to television and thus could afford to abandon AM.[52]

It would be reductive, however, to suggest that the relation between property creation in broadcasting and broadcast markets is a simple, mechanical one. Paley's plea for "orderly progress" is revealing not only because it demonstrates the importance of political struggles over the creation of properties in the spectrum, but because it expresses a habit of thought characteristic of industry practice in general: the ideal of "orderly progress."

Paley, like most corporate heads, was continually faced with pressures from three sides: pressures toward growth and expansion, pressures toward stability, and political pressures of legitimation. Too much emphasis on stability, and a business can stagnate and be left behind; this is what happened to Marconi when he ignored the importance of continuous-wave and voice-transmission technologies. Conversely, too much expansion, and things become uncertain and risky, and open to unexpected forms of competition; this is what happened during the broadcast boom in the early 1920s, and what Paley and Sarnoff successfully circumvented by delaying FM's introduction in the late 1930s. And if political legitimacy is neglected, favorable government regulation becomes more difficult to obtain, and the threat of unfavorable regulations is raised; this is largely why the industry occasionally engages in small forms of unprofitable, "altruistic" behavior such as running public-service announcements or covering major news events live and uninterrupted by commercials. Effective management of corporations requires an artful negotiation of these conflicting pressures, not just on the part of individual managers, but on the part of the community of management overall: it requires collective creation, negotiation, and maintenance of shared behaviors and habits of thought.

Market Limitation

The order-maintaining social conditions that undergird markets in electronic media are complex, but it is possible to identify a few common

52. Vincent Mosco, *Broadcasting in the United States: Innovative Challenge and Organizational Control* (Norwood, NJ: Ablex, 1979).

patterns of social behavior in the industry. One familiar pattern involves what some call market limitation, that is, building institutional structures that effectively limit external competition without engaging in practices such as price-fixing that would be easily actionable under the antitrust laws. Corporate structure itself, with its tendencies toward vertical integration, effectively limits competition by, in Chandler's words, "internalizing markets," and thereby creating barriers to market entry simply in terms of initial capital investments. Potential competitors must be large, well-established, and wealthy enough to spend the huge sums necessary to build similarly vertically integrated enterprises, a prerequisite that automatically limits the field. This is particularly true in capital-intensive industries, of which broadcasting is a classic example.[53]

Compared to some of the classic examples of capital-intensive industries such as steel, railroads, or automobiles, however, radio and many other forms of electronics are relatively simple, cheap, and accessible. In the fields of high technology, therefore, a second set of practices have played a key role in limiting market entrance: patents. Although one administrative agency in Washington is devoted to maintaining competition (the FTC), another is devoted to creating monopolies: the patent office.

The history of technological innovation in electronics in this century is heavily colored, to some degree defined, by a series of titanic struggles over patents. Many of the major modern American electronics empires were originally launched from the platform of patent-created monopolies: AT&T and the patents on the telephone, RCA and radio patents, Westinghouse and alternating current (and later radio). When someone obtains a patent on a valuable technology outside of corporate control, as was the case with De Forest's Audion, Armstong's FM, and Farnsworth's television, every effort is made to obtain it, typically through purchase (e.g., the audion, television), though sometimes through protracted legal assault (e.g., Armstrong's FM). Patents build strong legal walls against competition, particularly in the early stages of an industry. Later, as an industry and its dominant corporations grow (and as patents begin to expire), traditional economies of scale and vertical integration often develop and take the place of patents as barriers to competition. This is the history of several of the major American electronics corporations, including AT&T and RCA.

But if the life of individual patents has limits under law, corporations have another strategy for holding off the onset of competition: cultivat-

53. Alfred D. Chandler, Jr., *The Visible Hand: The Managerial Revolution in American Business* (Cambridge: Harvard University Press, 1977), 6–7.

ing patent libraries by means of investment-guided research and development, that is, in-house research carefully geared to incrementally improve on a corporation's existing technologies with small but patentable improvements. Corporations have been known to "sit" on patents for technologies that would compete with existing systems, but often investment- guided research and development is guided in such a way as to discourage the invention of competing technologies altogether.[54]

Closely linked to patent strategies is the process of technological standards setting. Although in the electronic media many aspects of technological standards are set directly by international bodies and the FCC in the frequency-allocation process, standards are also set by trade associations and other "neutral" bodies, and sometimes in a process of struggle among dominant corporations (e.g., VHS videocassettes). Although standards setting does involve purely technological considerations, the scientific neutrality of the process is easily exaggerated. History shows that standards setting generally involves the same kind of massive turf considerations, and thus the same kind of struggles, as the patents process. The battles over television standards are a famous example. The current U.S. National Television Systems Committee (NTSC) system for color television, for example, was created in a rush by RCA in the late 1940s, largely to prevent the adoption of a competing (and in some ways superior) CBS-controlled system; to this day, American broadcast engineers like to joke that NTSC stands for "never the same color."[55]

In most cases, standards setting inevitably rewards some at the expense of others; it involves the creation and distribution of power, and thus the building of walls against competitors. But standards setting should not be viewed as simply another kind of struggle among corporations, as marketplace competition by other means. For at its base, setting a technological standard creates a shared extramarket social framework, a baseline for industry behavior, a common industrial language. Standards are a (sometimes brutal) means of social coordination, not of competition. When, as is typically the case, they do involve struggles among corporations, those struggles are better understood as political

54. Senate Committee on the Judiciary, Subcommittee on Patents, Trademarks, and Copyrights, "An Economic Review of the Patent System" (1958), a study prepared for the subcommittee by Fritz Machlup.

55. For a discussion of technological standards and the media in general, see Brian Winston, *Misunderstanding Media* (Cambridge: Harvard University Press, 1986). For an important case study, see Brad Chisholm, "The CBS Color Television Venture: A Study in Failed Innovation in the Broadcast Industry," Ph.D. diss., Department of Communication Arts, University of Wisconsin, Madison, 1987.

struggles, as struggles over territories, struggles over the creation of shared frameworks of behavior.

Consensual Behaviors

Perhaps more than most industries, broadcasting has evolved a large and crucial body of shared frameworks of behavior that are maintained largely by the weight of tradition and relied on by participants that are otherwise competitors. Some are as innocuous and obviously arbitrary as the old tradition of giving call letters beginning with *K* to stations west of the Mississippi and *W* to stations to the east. But many are less innocuous.

For example, the American broadcast schedule is organized into an extraordinarily rigid grid. While it is not uncommon in European public-service systems to broadcast films and other programs of odd lengths (e.g., seventy-three and a half minutes), in the United States all programs are required to fit into an unalterable lattice of half-hour blocks (twenty-two minutes minus commercials). The reason for this is obvious enough: it allows for commodification, for exchangeability; station managers, program producers, and so forth can easily buy, sell, and schedule programs that are constructed like replaceable parts. Made-for-television programs are thus produced according to these specifications, and theatrical films are chopped to fit. The result is often aesthetically absurd: dramatic tension is greatly reduced when a glance at our watch can tell us whether or not the day will be saved at the end of a scene, or whether or not the romance will be consummated.

Series programming (as opposed to single-episode programming) is similarly convenient for the industry: it allows for repetitive and thus cost-effective use of sets and actors as well as the cultivation of more predictable and thus more measurable audiences. Yet it also creates artistic problems, such as the certain knowledge that life-threatening situations are never really life-threatening for series characters: when the crew of the starship *Enterprise* beams down to a dangerous planet, everyone knows that the only truly doomed characters are the unknown ones.

The point here, however, is not to argue the issue of the aesthetic value of industry practices. Certainly, rigid time constraints or series programming can also be the occasion for new aesthetic possibilities, such as the genre of the daytime soap opera, which arguably has made a virtue out of the constraints that seem to plague prime-time series programs. The point is simply that such practices embody industrywide cooperative patterns of behavior that cannot be attributed solely to audience desires or to aesthetic demands of the medium. The business of broadcasting takes place within a frame of these and many more such traditions. The broadcast week is divided into demographically targeted zones (daytime, prime time, the Saturday-morning children's ghetto, etc.); program production is systematized by the use of generic traditions (action, family drama, sitcoms); and (as we will see in chapter 8) the statistical ambiguities of the ratings are circumvented by a tacit collective agreement to take the numbers at face value.

These behaviors are neither a product of competition between industry members nor a product of externally imposed government regulation. They are not competitive behaviors. Yet it would make little sense to call them monopolistic: they are not created by one company and forced on others, and they can coexist with, and are sometimes prerequisites to, highly competitive conditions. They are constitutive of but outside the marketplace, and thus difficult or impossible to speak of within the traditional language of markets, competition, and monopolies. A discourse that can speak only of competition and monopoly is incapable of speaking about such practices directly.

Beyond the Government/Market Dichotomy

Lessons from Economic Sociology

With what vocabulary and theory, with what language, are we to describe and discuss the social forces that shape broadcast industry behavior? Among the many competing scholarly approaches to economics, the tradition sometimes called economic sociology is arguably the most useful in this context. In a tradition of thought going back at least to Veblen and extending through Karl Polanyi to contemporary scholars like Fred Block and Marshall Sahlins, economic sociology has marshalled substantial evidence and argument in support of the notion that markets and economic conditions generally need to be understood as socially constructed. If neoclassical economists tend to begin with the assumption that the economy is an analytically separate realm of society that can be scientifically understood on its own terms, economic sociology argues the reverse: what economists call "externalities" (conditions that influence market relations that are outside markets proper) must serve as a starting point of analysis, not as something to be factored in in the later stages. Externalities make markets possible and often profoundly shape the character of specific markets. Externalities, in other words, are external only to a certain type of abstract economic theory; they are *internal* to real-life economic relations.[56]

The problem with the claim that commercial broadcasting in the United States operates according to the dictates of the natural marketplace, then, is not that there is no marketplace but that the marketplaces that do exist are neither natural nor apolitical. Economic sociology does not argue that traditional discussions of markets and economic efficiency are inherently wrong, worthless, or mere ideological smokescreens for the ruling class. Rather, the argument is that the conditions that economists analyze, though often quite real, are what they are because of extramarketplace arrangements. People do often act rationally to satisfy needs and to maximize their self-interest in a marketplace, as when they buy a television set. Needs, self-interest, and rationality, however, are neither self-explanatory nor invariable; they are constituted by social conditions and vary according to the specifics of those social conditions, such as the socially constructed production/consumption boundary characteristic of a consumer culture. Similarly, businesses do compete in measurable marketplace conditions with one another, often with relatively economically efficient results, as in the case of the markets for advertising time or broadcast stations. The conditions that make it *possible* for businesses to buy stations or advertising time, however, rest on political and social conditions that cannot be accounted for in classical economic terms.

The word "market" can be used in a bewildering variety of ways. In many cases a market is something precise and measurable, such as the farmer's market in the town square with numerous buyers and sellers competing on equal footing in such a way as to drive prices down while generating incentives to enhance efficiency. It also can be used in a political sense, to suggest an absence of government interference in business affairs (which, of course, does not always lead to a market in the measurable economic sense). The word is also used highly metaphorically, as in "the marketplace of ideas," where what is being discussed is not really economic in the conventional sense at all. And in many contexts the word "market" takes on several of its possible meanings at once. It is when these multiple meanings are left unstated that the word can be at its most ideological.

Sociologist Fred Block has coined the inelegant but useful term

56. This summary of economic sociology owes a good deal to chapters 2 and 3 of Fred Block, *Postindustrial Possibilities: A Critique of Economic Discourse* (Berkeley: University of California Press, 1990), 21–74.

"marketness" as a way to analyze and compare market conditions without lumping them together in a single category and without eclipsing their socially constructed character. Markets, Block argues, can be classified along a continuum ranging from social relations with very high levels of commodification and open competition, that is, a high degree of marketness, and relations heavily shaped by extramarket conditions, that is, a low degree of marketness. Block's continuum is not exactly the same as the continuum between highly competitive markets and monopoly well known to traditional economists: monopoly is typically understood in strictly economic terms, whereas a low degree of marketness (as in, e.g., homemaking) is lacking not just in competition but in the social construction of activities as marketable labor and commodities. But for comparative purposes, Block suggests that degrees of marketness can be specified in terms of the prevalence of the price mechanism. In Block's words,

> High marketness means that there is nothing to interfere with the dominance of price considerations, but as one moves down the continuum to lower levels of marketness, nonprice considerations take on greater importance. It is not as though prices are irrelevant under conditions of low marketness, it is just that they compete with other variables, so that one would expect price differences to be much larger before they led actors to respond.[57]

The commercial broadcast industry in the United States can be characterized as a rich mix of structures from points all along the marketness continuum. The buying and selling of advertising time, for example, is highly volatile, complex, and competitive, particularly on the level of national spot ads; it is characterized by a high degree of marketness.[58] Relations between the major television networks are characterized by a medium degree of marketness; though they compete with one another for advertising dollars, they also enjoy considerable "market power" that insulates them from competition, and are subject to extramarket forces such as FCC regulation and the strictures of industry convention. Viewer program choices, on the other hand, are characterized by a relatively low degree of marketness: although price issues such as the cost of sets, the economic value of a viewers' labor time, and so forth do play a role in viewing or listening choices, nonpriced variables such as the culture of leisured domesticity and the gendered division of labor play more im-

57. Ibid., 51.

58. Willard G. Manning and Bruce M. Owen, "Television Rivalry and Network Power," *Public Policy* 24 (winter 1976): 36.

portant roles in shaping how and what audiences choose. Calling the relation between commercial broadcasters and their audience a "marketplace," therefore, is accurate only in a highly metaphorical sense.

One implication of this is that the world of markets and competition as we know it has limits, that there are boundaries of acceptable corporate behavior beyond which free markets are not allowed to go. The corporate community, with various degrees of deliberateness, works to construct institutions, rules of behavior, and other order-maintaining devices that fulfill its imagined and real needs for stability and predictability. This is not to say that corporations do not compete with one another, but merely that the terms of the competition are set by collectively constructed boundaries. Nor are those boundaries rigid or impermeable: they shift from time to time, and are generally a matter of interpretation. Many squabbles within the industry can be understood as the result of competing interpretations of boundary rules, and in some cases the squabbles can be fierce and diverse. The point, however, is that squabbles over boundaries are characterized by low marketness: even though often treated by observers as if they were a form of market competition, they involve political and definitional struggles, not struggles over price.

Ideologies of Property

Property creation is just one of the social arrangements embodied in broadcasting that theoretically might be subject to political intervention without being inherently anticompetitive or involving a form of censorship. But property creation in electronic ephemerals is useful to focus on because it is both central to the creation of industry power and a centerpiece of contemporary systems of belief.

So what is property? The concept of property is properly described as ideological, not in the naive sense of "ideological" as illusory, as more imaginary than real (property relations are hardly an illusory element of contemporary capitalist societies), but in the more interesting sense of something that is *both* very real and profoundly imaginary at the same time, as something whose reality is conditioned on (though not simply reflected in) its imaginary quality.[59]

59. For all the revisions, criticisms, and qualifications that Althusser's scholarly work has been subject to since the "Althusserian moment" of more than a decade ago, his definition of ideology as "systems of representation in which men [*sic*] live their imaginary relation to the real conditions of existence" is still pertinent, not as a formula for unlocking the secret of ideologies, but as a description of a difficult but crucial intellectual and political *problem* faced by social thought today (Louis Althusser, "Ideology and Ideological State

On a broad popular level it is probably safe to say that in our time the prevailing way of imagining property remains, loosely speaking, classically liberal: property is a natural right of ownership, a kind of absolute sovereignty, over physical things. The poor fit between this common understanding of property and electronic intangibles helps explain why it is so often suggested that software and other intangible goods of the information age present challenges to legal systems of ownership quite unlike the blunt and obvious material forms of property characteristic of the nineteenth century. It also may help explain why professors and other professionals—people who would recoil in horror at the thought of, say, using a stolen $400 television set—calmly use illegally copied $400 software without giving it a second thought. Because whirring electrons evoke little sense of anything either natural or thinglike, it's hard to think of them as subject to theft.

The belief that property involves a natural right over physical things has important roots in our legal and philosophical tradition. For John Locke, the authors of the U.S. Constitution, and early American legal theorists, private property was the most fundamental of the natural rights, a precondition for liberty and human fulfillment far more important than, say, free speech or electoral politics. Property was understood as both an essential fact of human existence and as a prerequisite to a free society and a cornerstone of capitalism, a sacred moral value, the central measure of fairness and freedom. For utilitarians and their successors, the system of private property has been understood less as natural than as an artificially achieved but, on grounds of utility or efficiency, nonetheless highly desirable condition. In that murky but crucial world of contemporary common sense, property seems to be conceptualized largely through a shifting mixture of natural rights and utilitarian conceptualizations.[60]

However, as the legal realists first demonstrated early in this century, and as political scientist C. B. Macpherson has since elaborated, the popular views of property are neither philosophically coherent nor empirically accurate.[61] Property of any sort is neither blunt nor obvious.

Apparatuses: Notes towards an Investigation," in *Lenin and Philosophy and Other Essays*, translated by Ben Brewster [New York: Monthly Review, 1971]). See also Stuart Hall, "The Problem of Ideology: Marxism without Guarantees," in *Marx 100 Years On*, ed. B. Matthews (London: Lawrence & Wishart, 1983), 56–90.

60. For a pragmatist approach to property law and its place in broad streams of both scholarly and popular thought, see Margaret Jane Radin, *Reinterpreting Property* (Chicago: University of Chicago Press, 1993).

61. C. B. Macpherson, introduction to *Property: Mainstream and Critical Positions*, ed. C. B. Macpherson, 1–14 (Toronto: University of Toronto Press, 1978); Morris R. Cohen, "Property and Sovereignty," *Cornell Law Quarterly* 13 (1928): 8–30.

Ownership of a stock, for example, merely confers a narrow set of rights to income under certain circumstances, not exclusive control over anything physical. Similarly, the right to private property, subject to shifting interpretations and based as it is on state enforcement, is logically indistinguishable from a government-granted privilege. Property, the legal realists concluded, is neither a thing nor a natural condition of human existence, but is rather a shifting, flexible bundle of rights, a set of contingent political decisions about who gets what in what circumstances. As Macpherson puts it, property is an "institution which creates and maintains certain relations between people," and the relations created and maintained, and thus the meaning of the word "property" itself, are not constant but rather constantly changing from context to context.[62]

There are those—legal scholar Richard Epstein, for example—who would deny this argument entirely, and overtly apply a revised view of the classical liberal view of property to broadcasting.[63] Broadcast property, they would argue, like all property, is indeed natural to human life and is apolitical in its nature, and law and policy should be designed accordingly. But, as we will see, this is not the mainstream view. Broadcast policy discussions in the United States are more likely to ignore the question of property than to approach it from one or another point of view. When questions of property do come up, it is in the context of new technologies like today's digital sampling and computer software; the question is how to extend the regime of property into new areas, not what to do about property in already existing institutions, nor the idea of property itself. The coherence and legitimacy of the liberal idea of property in commercial broadcasting in the United States, then, is assumed to be part of the natural order of things, less because it is overtly asserted as such than because it is assumed to be so obvious as to be not worth talking about. It is taken for granted, both as an idea and as a set of practices and conditions.

The argument that the meaning of property is never constant, it should be pointed out, need not be taken to mean that property is meaningless, that it is *merely* a bundle of rights, that as a concept it is "disintegrating."[64] That would be to mistake something that is ideologically

62. Macpherson, *Property,* 1.

63. See, for example, Richard A. Epstein, "Property and Necessity," *Harvard Journal of Law and Public Policy* 13, no. 1 (1990): 2–9: "the grand idea of property and its principled necessity limitations provide the best guide for dealing with the complex modern issues that dominate our collective agenda today" (9).

64. Thomas C. Grey, "The Disintegration of Property," in *Property: Nomos XXII,* ed. J. R. Pennock and J. W. Chapman (New York: New York University Press, 1980), 69–70. In conversation in Palo Alto in July 1990, Professor Grey told me that today he would qualify

imaginary, that operates in the contingent realm of metaphor and representation, for something that is *merely* illusory. The illuminating oxymoron "property is theft" raises questions, but does not answer them. It is a useful trope because it discomfits traditional common sense, but it does not resolve the problem of the nature of property; it only opens it up.

Far from being meaningless or illusory, any definition or assertion of property rights is an action in the real world with potentially crucial material consequences. Any assertion of property involves a particular vision of how society is or ought to be. As Macpherson puts it, property changes in ways "related to changes in the purposes which society or the dominant classes in society expect the institution of property to serve." It also involves social struggle and power: property is an attempt to get social organizations (typically the state) to enforce and legitimize particular claims embodying particular social relations. Property is thus foremost "a political relation among persons."[65] The historian Martin Sklar summarized the scope of property nicely: "Property . . . is not simply a thing, nor simply an economic category, but is a complex social relation—of an intra- and inter-class character—that involves a system of authority inextricably interwoven with the legal and political order as well as with the broader system of legitimacy, the prevailing norms of emulative morality and behavior, and the hierarchy of power."[66] The fact that information, the radio spectrum, the broadcast audience, and electronic software are resources, therefore, should not be conflated with the treatment of these resources as commodities. As Dan Schiller has remarked, "A resource is anything of use, anytime, anywhere, to anyone; but a commodity . . . bears the stamp of society and of history in its very core."[67]

Bureaucratic Simulation of Property and Markets

So how can one approach property in a way that acknowledges its pervasive, constitutive, and powerful character without assuming it to be a thing or a universal economic category? Most of the answer lies in

---

his argument with the caveat that property retains an important "metaphorical" power in the American legal system, even as it has lost its coherence on the level of concrete application. Although one might wonder about the distinction between "metaphorical" functions in legal language and more concrete functions—isn't all legal language metaphorical on some level?—the following analysis is consistent with, and perhaps elaborates, Grey's current view.

65. Macpherson, *Property,* 1-14.

66. Sklar, *Corporate Reconstruction of American Capitalism,* 7.

67. Dan Schiller, "How to Think about Information," in *The Political Economy of Information,* ed. Vincent Mosco and Janet Wasko (Madison: University of Wisconsin Press, 1988), 33.

detailed analysis of specific forms of property in social and historical context—the substance of the next three chapters. But in order to foreshadow some of the general patterns revealed in those chapters, a hypothesis is useful: property may not be eliminated or disintegrated by contemporary conditions as much as it is simulated.

I borrow the word "simulation," of course, from Baudrillard. But there are echoes of the notion in otherwise mainstream discussions of broadcast policy. For example, in what is probably the most elaborate and respected application of neoclassical economic theory to spectrum regulation, economist Harvey J. Levin, after concluding that a full-fledged spectrum market would face serious economic difficulties because of high transaction costs, makes a curious suggestion. In a section titled "Market Simulation and Shadow Prices," he argues, roughly, that administrative means might be used to create a spectrum system whose behavior approximates that of what a "real," but unachievable, marketplace in spectrum would be.[68] Coming from someone who presents himself as scientifically concerned with hard empirical realities, this is a curious proposal: what is called for is a "simulation" of a marketplace that according to his own analysis *does not and cannot exist.* Levin, in other words, sets out speaking the language of Adam Smith and ends up speaking a language reminiscent of postmodern theory.

Such curious argumentative displacements occur with remarkable frequency in mainstream policy discussions. Perhaps the central example involves the discourse of consumer sovereignty, wherein the audience begins as the customer, but then turns out to be the thing sold. Even the more brazen of the Reagan-era deregulators had to qualify their visions of marketplace democracy with the observation that broadcasters are able to "determine the wants of their audiences" only by way of broadcasting's "indirect market mechanism [wherein] the advertiser acts as the representative for consumers."[69] In other words, the effort to replace indirect government regulation with direct market relations nonetheless must fall back on another indirect mechanism: the advertising system serves as a stand-in for the audience.

Similarly, we think of the battles between cable, broadcasters, phone companies, and so forth as simple expressions of marketplace competition, but it is more complicated than that: regulatory struggles are first of all political, not market driven. As I will show in the following

68. Harvey J. Levin, *The Invisible Resource: Use and Regulation of the Radio Spectrum* (Baltimore: Resources for the Future and Johns Hopkins University Press, 1971), 118.

69. Mark S. Fowler and Daniel L. Brenner, "A Marketplace Approach to Broadcast Regulation," *Texas Law Review* 60 (1982): 210, 232.

chapters, regulatory pie-sharing struggles in general displace actual marketplace competition over price and quality of actual commodities with political struggles among industry factions that vie for regulatory advantage in front of Congress, the FCC, and the courts. Copyright collectives, in the name of upholding a system of market exchange, create a bureaucracy in which statistical abstractions and formulaic procedures displace actual exchange of money for goods.

Most ways of accounting for these patterns of displacement are functionalist. Some forms of Marxist state theory, for example, lean toward accounting for the "bureaucratization" of business behavior in functionalist terms: capitalism "needs" the stability that comes with vertical integration, government regulation, and so forth in order to overcome internal contradictions such as the crisis of overproduction and class conflict. Chandler, for his part, explains the relation of corporate bureaucratic practices to the market with the concept of "internalization." By coordinating the activities of what had previously been numerous independent suppliers, distributors, manufacturers, and retailers, Chandler argues, corporations "internalized" the relationships that had previously been governed by market mechanisms. The diffuse and helter-skelter market patterns of allocation and distribution of the nineteenth century, he suggests, were thus reorganized and brought under the centralizing umbrella of the more coordinated, rational, and efficient corporate administrative system. Chandler's functionalism is evident in his assumption that the success of the "managerial revolution" in American business proves its inherent productivity, rationality, and superiority. Why did the managerial revolution succeed? Because it was efficient. How do we know it is efficient? Because it succeeded.[70]

In order to account for these patterns without falling back into functionalism, it is useful to borrow the concept of simulation. It's been said that we live in a time in which "all that is solid melts into air," in a "postmodern condition" in which life seems to be characterized more by the dizzying manipulation of words, signs, and symbols than by the iron necessities characteristic of nineteenth-century industrial society. We no longer deal with things themselves, the consensus seems to be, but with what Baudrillard calls simulations.[71]

70. This assumption is bolstered by Chandler's use of case studies such as the railroads, where consolidation and introduction of administrative logic in the nineteenth century indeed allowed for a much more efficient and effective coordination of scheduling and pricing. The fact that bureaucratic methods help make the trains run on time need not, however, stand as proof that administrative practice should be extended to all areas of life.

71. Marshall Berman, *All That Is Solid Melts into Air: The Experience of Modernity* (New York: Simon & Schuster, 1982); Jean-François Lyotard, *The Postmodern Condition:*

For the purposes of this analysis, simulation can be taken to mean, more or less, a representation once removed, a representation that has taken on a life of its own, divorced from its referent. Without subscribing to Baudrillard's entire intellectual framework, it is possible to suggest that liberal market relations are not so much eliminated by contemporary conditions as they are simulated. The postmodern experience, in other words, might be related to the often tense relations between the liberal exterior and bureaucratic interior of our corporate liberal political economic system.

Bureaucracy invariably defines itself as the rational and efficient means to achieve a collective end, as a neutral and transparent tool. An alternative approach might be to think of bureaucracies as systems of signification or representation, as means of *simulating* aggregate goals or purposes. This would then dislodge bureaucracy's self-definition by loosening the mechanistic link between the bureaucratic "signifier" (administrative means) and the bureaucratic "signified" (collective goals). Combine the notion of bureaucracy as a means for simulating goals with the nominally liberal political system that is a condition for the existence of the corporate system, and you arrive at the notion of bureaucratic market simulation. When faced with the absence or breakdown of traditional market relations, our bureaucratically structured business world sometimes sets out to establish an administrative counterpart to the market, a *simulation* of the market using the language and procedures of bureaucracy.

None of this is to suggest that simulated property and markets do not have complex and important economic significance. Many, if not all, nonmarket bureaucratic systems (e.g., income tax or welfare) have economic consequences, but we do not therefore call them "markets." Nor does the fact that the bureaucratic structures of simulated markets can be guided by "marketplace policies" change matters. Market rationales are *rationales;* persuading a bureaucracy to reduce your costs on the grounds that demand is down is not the same thing as responding to a market.

Nor is the point here that commercial broadcasting is merely a colossal ruse, that corporations essentially dupe themselves and the general populace into falsely believing they are engaged in market relations

---

A Report on Knowledge (Minneapolis: University of Minnesota Press, 1985); Jean Baudrillard, "Simulacra and Simulations," in *Selected Writings,* ed. Mark Poster (Stanford: Stanford University Press, 1988), 166–84. Berman uses the word "modernism" whereas Lyotard and Baudrillard are associated with "postmodernism." While there are important differences between the two approaches, the modernism/postmodernism distinction is often exaggerated, and the processes and patterns of life they refer to are at least similar, if not identical.

when they are not. The corporate environment is complexly structured in a way that encourages some procedures and strategies, discourages others, and generally sets boundaries to what can and can't be done. Mastering the structure of that environment, its grammars and codes, its pressures and limits, is a large part of what managerial skill is all about. The structure of that environment, its "discursive economy," is such that bureaucratic practices are favored in day-to-day procedures, yet on a broader level pressures are exerted and limits are set by the basic terms of liberal capitalism. Simulating markets, in other words, is the product of intelligent and skilled managers steering a course through the treacherous shoals of the corporate environment. It is an accomplishment, not a falsehood.

Hence, as we will see, to solve the dilemmas of copyright in broadcasting, the managerial community creates copyright collectives, bureaucracies that statistically simulate market exchange between program owners and broadcasters. In order to legitimate aggressive government intervention on behalf of private businesses, licenses to broadcast are not defined as property under law, but serve the functional equivalent of property within the industry: they are simulated property. For producers and distributors of broadcast programs, the market breakdown problem is obvious: the absence of a ticket booth in broadcasting precludes a genuine, direct market relation between audience-buyers and producer-sellers. Advertising-supported television *simulates* the ticket booth with a bureaucratically organized triangle of exchange where advertisers pay broadcasters and the audience, in theory, pays advertisers through the purchase of consumer products. Skepticism might exist about the accuracy of the ratings or television advertising's overall effectiveness in selling products to consumers. But such doubts are secondary as long as this triangle of exchange ensures a steady enough flow of income for system maintenance and simulates a market relation, at least as represented on a simple organizational flow chart; money flows from consumers to advertisers to networks to program producers, and programs flow the other way.

The other link in the chain, consumer-product manufacturers, are faced with a situation arguably, though less obviously, similar to that of program producers. A massive institutional system of manufacturing and distribution like Procter and Gamble, for example, cannot afford the risk of subjecting its elaborate operations to the whims of a vigorously open marketplace for consumer goods; that would fall outside the bounds of "orderly progress." At the same time, however, P&G can hardly resort to a monopolistic, centralized system of planning along the lines of state socialist enterprises; its legal and political legitimacy, its right to existence, rests on the belief that it is a private enterprise selling goods in a marketplace fashion. While P&G's strategies for dealing with the problem are numerous—for example, domination of distribution channels, product differentiation as a means to monopolize shelf space in stores—the system of commercial television advertising is clearly one of its most important tactics. From P&G's point of view, the importance of television advertising is that it simultaneously allows two things: an administratively formalizable and predictable set of procedures (fixed program schedules, the data of market research and ratings, etc.) that can be safely incorporated into its corporate system, and a set of procedures that approximate or simulate actual marketplace barter over goods with consumers. One of the primary incentives for regularly purchasing massive amounts of advertising time on television, therefore, may be simply that the practice, within the grammar of bureaucratic systems of representation, simulates a seller offering wares to a buyer in a marketplace. P&G's career managers, of course, also generally believe that purchasing advertising increases product sales. Their motivation to maintain smooth relations *within* the system of manufacture and distribution, however, is more immediate than the desire to increase sales to consumers over the short term, and the advertiser-supported system of broadcasting might be just as, or even more, useful for the first task as it is for the second.

Woven into the generally lumbering, bureaucratic behavioral habits of broadcast corporations are strands of high drama: the network ratings race, the constant manipulation of program schedules, the continuous struggles between producers and networks over scripts and other program details, the rise and fall of series, rocketing and plummeting careers, and so forth. Clearly, this curious mix is shaped by numerous different forces: classical market constraints sometimes play a part, as do the drives to expand turf and rise through the ranks. Bureaucratic market simulation, however, suggests at least one way that the different strands of this mix can be related. Corporate practices such as the ratings race may not construct a real marketplace, but may instead serve as administrative stand-ins for the market.

A few industry heretics provide support for this view. For example, industry executive Deanne Barkley is a skeptic who thinks that network scheduling strategies are generally irrelevant: "[B]ut don't you see, how does Fred Silverman justify his existence, or any of those people who are there. I maintain you could run a network with ten people. All these people justify their existence by making decisions, pretending to make decisions as to what's on."[72] Of course, this is not a view of things likely

72. Quoted in Todd Gitlin, *Inside Prime Time* (New York: Pantheon, 1985), 62.

to be popular within the industry, at least not on the network end of things. Taken to its logical conclusion, it would suggest that in classical economic terms operating a network is largely a matter of keeping undifferentiated contents flowing through electronic plumbing, and that network executives are hardly different in function from the nameless technicians who maintain municipal water supplies and sewer systems.

Whether or not network executives are overpaid, much about industry behavior suggests that what they do is related to the market ritually rather than directly. Program producers as well as advertising, network, and broadcast station executives are in most cases people with careers, not invested capital. They thus share an interest in maintaining and expanding the bureaucratically organized, smoothly operating system of production and distribution that employs them, an interest sometimes at odds with the interest in short-term maximization of profit for their employers. The ground rules of the system in which they operate, however, are circumscribed by the terms of liberal capitalism, central to which is a belief in the efficiency and democratic character of the market. Paradoxically, therefore, one of the functions that managers must perform for system maintenance is the creation of an appearance of market behavior. Industry managers are constantly caught between the drive toward stability and the political or ideological pressure to maintain something that looks like a market.

So in response to this dilemma, the industry uses ratings, but in a way that does more to ensure internal industry stability than a clear understanding of its audience. And the industry regularly drops low-rated programs and upper-level executives, in an apparent show of competitive bravado, but it just as regularly rehires the same executives and contracts with the same program producers; organizational reshuffling thus comes to stand for competition.[73] By means of such procedures, the circle is squared, and the bureaucratic corporation is symbolically reconciled with liberal capitalism.

The people who make broadcasting the way it is, in sum, are caught between the drive to manage and maintain a series of complex corporate systems, and the drive to uphold a liberal system of property rights and

73. Some readers may point to the dramatic layoffs at CBS and elsewhere in the 1980s as counterexamples, as a case where marketplace pressure did assert itself in a direct and drastic manner. The exceptional character of this case alone may be enough to prevent it from damaging my general argument. But even in the case of the CBS layoffs, the trade press reports that almost all upper-level management are still happily employed within the industry, either with competing firms or as consultants, university professors, and the like (Diane Mermigas, "Where Are They Now?" *Electronic Media,* September 19, 1988, 1).

market relations without which their employers, and hence their jobs, would have no justification for existence. One response to this dilemma has been the construction of a system that bureaucratically simulates market relations between program producers and the broadcast audience, and between consumer-product manufacturers and consumers. As the middlemen in these relations, broadcasters, networks, and advertisers may thus serve, not so much as intermediaries in a chain of actual market relations, but as providers of bureaucratic simulations of market relations between consumers and corporations.

Conclusion

The following three chapters elaborate on the notion of bureaucratic market and property simulation by exploring three key forms of property creation constitutive of commercial broadcasting in the United States: licensed stations, copyright, and the audience commodity. The analysis proceeds from questions of property to broader issues, not by reducing social complexities to matters of property. This argument toward complexity is present both within each chapter and in the relation of the chapters to each other: each successive chapter casts its net more widely in the social formation. Chapter 6, on broadcast licenses, stays fairly close to traditional legal analysis, tracing the legal and legislative strategies that created the present system. Chapter 7 expands its scope into sociological questions of corporate bureaucratic institutional structures. Chapter 8, on the creation of property in viewership, finally takes the analysis to the level of one of the basic features of the contemporary social formation, namely, the division of life into spheres of production and consumption, of work and home.

Broadly, it will be shown, these acts of property creation are conditioned in part on a liberal culture in a corporate era, on a deliberate attempt to implement the principles of private property and competition in the world of electronic communication and bureaucratically centralized businesses. And these acts require considerable political and ideological effort. The creation of marketable, privately owned broadcast stations involves a permanent system of federal licensing, systematic government elimination of small private entities from broadcasting in favor of large corporate and government institutions without compensation to the former or payment from the latter, and some tricky efforts at legitimacy that include a tenuous legal regime that simultaneously forbids ownership of the airwaves and invites their treatment as private property. Ownership of broadcast programs requires the approximation of property by statistical abstractions within the context of bureaucratic structures of copyright collectives, as well as elaborate webs of contractual relations between program suppliers and distributors that are in turn heavily contorted by fluctuating government regulations, and regular systems of industry dispute resolution that emerge from the inevitable wrangling between industry factions. And the economic lifeblood of the system, the audience commodity, rests on the social construction of a complex boundary between consumption and production, articulated with the methodologically tenuous systems of the ratings and ideologically tenuous practice of understanding listeners and viewers, not as the free, active, rational individuals of liberal anthropology but as themselves a form of property, as audiences that are sold to advertisers. On these elaborate foundations the institution of commercial broadcasting is built: they are prerequisites to commercial relations of any kind, whether competitive or not.

Selling the Air: Property Creation and the Privilege of Communication

"But Not the Ownership Thereof": The Peculiar Property Status of the Broadcast License

A license provides for the use of such channels, but not the ownership thereof.
RADIO ACT OF 1927

Secretary Hoover's signature in New York City sells for $150,000 to $200,000, and the applications are now being picked up as for sale.
SENATE TESTIMONY, 1926

Introduction: The Broadcast Station as Legal Creation

This chapter focuses on the form of ownership that constitutes the most fundamental kind of power in broadcasting, the power of access to the airwaves, which is also the power to speak, to have free speech rights, in radio and television. The principal organizational unit in American broadcasting is the station, and a station is something that is owned, bought and sold. By law, station ownership grants power over and responsibility for what is broadcast; whatever free speech rights exist in broadcasting exist foremost for station owners. The principal way to gain the right to exercise free speech in radio and television, the principal way to gain access to the broadcast airwaves, is to buy a station.

A broadcast station, however, is not self-evidently an object. It is not just some equipment and studios. It requires a legally enforced boundary in the formless continuum of the radio spectrum—in other words, a channel. There is nothing inherent to the spectrum that indicates exactly where boundaries within it should be drawn. The spectrum itself is only a potentiality; unless a signal is generated, nothing exists, not even the "ether" that was once imagined to be the omnipresent substance in which electromagnetic waves propagated. The bands, channels, and technical standards that make up frequency-allocation charts are for the most part human projections onto that continuum, not scientific descriptions of natural objects.

In a sense, the charts are less like topographic maps of natural areas than they are like the street map of a city. But even this analogy is imperfect. Once built, city streets have an existence more durable than the city planner's imaginary scheme that created them. But if the spectrum's "street map" were to disappear, so would the streets. Roadways in the airwaves exist only as long as they are imagined to exist; their imaginary status is integral to their reality.

True, different frequency ranges are characterized by different propagation and information-capacity characteristics; for example, medium and shortwaves can travel over the horizon, higher frequencies cannot. But these technical characteristics only provide the most general limits to social choice. AM radio channels were originally set to their present width of 10 kilohertz, for example, not so much for technical reasons but because humans have ten fingers: although 1 kilohertz channels would be unworkable and 100 kilohertz channels impractical, the choice of 10 instead of 9 or 11 (or 10.0267) is basically because ten is a convenient number for a species that does most of its arithmetic in the decimal system.[1]

This is not to say that the boundaries drawn in the spectrum are whimsical. Like the borders between nations, borders in the spectrum are often coded records of past political struggles. The location of the FM band between channels 6 and 7 on the television dial, for example, is largely the product of the David and Goliath struggles that took place in the 1930s and 1940s between upstart FM advocates and the RCA Corporation.[2] The mediocre NTSC technical standards for color television in the United States grew out of a similar struggle between RCA and CBS, which was resolved as much by RCA's deep pockets and manufacturing base as by questions of broadcast quality.[3]

In the broad view, then, the relation of a license to a channel is not merely one of granting access to something that already exists, but one of creation. When the government creates a legal regime that regulates access to the spectrum, the statement "We grant you a license to channel 6" is what language theorists call a performative; like the statement "I

1. In the early 1980s AM channels were shrunk to 9 kilohertz in most of the world outside the United States as a way to open up more channels. The U.S. AM broadcast industry successfully kept the FCC from going along. Even if one accepts the industry's argument at the time that 9 kilohertz is technically inferior to 10 (many suggested the industry was worried more about increased competition in the band than decreased signal quality), this does not mean that 10 kilohertz is technically ideal; by the industry's reasoning, 11 kilohertz would be even better.

2. Erwin G. Krasnow and Lawrence D. Longley, "Smothering FM with Commission Kindness," in *The Politics of Broadcast Regulation,* 2d ed. (New York: St. Martin's Press, 1978), 107–17.

3. Brad Chisholm, "The CBS Color Television Venture: A Study in Failed Innovation in the Broadcast Industry," Ph.D. diss., Department of Communication Arts, University of Wisconsin, Madison, 1987.

pronounce you man and wife," it not so much describes an existing situation as it creates a new one. The practice of licensing is the primary means by which we create, enforce, and maintain the socially defined boundaries in the spectrum. When the 1927 Radio Act stated that its intention was "to maintain the control of the United States over all the channels of . . . radio transmission," it was *creating* channels, not, as the phrasing suggests, simply taking charge of channels that preexisted the law in a state of nature.

A station, then, is not simply a building with a transmitter in it. It is a combination of a particular channel with a particular transmitting facility, legally constituted and protected with a federally issued license. Scientifically speaking, there is nothing natural or technologically necessary about this combination: transmitters, channels, and conditions of control and ownership are all independent variables. Transmitters can be adjusted to different channels, and licenses need not be tied to either frequencies or facilities.[4] The station is an arbitrary social creation, as much imaginary as it is real; and it is that act of creation that makes it possible for the station to become a commodity, an "object" available for purchase and sale on a for-profit basis. The government's role in this act of commodity creation has been elaborate and ongoing: as technologies and industrial practices have evolved over the years, the number and character of government-created channels has continually expanded and mutated, requiring constant regulatory involvement with the maintenance of broadcast channels and the creation of new ones. Arguably, the single most important government intervention associated with commercial broadcasting is the legal creation and maintenance of that marketable entity we call a broadcast station.

This situation has created something of a quandary for liberal ways of thinking, in which property and the marketplace are thought of as autonomous and in need of shielding from government and politics. In this case they are thoroughly entwined with and dependent upon ongoing government intervention. Squaring this government intervention with liberal principles, therefore, has proven ideologically awkward. The most glaring problem concerns the "nonownership" clause of the Communications Act: the law simultaneously forbids ownership of the airwaves and invites their treatment as private property. But there are other oddities as well. In what would appear to be a violation of fundamental liberal principle, the creation of marketable, privately owned

4. It is possible, for example, to link the license to a person instead of to equipment or frequencies, as is the case with ham or amateur radio operators, whose licenses allow them to transmit on a variety of frequencies with a variety of equipment. See 47 C.F.R. pt. 97.

broadcast stations historically involved systematic government elimination of small private entities from broadcasting—amateurs, nonprofit broadcasters—in favor of large corporate and government institutions without compensation to the former or payment from the latter. And throughout its history, the licensing system has been characterized by a doctrinal instability, wherein the depiction of licensees in legal argument oscillates wildly between descriptions of them, on the one hand, as business entrepreneurs fundamentally autonomous from government and, on the other, as recipients of government privilege and thus fundamentally different from traditional entrepreneurs.

This chapter investigates both the processes by which the broadcast station has been commodified and the residual tensions that act of commodification has created. The first part of the chapter outlines the creation of the existing system of regulation in the early part of this century. The key features of that system, such as the simultaneous creation of marketable broadcast stations and the regulation of those stations "in the public interest," are shown to be the result of a distinctly ideological pressure: the need, central to the liberal imagination, to maintain a boundary between private property and government in the face of the government's helpful reach across that boundary when it creates broadcast stations.

The second part discusses the variety of political and regulatory responses to the tensions inherent in the existing system that have surfaced over the history of broadcasting. The responses have involved two mutually antagonistic legitimatory tactics. One tactic attempts to preserve the industry/government boundary by shielding government-created broadcast licenses from private control, that is, by limiting licenses' propertylike character. The other tactic attempts to uphold the same boundary by enhancing the propertylike character of licenses and by limiting government interference.

The effort to pursue the principle of private property in broadcasting, as a result, has been beset by a new version of an old tension: the regulation of the broadcast spectrum involves a paradoxical effort to use elaborate political intervention to achieve the goal of limiting political intervention. Most of the effort that has gone into the regulation of broadcast channels, this chapter will argue, has been directed toward negotiating this contradiction.

Enclosing the Spectrum, 1900–1920

To many of the first entrants into the world of radio communication, legal regulation of the spectrum was not a priority, and was for a period actively opposed. Entrepreneurs were interested above all in developing and manufacturing individual pieces of equipment, physical things, for sale. Government regulation of radio signals was seen only as a restriction on entrepreneurial activities. And as we have seen, the amateurs were even more adamant about maintaining "open" access to the airwaves; radio's wide-open character was precisely what made it intriguing.

If any one individual deserves credit for inventing the notion that the radio spectrum might be usefully "bounded," and thus given some of the characteristics of property, it is probably Marconi. Marconi's managerial "exclusivity policy" was designed to extract profit, not by manufacturing devices for sale, but by regulating access to a communications system, which meant controlling access to the radio spectrum that made it possible. Implicit to the Marconi strategy, then, was a vision of the spectrum as a space or territory to be conquered and cordoned off, as something analogous to property.[5]

Yet at first Marconi was alone in his vision. The enclosure of the radio spectrum ultimately emerged in a process of interaction between military, corporate, and government interests that led to the first international treaty regarding radio in 1906. It is here, in the relations among large national and international institutions, not in the activities of the private individuals of Lockean fable, that a vision of the radio spectrum as something with propertylike characteristics crystallized.

The spectrum was not explicitly spoken of as a kind of property at first. Yet, as the navies of the United States and the European powers coaxed their governments into establishing legal powers over access to radio between 1903 and 1912, an understanding does seem to have emerged of the spectrum, if not as a commodity, then at least as a kind of *territory*. The leaders of the United States and the European powers came to assume that the radio spectrum was an unsettled, strategic territory analogous to the foreign lands they were then competing to colonize.

Within the United States, the principal form of resistance to these efforts came from the amateurs with, it seems, some help from entrepreneurs. Together, they successfully lobbied against attempts to bring U.S. law into line with the 1906 treaty for four years, arguing that the 1906 international rules were restrictive, premature, and technically naive.[6] There is little evidence that they based their resistance on any kind of sophisticated political or legal analysis. Yet in the amateurs' organiza-

5. Susan J. Douglas, *Inventing American Broadcasting, 1899–1922* (Baltimore: Johns Hopkins University Press, 1987), 101.

6. Douglas, *Inventing American Broadcasting,* 216.

tional practices one can detect the outlines of an alternative vision to the one that was driving the legal enclosure of the airwaves by corporate-military coalitions. The amateurs quite successfully developed extralegal, grassroots means of creating order in the airwaves, such as time- and channel-sharing arrangements in populous areas, informal traditions and codes of etiquette for on-air behavior, and, eventually, a volunteer network for relaying messages that spanned the nation.[7] If there is a legal precedent to the vision implicit in amateur activities, it is the medieval tradition of the commons, which, in its ideal form at least, functioned as a common public space open to all, owned neither by individuals nor by the state, and maintained as much by shared traditions as by legal policing.

The possibility of a treatment of the spectrum as a commons, however, was eliminated with the assertion of the principle of legally enforced limitations on spectrum access in the Wireless Ship Act of 1910 and the Radio Act of 1912. Property rights were not discussed at the time. If they had been, it might have raised troubling questions. In the 1912 act, after all, private individuals—the amateurs—were forcibly ejected from their place in the spectrum without compensation, while others, notably the Marconi Company, were granted a place of privilege by what amounted to a government bequest. The corporate liberal tropes of technological necessity, expertise, the national interest, and overriding public purpose were relied upon instead. The aura of technological complexity and public urgency surrounding the 1912 act, in sum, thoroughly overshadowed potential concerns about political and social issues in general, and property in particular.

One can detect in the logic of the 1912 act some implicit answers to questions of property, however, based not in explicit principles but in corporate liberal habits of thought. Was it legitimate to eject the amateurs from their established places in the airwaves? Was it fair to grant Marconi such a large protected chunk of the spectrum free of cost? Yes, according to the logic of the act: these actions were legitimate and fair because of the complexities of radio, public safety, and the national interest as determined by the experts, that is, by Marconi engineers and navy officers. Were not some questions still left unresolved? For example, was Marconi accruing private property rights in the spectrum by dint of his investments after 1912? Yes, many questions were left unresolved: this is inevitable in complex, evolving technologies, which is why the act established a mechanism for dealing with such contingencies, the administrative power of the secretary of commerce and labor.

7. Ibid., 209.

Significantly, however, as long as the corporate liberal establishment imagined radio as a strictly point-to-point, strategic communications technology, the airwaves themselves were conceived merely as a means toward the ends of profit through the sale of other goods: equipment, patents, parts, and services. Property relations were extended only to relatively traditional realms, and the airwaves were legally regulated purely in terms of their role in systems of communication and manufacture. The idea that the airwaves themselves might be subject to commodity exchange was not seriously broached.

Commodifying the Spectrum, 1920–1934

The commodification of the spectrum, turning it into something that could be bought and sold, came during the first years of the broadcast boom, the years when radio became an instrument of broad-based popular communication and a key element in the consumer society. The marketable broadcast station seems to have been a casual and relatively uncontroversial outgrowth of the process by which the Commerce Department under Hoover's direction gradually established broadcasting as a corporate activity. Once Hoover established the principle of regulating broadcasting in terms of channel allocations that established different classes of service according to transmitter power, corporate affiliation, and broadcast content, it occurred to businesses interested in selling their broadcast equipment that they might include broadcast licenses as part of the package. At their request, the Commerce Department began to transfer their licenses along with the equipment. The license thus came to be understood as attached to the equipment rather than to the individual broadcaster, and the institution of the marketable broadcast station was born.

It says something about our culture that, like the erection of a barrier between two-way amateur radio and one-way broadcasting, this profoundly definitive policy was undertaken with almost no discussion. The only recorded discussions of the matter that do exist are buried deep inside the records of congressional hearings that occurred several years after the policy was initiated. During Senate hearings on the pending 1927 Radio Act, for example, Department of Commerce solicitor Stephen Davis testified,

> We have felt this way about it. . . . that the license ran to the station rather than to the individual. In other words, we have never felt it wise to adopt a policy under which we would say to an

> individual, "Yes; go in and build this station at whatever cost there may be. If you die it is worth nothing. If you change your mind and want to quit broadcasting it is worth nothing. If you get into business trouble it is worth nothing to your creditors. It has got only a refuse value."[8]

The policy of transferring licenses when stations were sold, then, was thought of as a means to increase the security and likelihood of profit for investors by extending the power to gain returns on investment beyond the profits from broadcasting itself and beyond traditional forms of property to the station itself. That it also helped to *create* that economic value went largely unremarked, perhaps because of the blurring of description and prescription characteristic of popular functionalism: the system simply existed "out there," and the government fulfilled its function of serving it. The general principle was the same that governed licensing procedures overall: the functional goal of nurturing, not formal private rights, but the autonomy and power of private capital and the "system" of broadcasting "necessary" to progress. To Hoover and others like him, in sum, the positive value of encouraging the corporate development of broadcasting was obvious; if this meant using government to transfer licenses when private individuals contracted to sell stations, Hoover's Commerce Department saw no reason to object.

Again, a mystified sense of technology that conflates social with technological choice seems to have helped legitimate the practice. During hearings, one senator remarked with regard to the structure of the policy: "I understand the policy of giving the licenses to the machine rather than to the individual. . . . I do not offhand see any fault with that, because I can see sound reason for not liquidating equipment all over the country. . . . There is no justification for abandoning this apparatus because the license expires."[9] What is odd about this comment is that, strictly speaking, "equipment" or "apparatus," that is, radio transmitters, physical plant, and so forth, would *not* be "abandoned because a license expires." Transmitters and the like can be and are regularly sold on a marketplace basis without licenses, just like any other device whose use is regulated, like a used car, an airplane, or ham radio equipment. Just because they require licenses to operate does not mean that they can't be sold. What was threatened with being worth "nothing" or with being "abandoned" in the absence of a license was not the equipment per se

8. Senate Committee on Interstate Commerce, *Radio Control: Hearings before the Committee on Interstate Commerce,* 69th Cong., 1st sess., 1926, at 42 (hearings on S. 1 and S. 1754), January 8 and 9, 39.

9. Statement of Senator James Couzens, ibid., 44.

but the station. What the senator seems to have been referring to, therefore, was the "station" as a whole, which he mistook for "apparatus." To him, and one suspects to some other politicians, the social and political underpinnings of a station were hidden behind a mystified technology.

In any case, by the early 1920s access to the spectrum was being controlled by two distinct mechanisms: licenses to broadcast could be had either from a government office or from private individuals, the former limited by administrative fiat, the latter largely by price. When the Commerce Department declared the airwaves full and ceased issuing licenses in 1925, the only way to gain access to the airwaves within the existing framework was by purchasing an existing station with a license. The market for broadcast stations has been brisk ever since.

This mixed bag of private and public means of access and regulation required some ingenuity to be rendered legitimate. The classical liberal faith in formal, bright-line property rights had only recently begun to lose its centrality in legal and political discourse, and so muted versions of it applied to radio did surface in the 1920s. Most significantly, in his memoirs Hoover hinted that in the early 1920s some commercial radio manufacturers were "insisting on a right of permanent preemption of the channels through the air as private property," which prompted him to organize the First Radio Conference as a means to resolve the conflict between these groups and other claimants to the spectrum.[10] The threat of individual private parties using classical liberal principles to stake claims in the spectrum against the designs of both the military and corporations thus may have been a key motivation lurking behind the entire range of Hoover's associational efforts in the twenties.

There were other classical liberal efforts as well. The American Bar Association, for example, took a short-lived stand in favor of formal property rights in the spectrum. In December 1926 an ABA committee released an interim report on radio legislation, which argued that when licenses are refused to existing stations, the stations are legally entitled to compensation.[11] Simply by virtue of the fact that broadcasters profited from the use and sale of their stations, in other words, they had a naturally existing property right in their channels, and any government usurpation of that right amounted to takings requiring compensation. Private use and sale for profit, the argument seemed to be, automatically created private property rights subject to legal protection from govern-

10. Herbert Hoover, *The Memoirs of Herbert Hoover: The Cabinet and the Presidency, 1920–1933* (New York: Macmillan, 1952), 139–40.

11. Stephen Davis, *The Law of Radio Communication* (New York: McGraw-Hill, 1927), 66–67.

ment interference, public-interest clause or no. Similarly, in the legal vacuum that ensued after the collapse of the Commerce Department's authority, a circuit court case tried to resolve a dispute between two broadcasters by turning to classical property rights; *Tribune Co. v. Oak Leaves Broadcasting Station* upheld a licensee's right to enjoin an interloper on the licensee's channel.[12]

The *Oak Leaves* case, however, was not understood at the time so much as a competing method of ordering broadcasting as it was a stopgap, an action taken using familiar tools to solve a local dispute until more comprehensive solutions were worked out at the federal level.[13] Hence, although Department of Commerce solicitor Stephen Davis took it as obvious that "[r]adio communication is a natural right" in some sense of that phrase,[14] he did not interpret this to mean that there were common law property rights in the spectrum. "[T]here is no absolute right of transfer," he told a Senate committee.[15]

In the years that followed, both Congress and the courts upheld Davis's view. As early as 1922, draft bills before Congress suggested that broadcasters be required to get permission before selling or otherwise transferring licenses.[16] A fear that common law property rights might be used against government efforts to regulate lay behind a proposed 1924 joint resolution of the House and Senate, "affirmed that the ether was a public possession and provided for limited grants for its use."[17] A Senate

12. *Tribune Co. v. Oak Leaves Broadcasting Station,* 68 Cong. Rec. 216 (1926; reprint of Circuit Court, Cook County, IL, decision of November 17, 1926). See also Harry P. Warner, "Transfers of Broadcasting Licenses under the Communications Act of 1934," *Boston University Law Review* 21 (November 1941): 585, 591; and Matthew Spitzer, "The Constitutionality of Licensing Broadcasters," *New York University Law Review* 64 (November 1989): 990, 1046.

13. "This court is of the opinion, from its interpretation of the act of August 13, 1912, that Congress did not intend to undertake to assume the right to regulate broadcasting under its powers given it to regulate commerce and that, *until such time as it does,* litigants may enforce such rights as they may have by reason of operating broadcasting stations in the State courts having jurisdiction" (*Tribune Co. v. Oak Leaves Broadcasting Station,* 218, emphasis added).

14. Davis, *Law of Radio Communication,* 14.

15. Senate Committee, *Radio Control,* 43.

16. The first version of this requirement appeared in a draft bill on April 20, 1922, which stated, "Such station license, the wave length or lengths authorized to be used by the licensee, and the rights therein granted shall not be transferred, assigned, or in any manner either voluntarily or involuntarily disposed of to any other person, company or corporation without the consent in writing of the Secretary of Commerce" (S. 3694, 67th Cong., 2d sess. [1922]). See also H.R. 13733, 67th Cong., 1st sess. (1923); and Warner, "Transfers of Broadcasting Licenses," 594.

17. H.R. 7357, 68th Cong., 2d sess.; Marvin R. Bensman, "Regulation of Broadcasting by the Department of Commerce, 1921–1927," in *American Broadcasting: A Source-*

resolution, passed only days before the Radio Act itself, was more assertive: "the ether and the use thereof for the transmission of signals, words, energy and other purposes, within the territorial jurisdiction of the United States is hereby reaffirmed to be the inalienable possession of the people of the United States and their government."[18]

In a similar effort to protect the licensing system against private property claims, Congress hit upon the idea of requiring licensees to sign waivers relinquishing any such potential rights against the regulatory body. On July 3, 1926, the Senate passed joint resolution 125, which required licensees to "execute in writing a waiver of any right or of any claim to any right, as against the United States, to any wavelength or to the use of the ether in radio transmission because of previous license to use the same or because of the use thereof."[19]

The ground was already well prepared, then, when Congress passed the 1927 Radio Act. Of course, the crucial phrase that divides the licensing mechanism from common law property is the nonownership clause, which provides for "the use of such channels, but not the ownership thereof," and which specifies that "no such license shall be construed to create any right, beyond the terms, conditions, and periods of the license."[20] The waiver requirement also made it into the 1927 act, though not without modification under pressure from concerned broadcasters: the references to "rights" were replaced with the vaguer "claim," and "as against the United States" was narrowed to "as against the regulatory power of the United States."[21] Section 12 of the act quietly wrote into law the Commerce Department's policy of transferring licenses along with stations by specifying that licenses to broadcast "shall not be transferred . . . to any person, firm, company, or corporation without the consent in writing of the licensing authority." In the years following the 1927 act, the constitutionality of license revocations without compensation was upheld by the courts.[22] A series of court

book for the History of Radio and Television, ed. Lawrence W. Lichty and Malachi C. Topping (New York: Hastings House, 1975), 545.

18. 67 Cong. Rec. 4152 (February 18, 1927).

19. S. Res. 47, 69th Cong., 1st sess., signed into law December 8, 1926.

20. Preamble to Public Law 632, 69th Cong., 1st sess. (February 23, 1927).

21. Warner, "Transfers of Broadcasting Licenses," 592. The full text of section 5(H) of the 1927 Radio Act reads, "No station license shall be granted by the commission or the Secretary of Commerce until the applicant therefore shall have signed a waiver of any claim to the use of any particular frequency or wave length or of the ether as against the regulatory power of the United States because of the previous use of the same, whether by license or otherwise."

22. *United States v. American Bond and Mortgage Co.,* 31 F. 2d 448 (N.D. Ill. 1929), affirmed 52 F. 2d 318 (7th Cir. 1931): regulatory authority does not violate the Fifth

cases in the early 1930s further affirmed the legitimacy of the regulatory practices established in the early 1920s.[23] And in 1934 section 12 of the 1927 act was simply folded into section 310 of the Communications Act, where it remains to this day.

Wealth through Regulation: On the Value of Stations

There is no doubt that the policy of using licenses to create and protect transferable stations has had the effect of establishing the broadcast spectrum as its own kind of real estate. The thriving and highly lucrative marketplace in broadcast stations has formed one of the key underpinnings of commercial broadcasting overall. Immediately following the passage of the 1927 act, the market value of stations went up dramatically, largely because the FRC reduced the supply of channels—between 1927 and 1929 the commission reduced the number of broadcast stations from 681 to 606 to reduce interference[24]—and because its new powers brought higher levels of stability and confidence to the broadcast business.

Since then, the regulatory system has continued to create new allocations at regular intervals, and therewith the conditions for new marketable broadcast stations. Since the late 1930s, technical improvements have allowed the number of AM radio allocations to grow from under one thousand to roughly five thousand. In 1941, furthermore, FM radio and VHF television bands were opened up, and the UHF band was made available for television in 1953.[25] In the 1980s the FCC took applications for nearly two thousand newly allocated low-power television (LPTV) channels.[26] And in the last few years, the FCC has initiated efforts

Amendment and "is not an unconstitutional taking of property without compensation or without due process of law," 31 F. 2d at 455. See also *General Electric Co. v. Federal Radio Commission*, 31 F. 2d 630 (D.C. Cir. 1929); *KFKB Broadcasting Association, Inc., v. Federal Radio Commission*, 60 U.S. App. D.C. 79, 47 F. 2d 670 (1931); *Journal Co. v. Federal Radio Commission*, 60 U.S. App. D.C. 92, 48 F. 2d 461 (1931); and *Federal Radio Commission v. Nelson Brothers Bond and Montage Co.*, 289 U.S. 266 (1933).

23. That Congress has the power to regulate the use and operation of radio stations under the "commerce clause" of the Constitution was affirmed in *Technical Radio Laboratory v. Federal Radio Commission*, 59 U.S. App. D.C. 125, 36 F. 2d 111 (1929); *General Electric Co. v. Federal Radio Commission; KFKB Broadcasting Association, Inc., v. Federal Radio Commission;* and *Journal Co. v. Federal Radio Commission.*

24. Christopher H. Sterling and John M. Kittross, *Stay Tuned: A Concise History of American Broadcasting*, 2d ed. (Belmont, CA: Wadsworth, 1990), 632.

25. Ibid., 632–33.

26. LPTV channels operate at transmitter power levels small enough to allow their

to create channels for the new broadcast technology of high-definition television (HDTV).

Although the government plays a necessary role in *creating* this supply of marketable broadcast stations, however, the principal hurdle to access to the broadcast marketplace facing the majority of the population is not the FCC but the ability to buy a station. As a general rule, at any given time the supply of stations for sale greatly exceeds the supply of unclaimed licenses. This is especially true if one thinks in terms of access to broadcast audiences instead of access to channels: available unclaimed channels typically provide access to relatively small audiences because they tend to be of lower power, to be in less populated areas, or to involve technologies that have not yet established themselves among consumers (e.g., FM and UHF television in the 1950s). The number of those who enter the spectrum via new channels, therefore, is typically exceeded by those entering by station purchase.[27]

It is an undeniable feature of the existing system of regulation, furthermore, that broadcasters as a rule are able to sell their government-licensed stations to just about anyone for just about any price. In principle the FCC retains the power to revoke licenses without compensation, and to interfere with or even forbid the sale of a license. Nonetheless, over the years the FCC has stuck to the broad policy of maintaining a broadcast system based on the free exchange of capital and maximum autonomy from government interference, and has thus been extremely reluctant to invoke its theoretical powers. Only two television licenses have been revoked in the forty-year history of the medium, and fewer than 150 licenses overall have been revoked or denied renewal in the history of regulation, most of them involving technical problems in small radio stations. Radio and television licenses have changed hands by sale with FCC approval, on the other hand, in more than six thousand trans-

---

introduction into areas already saturated with standard high-power channels. As of 1988, 455 LPTV stations were on the air, and the FCC had granted construction permits for another 1,359 (ibid., 467).

27. For example, in 1986, a period of heavy activity in the market for broadcast stations, 1,558 commercial radio stations changed hands by sale, compared to 123 stations that were new to the airwaves that year. Similarly, 37 new commercial television stations went on the air while 128 existing television stations changed hands. For new stations, see ibid., 633; for station transfers, see Joseph M. Foley, "Value and Policy Issues in the Marketplace for Broadcast Licenses," in *Telecommunications, Values, and the Public Interest,* ed. Sven Lundstedt (Norwood, NJ: Ablex, 1990), 273–74. These numbers, moreover, reflect numbers of channels but not audience size represented by each channel; in terms of audience size, the ratio of market entry by purchase versus entry by new licenses is likely to be much greater.

actions, many of which involved more than one station.[28] The overwhelming majority of applications for transfer of control of licenses have been approved, and the majority of existing broadcast licensees obtained their licenses through the purchase of stations.[29] The easy and most common mode of access to the broadcast airwaves, in sum, is by purchasing a station, and once a station is obtained its continued possession is very nearly guaranteed. If, as a few property-rights purists have darkly suggested, the FCC's power to revoke licenses and deny transfers constitutes a slippery slope with government control and censorship at the bottom, it also must be acknowledged that the commission has managed to cling to the uppermost edge of the slope with nary a slip for more than seventy years.

Making Sense of Spectrum Regulation in a Liberal Universe

The idea of property as a natural right is so deeply ingrained in American consciousness that it cannot be said to have ever completely disappeared from the discourse surrounding broadcast channels, and has resurfaced in small ways over the years. In a few cases, property rights have been invoked explicitly. For example, when a few liberal activists managed to reserve a handful of the newly opened FM and UHF television channels for nonprofit broadcasters in the 1940s, the militantly antiregulation industry trade magazine *Broadcasting* attacked the action with the suggestion that the reservations somehow constituted a violation of the industry's property rights. More frequently, however, the intimacy of government-business relations evidenced in licensing has generated vaguer forms of ideological uneasiness. There has always been some grumbling about government red tape in the licensing process, particularly when broadcast executives find themselves faced with the inconvenience of hiring lawyers to file lengthy license-renewal and station-transfer applications with the FCC. And when an FCC action makes the government-business linkage overly transparent, complaints inflected by property-rights ideology are heard. For example, when the

28. For denials, see Richard Ellmore, *Broadcasting Law and Regulation* (Blue Ridge Summit, PA: Tab Books, 1982), 114–15. For station sales, see Christopher H. Sterling, *Electronic Media: A Guide to Trends in Broadcasting and Newer Technologies, 1920–1983* (New York: Praeger, 1984), 45.

29. Note, "Radio and Television Station Transfers: Adequacy of Supervision under the Federal Communications Act," *Indiana Law Journal* 30 (1955): 351.

FCC recently guaranteed existing broadcasters HDTV channels, competitors complained about the injustice of this government "giveaway."[30]

In general, though, the system worked out in the 1920s has been a roaring success. The profits to be made in broadcasting are large, so the complaints have generally been muted. The FCC's theoretical powers to interfere with the ownership, sale, and control of broadcast stations have been rendered acceptable by the argument that broadcasting is technically unique, coupled to a policy of extreme constraint in invoking those powers. Some government intervention is necessary, it is thought, to build a free enterprise system that is free of government intervention—in the special case of broadcasting.

In the netherworld in which broadcast law and policy experts operate, however, a nagging unresolved question has remained: What is too much government intervention, and what is too little? At what point does government intervention cease to help and start to interfere with the free enterprise system it is supposed to protect? How are the "experts" that run the FCC supposed to find a politically neutral, objective way to draw a boundary between appropriate and inappropriate government intervention?

These are not just abstract questions. They are rendered "practical" in the policy world because lobbyists, in their search for ways to translate the designs of their clients into "neutral" terms, regularly exploit the ambiguity of the questions and manipulate them to their advantage. Because the exact location of the appropriate boundary between government and business is uncertain, it is easy for an industry faction to argue that desired FCC actions uphold the boundary and undesired actions inappropriately blur or cross it, and just as easy for that faction's opponents to argue the reverse. Over the years, then, industry squabbles have generated a series of opposing arguments that draw the boundary in different places, and thus put the FCC in the role of resolving those disputes by deciding where, for the moment, the line lies.

The different argumentative strategies that have been advanced over the years for drawing the appropriate boundary between government and business in the licensing mechanism can be placed in two broad categories. One constructs the principal threat as private interests gaining unfair advantage from the fact of government involvement, and thus interprets the public airwaves as a bulwark against private privilege. The other takes an opposite approach, seeking to minimize FCC intervention in licensing as a means to reduce government interference in private affairs.

30. Doug Halonen, "FCC Offers New Channels to TV Stations," *Electronic Media*, April 13, 1992, 1.

Public Airwaves as a Bulwark against Private Privilege

Since its origins in the 1920s, a constant theme of regulatory decision making has been the idea that the use of government-created channels by private businesses is justified only if private discretion is carefully limited. If a private broadcaster is able to get a license for free from the FCC and then turn around and sell that license with her station for a profit, is the broadcaster not profiting unfairly from a government bequest? If licensees select both their successor and the price paid for their stations, does this give them dramatically more control than the FCC over selecting entrants to the spectrum resource, and thus undermine the public-interest principle which justifies broadcaster's power in the first place? By this logic, appropriate FCC actions should seek to uphold the boundary between public and private interests by restricting the control of private interests over licenses. It is necessary, the argument goes, to carefully limit the powers of private interests in the public broadcast spectrum to prevent unfair advantage.

This pattern of thought appeared repeatedly in the years leading to the passage of the 1927 Radio Act, and left its mark on the legislation. During the Fourth Radio Conference, concerns were raised that some individuals were obtaining broadcast licenses solely for the purpose of resale at a profit, and suggestions were made that the Department of Commerce take action to prevent such "trafficking."[31] Shortly thereafter, during the hearings for the Radio Act, objections were raised when it was revealed that a station had sold for $50,000, considerably more than the value of its tangible assets.[32] Concerns were also expressed about the loss of regulatory control implicit in the practice of allowing license transfers at times when direct applicants were being refused licenses on the grounds of spectrum scarcity, and about the propriety of creating a de facto franchise through licensing.[33] These concerns helped ensure the presence of the nonownership clause and the requirement of FCC approval of transfers.

The nonownership principle, however, raised as many questions as it answered—questions that go right to the heart of what we mean by "ownership" and "value." For although the law did not grant full-fledged property rights in the spectrum and gave the FCC theoretical powers to intervene in station sales, the 1927 act and its successor nonetheless

31. Warner, "Transfers of Broadcasting Licenses," 595.
32. Senate Committee, *Radio Control*, 46.
33. Ibid., 45–47.

were clearly intended to underwrite Hoover's initial policy of upholding private ownership and exchange of stations in the name of free enterprise. After all, does not the ability of broadcasters to transfer licenses by sale create economic value in the license, and thus a form of de facto private property, in spite of the waivers and public declarations to the contrary?[34]

The dominant response to this question has been to resort to a common formalist legal strategy: turning a blind eye in the name of neutrality. The price paid for stations, the argument goes, is none of our business. The general policy was articulated in 1926 by the Commerce Department solicitor Stephen Davis, when he said, "We have never felt . . . that it was any part of our concern as to what price a man received for his broadcasting apparatus. . . . I have no doubt that the broadcasting privilege is going to be of very considerable value, the same as any other franchise becomes of value."[35] The same attitude was reflected in the following decade. In the *Seitz* case, for example, the FCC opined that "our primary consideration, from the standpoint of the public interest, deals not with the prevailing relationship between contract price and the items to be transferred, but rather with the qualifications of the proposed transferees and their ability to provide the public with an improved broadcast service."[36] This remains the policy today; prices paid in station sales are not considered particularly relevant to FCC decision making.

Yet in a corporate liberal environment, formalist limitations on legal and administrative inquiry are hard to maintain. Corporate liberal experts are expected to take into consideration entire systems. By a corporate liberal logic the structures and patterns underlying systems of market exchange are appropriately within the purview of decision making. So it is not surprising that the question of the value of station licenses, though not at center stage in the policy arena, has returned to haunt policy discourse at odd intervals over the years.

34. In the 1920s unease with this practice was sometimes expressed in terms of fears of monopoly. One senator, for example, argued that "[f]reedom to barter and sell licenses threatens the principle that only those who render a public service may enjoy a license. It would make possible the acquisition of many stations by a few or by a single interest. . . . this [is] a possibility to be guarded against." The senator was recommending the enactment of H.R. 9971, 69th Cong., 1st sess., an amendment restricting license transfers (*In the Matter of Powel Crosley, Jr.*, Docket 6767, in 11 F.C.C. 3 [1945], 40 [hereafter *Crosley*]).

35. Senate Committee, *Radio Control,* 43; Warner, "Transfers of Broadcasting Licenses," 600.

36. *In re Seitz,* Docket 5313, decided June 27, 1939, cited in Warner, "Transfers of Broadcasting Licenses," 612.

Bright Lines from the Left: The Bare-Bones Theory

One form that question took was what became known as the bare-bones theory, a policy tactic that, although defunct today, gained attention in the twenties, thirties, and forties.[37] The idea was that, if licenses cannot be owned, broadcasters should not be able to make money by selling them. In a sense the bare-bones theory addresses the ambiguities of corporate liberal boundary blurring by trying to redraw a classical bright line in a narrow context. Tacitly invoking the classic liberal assumptions that property exists most of all in physical things and is distinct from government-granted privileges, the bare-bones approach suggests that the public character of the spectrum should be preserved by drawing a sharp line between selling the tangible assets of a station—its equipment, buildings, and related "things"—and selling the ephemeral license itself, with the former allowed but the latter prohibited.

The Senate draft of the 1927 act did just that: it prohibited license transfers "if the consideration be greater than the reasonable value of the apparatus for which said license has been issued, and said exchange value shall in no case exceed the original cost of the apparatus."[38] Such a restriction, it was presumed, would eliminate any economic value in the license itself, preventing both trafficking and the accrual of any legally protected property rights in the spectrum. The clause was removed in conference committee,[39] but the practice suggested by the bare-bones theory was not expressly prohibited, thus leaving open the possibility that the commission could adopt a bare-bones approach in the future as an administrative rule, logically supported, perhaps, by the legislatively explicit public character of the spectrum.

After the 1927 act was passed, members of both Congress and the FRC continued to express concern about the possibility that licenses should not be allowed to take on the character of private property by accruing exchangeable monetary value. On January 29, 1932, the FRC proposed that transfer applications include an itemized breakdown of the values of both the tangibles and the intangibles included as part of the sale, on the ground that "the information now required by the Federal Radio Commission is not complete enough to permit the commission to determine whether or not value is being placed upon the wavelength or

37. Warner, "Transfers of Broadcasting Licenses," 587.

38. Ibid., 596. This policy was recommended by, among others, the ACLU. In Senate testimony an ACLU representative argued that "Secretary Hoover's signature in New York City sells for $150,000 to $200,000, and the applications are now being picked up as for sale. . . . you should prevent the sale above the cost of the equipment or the cost of the plant" (Senate Committee, *Radio Control,* 127).

39. *Crosley,* 23.

license, and as a result there is considerable commercializing and trafficking in wavelengths and licenses," which at least some members believed to be contrary to the Radio Act.[40] Continuing concern about the casual character of commission review of license transfers, particularly the paucity of information gathered, produced a change in the language of the transfer clause as it was transcribed into the Communications Act: section 310(b) of the original 1934 act stated that license transfers should be allowed only if "the Commission shall, after securing full information, decide that said transfer is in the public interest." The fact that this was one of the very few original pieces of language introduced in the broadcast portion of the 1934 act suggests that Congress continued to cast a nervous eye on the practice of license transfers.

For the remainder of the 1930s, FCC decision making struggled to reconcile the broad policy of minimal interference in business affairs with the belief that licenses should not take on the character of private property. On the one hand, the commission had been directed to gather detailed information about transfers for the purposes of preventing licenses to gain economic value in and of themselves. On the other, its broad policy mandate was that it should provide the conditions for the free exchange of stations on a marketplace basis. Not surprisingly, when it did scrutinize the accounting details of transfers with great care, it brought upon itself accusations of meddling with management prerogatives.[41]

One of the FCC's more common responses to this dilemma was euphemism: as it became clear that stations regularly changed hands at values far in excess of their tangible assets, the commission described the intangible assets in terms of "earning capacity," network affiliation contracts, the existence of established audience habits of listening to a station, and so forth—in terms of anything but the possession of a license.[42] If stations involved property in intangibles, the thinking seemed to be, at least the intangibles ought to be nongovernmental. Of

40. Ibid., 41 n. 12.

41. For example, *Travelers Broadcasting Service Corp.* (WTIC), 7 F.C.C. 504 (1939).

42. An FCC press release from July 25, 1944, stated that "[t]he Commission . . . has approved transfers that involve going-concern value, good will, etc. There remains, however, a serious question . . . on which the law is not clear, as to whether the Commission should approve a transfer wherein the amount of the consideration is over and beyond any amount which can be reasonably allocated to physical values plus going-concern and good will, even though the written record does not itself show an allocation of a sum for the frequency" (Murray Edelman, *The Licensing of Radio Services in the United States, 1927–1947: A Study in Administrative Policy Formation* [Urbana: University of Illinois Press, 1950], 98, reprinted in *Administration of American Telecommunications Policy,* vol. 1, ed. John M. Kittross [New York: Arno Press, 1980]). See also Warner, "Transfers of Broadcasting Licenses," 601.

course audience size, affiliation, and earning capacity are all quite closely tied to the fact of a station's government-protected channel. A government-granted television channel in New York City, for example, will inevitably have a larger audience, more desirability to a network, and a greater earning capacity than one in a small Nevada town, even if all other variables such as owner investment, physical plant, and so forth are held equal. But this was not frequently discussed.[43]

Residual unease about the property status of broadcast licenses persisted into the 1940s. A 1945 bill introduced to the House, for example, would have amended the transfer section of the Communications Act to limit the price of stations, if not to tangible assets per se, then to double the value of the tangible property—a kind of qualified bare-bones approach.[44] The most emphatic attempt to resolve the dilemma, however, came when the FCC attempted to draw a hard and fast line between private property and broadcast licenses in the form of what became known as the AVCO rule. In 1945, in the wake of a license-transfer decision involving a company called the Aviation Corporation (AVCO), the FCC grew concerned about the way that relatively unrestricted license transfers amounted to an apparent abrogation of the commission's duty to enforce the public character of the broadcast spectrum.[45] Perhaps overstating the clarity of its mandate, the FCC intoned:

43. For the relation of market size and audience to station value, see Benjamin J. Bates, "The Impact of Deregulation on Television Station Prices," *Journal of Media Economics* 1 (spring 1988): 5–22.

44. H.R. 4314, 79th Cong., 1st sess., introduced on October 9, 1945. The bill would have amended section 310(b) of the act by adding "No transfer or assignment shall be approved in which the total consideration to be paid for broadcast property, tangible and intangible, exceeds the fair value of such property: Provided, that such fair value shall not exceed double the depreciated cost value of the tangible broadcast property transferred or assigned." The bill died in committee.

45. AVCO had contracted to buy the bulk of the manufacturing empire of Powel Crosley, Jr., which included a number of broadcast stations, including one of the largest in the country. AVCO thus had to apply for permission to transfer the licenses. In the course of the proceedings, it was revealed that AVCO was buying the stations only because Crosley refused to separate them from AVCO's real interest, his manufacturing concerns. AVCO executives, as a result, were thoroughly unfamiliar with broadcasting and broadcast law and performed embarrassingly before the commission. The transfer was approved by a vote of four to three after the executives expressed a "commitment" to acquaint themselves with the details of broadcasting. The entire affair disturbed even the commissioners who approved the transfer, however, and prompted the promulgation of the AVCO rule (*Crosley*, 3–43). For a discussion of the case that reflects some of the views that were current at the FCC during the period, see Charles Siepmann, *Radio's Second Chance* (Boston: Little, Brown & Co., 1947), 167–83.

"Our opinion has remained steadfast that people who enter broadcasting must recognize their obligation to render a public service. They cannot operate a station as they would a department store or a steel mill—for purely financial benefits."[46] The resulting AVCO rule required broadcasters who had contracted to sell their stations to give notice of the deal and its price for a sixty-day period, so that others wishing to buy the station could apply to the FCC, who would then choose among the competing applicants. In theory the rule shifted the power to choose a successor from the station owner to the FCC, while allowing the owner a fair price for his or her station.

The AVCO rule proved practically and politically unworkable. The rule imposed serious delays on sellers—sixty days plus the time needed to review competing applicants—and new uncertainties on buyers—why go through the trouble of negotiating to buy a station if the FCC might give the station to someone else after the contract had been concluded? During its four-year existence, very few competing applications were filed, so the FCC was unable to exercise the discretion it had hoped would be created by the rule. As a result the commission repealed the rule in 1949.[47]

Nonownership as "Soft" Property

The demise of the AVCO rule marked the end of the bare-bones theory. After 1949 the strong interpretation of the nonownership clause of the Communications Act was replaced by a soft reading, and ever since the FCC has taken as a given the fact that broadcast licenses have economic value. Since then, the blurred character of the boundary between property and government licenses has been largely accepted as a fact of life (with the exception of the New Right's "marketplace approach" discussed below). The FCC has abandoned all pretense of trying to maintain a bright-line distinction between public licenses and private property, allowing the distinction to become a matter of degree rather than of kind. The nonownership clause has been interpreted in a fully corporate liberal sense as a functional guideline, not a boundary-drawing rule. To the extent that it is addressed at all, it is taken to mean merely that the purchase and sale of stations involves some special conditions that allow for slightly more legal restraint than the exchange of unregulated goods when those restrictions serve some functional purpose, such as enhancing competition, social diversity, or quality programming.

Between 1962 and 1982, for example, the FCC enforced the "three-

46. *Crosley,* 24.

47. Ellmore, *Broadcasting Law and Regulation,* 101.

year rule" as a means to reduce trafficking in licenses.[48] The rule required a hearing for transfer requests within three years of an original grant, in the hopes of thus discouraging at least the most obvious forms of speculation in broadcast licenses. During its existence, the rate of growth of license transfers slowed, suggesting the rule had some effect.[49] In 1962 the commission also invoked a minor version of the bare-bones theory when it limited the cost of transferred construction permits it requires of broadcasters seeking to build a new station. Since then, construction permits may be sold, but not for more than actual expenses invested at the time of sale.[50] Interestingly, the Reagan-era FCC that abolished the three-year rule could not see fit to do the same for the limits on construction permits. Since construction permits are typically given before any physical plant is purchased or built, the absence of "tangibles" in the value of the permits is absolute. While the practical implications of this are trivial—the intangible value of broadcast licenses is of a nearly identical nature to construction permits, and generally of much greater value—the ideological implications are not. The role of government in creating the value of construction permits is so obvious and thus troubling to the liberal desire to see property in terms of physical things that even the radically deregulatory FCC of the early 1980s could not bring itself to remove this one bit of government interference in business affairs.

It is significant that the post-AVCO liberal strategy shifted from trying to *prohibit* "ownership" of licenses to accepting the fact of ownership and trying to shape the character of station owners, through policies on diversity of ownership and control of broadcasting. Since the 1940s the FCC has prohibited ownership of more than one broadcast network and more than one station in a single market.[51] Over the years it has also prohibited cross-ownership of broadcast stations with cable systems and newspapers, and limited the total number of stations a single owner can control.[52] In the name of ownership diversity, it has also sought to encourage ownership of stations by minority-group members

48. "March 15, 1962: Applications for Voluntary Assignments or Transfer of Control," *FCC Annual Report* 32 (1962): 689. See also *FCC Annual Report* 28 (1962): 56–57. Exceptions were made for lack of finances and (obviously enough) death of the licensee.

49. Foley, "Value and Policy Issues," 281.

50. Ellmore, *Broadcasting Law and Regulation,* 101.

51. *National Broadcasting Co., Inc., et al. v. United States et al.,* 319 U.S. 190 (1943).

52. On FCC regulations limiting newspaper cross-ownership, see 47 C.F.R. 73.35, 73.240, 73.636, upheld in *Federal Communications Commission v. National Citizens Committee for Broadcasting,* 436 U.S. 775 (1978). On limitations of number of stations—the so-called rule of sevens—see 47 C.F.R. 73.35, 73.240, and 73.636. See also *Multiple Ownership of AM, FM, and TV Stations,* 18 F.C.C. 288 (1953), affirmed in *United States v. Storer Broadcasting Co.,* 351 U.S. 192 (1956).

by occasionally giving weight to minorities in comparative license hearings, allowing transfer of stations under FCC investigation to minorities (the "distress sale" policy), and allowing tax benefits to owners who sell their stations to minorities.[53]

None of the policies introduced since 1949 involve direct intervention in the majority of ordinary station sales; the basic fact of a marketplace in government-created broadcast licenses has been left untouched. To be sure, a few potential buyers and sellers of stations at various times have faced a few restraints because of post-1949 rules. Owners who wanted to sell stations within three years of purchasing them had to ask for special permission, potential buyers who already owned the maximum number of stations were prevented from buying more, and potential minority buyers have been given small advantages in the market for stations. Even when taken together, however, these cases directly involve only a small fraction of actual station sales and purchases.[54] None of the rules have substantially altered the general practice of freely buying stations at a market-determined price; throughout the rules' existence, most buyers and sellers of stations have been able to transact their business without any government interference beyond the filing of the appropriate forms.

Since 1949, in sum, regulatory interventions into the buying and selling of stations have been at most pale echoes of the AVCO rule and the "bare-bones" interpretation of the nonownership clause. After experimenting with efforts to fully insulate the government-issued broadcast licenses from the realm of private property between 1921 and 1949, Congress and the FCC have accepted, or at least acquiesced to, the principle that licenses confer economic value that can be bought and sold on an open market.

Private Airwaves as a Bulwark against Government Interference

Adding to the Bundle of Rights

Another theme of regulation over the years adopts an opposite strategy to that of shielding public licenses from private ownership. The appro-

53. FCC, *1978 Statement of Policy on Minority Ownership of Broadcasting Facilities, FCC Annual Report,* 2d ser., vol. 68 (1978): 979; *Metro Broadcasting, Inc., v. Federal Communications Commission,* 110 S. Ct. 2997 (1990).

54. In 1986, for example, nearly a decade after the minority ownership rules were adopted, minorities still owned just 2.1 percent of the nation's more than eleven thousand broadcast stations (*Metro Broadcasting, Inc., v. Federal Communications Commission,* 3002 n. 1, 3003.

priate response, it is argued, is to move in the other direction, enhancing the propertylike qualities of licenses and further limiting the government's ability to intervene in their purchase and sale. This has been the logic underlying a broad variety of policy initiatives over the years.

Critics from both the right and the left have often found it irksome that, in the case of new allocations, broadcasters receive their licenses from the government for free, but are then able to turn around and sell those licenses for a profit. Whereas the bare-bones approach sought to rectify this boundary blurring by trying to somehow eliminate the property character of the airwaves, it has frequently been suggested that the property character of licenses should be made more consistent: broadcasters should pay the government for new licenses just as they would pay other broadcasters for existing licenses. If private enterprises are going to make a profit on a public resource, the argument goes, they should pay for the privilege, not get it for free. On this theory, a draft bill was introduced in 1933 that would have required broadcasters to pay assessments on their licenses.[55] The idea of assessments for licenses resurfaced in the 1950s,[56] and has been experimented with in nonbroadcast portions of the spectrum in the last decade. And numerous proposals have been advanced for leasing or even auctioning broadcast channels.[57]

Most broadcast license transactions involve existing licenses, and thus are already obtained by purchase. Leasing or auctioning broadcast channels, therefore, would only affect newly assigned or unused frequencies, which typically involve only a minority of license transactions. Another strategy for enhancing the propertylike character of licenses, therefore, involves attempts to reduce the government's ability to interfere with the actions of existing license holders. This was the strategy, for example, behind a 1952 amendment to the Communications Act. Concerned that the AVCO procedure constituted "an unwise invasion by a Government agency into private business practice,"[58] Congress ensured

55. S. 5201, 72d Cong., 2d sess. (1933). See also Harvey Sarner, "Assessments for Broadcast Licenses," *Federal Bar Journal* 21 (1961): 245.

56. Senate Committee on Government Operations, *Adjustment of Fees of the Federal Communications Commission,* Staff Memo 85-1-70 (October 28, 1957).

57. Harvey J. Levin, for example, has proposed that the public trustee concept be replaced with a system of government-leased broadcast channels priced with "shadow prices," that is, prices calculated to simulate the cost of a "real" market, thus inducing economic efficiency while retaining government control over the long term (*The Invisible Resource: Use and Regulation of the Radio Spectrum* [Baltimore: Resources for the Future and Johns Hopkins University Press, 1971], 119-30).

58. Senate Committee on Interstate and Foreign Commerce, on S. 658, S. Rep. 44, 82d Cong., 1st sess., reprinted in 97 Cong. Rec. (1951), 967. Previous attempts to restrain the commission's ability to intervene in station sales included the 1942 Sanders bill, H.R.

that the AVCO rule could not be revived by prohibiting the commission from considering competing applicants in the case of a transfer.[59]

Although sometimes depicted as a radical departure from regulatory history, the many deregulatory efforts in broadcasting of the 1980s can be equally well understood as a continuation of the same logic used in the 1933 proposal and 1952 amendment. For example, in 1982, prompted by arguments that the three-year rule constitutes "a needless inhibition on normal business and marketplace forces in the radio and television industries," the FCC eliminated the rule.[60] For similar reasons, in the 1980s the ownership limit on broadcast stations was raised to twelve, license terms were extended from three to five years for television stations and from five to seven years for radio stations, and the licensing process was greatly simplified—many licenses can now be renewed by postcard.[61]

Significantly, in all of these cases the principle of government licensing was left intact. What was constrained was the power of the government to intervene in private business practices in certain circumstances. Though not always recognized as such, there is a decidedly corporate liberal slant to many of these regulations. The goal can be construed as not so much formal or ideological consistency but as prudent management of the regulatory structure—trimming some regulations here, removing some barriers to entry there—in order to enhance the efficiency of the system overall.

Bright Lines from the Right: Deregulation and the "Marketplace Approach"

In the late 1970s and early 1980s, the notion of property rights in the spectrum, after gestating for a few decades among neolibertarian economists and think tanks, took on a newly legitimized form and emerged as a

5497, 77th Cong., 2d sess., which limited the "public-interest" test to the transferee's ability to construct and operate a station, rather than applying it to the entire transfer proceedings; and the similar 1943 White-Wheeler bill, which would have added a requirement that the transferee's qualifications matched those of the original licensee. See Note, "Radio and Television Station Transfers," 352.

59. Public Law 554 (July 16, 1952), 66 Stat. 716. The text of section 310 was amended to its present form by, among a few other changes, adding "Any such [transfer] application shall be disposed of as if the proposed transferee or assignee were making application under section 308 for the permit or license in question; but in acting thereon the Commission may not consider whether the public interest, convenience, and necessity might be served by the transfer, assignment, or disposal of the permit or license to a person other than the proposed transferee or assignee."

60. FCC, "Amendment of Section 73.3597 of the Commission's Rules (Applications for Voluntary Assignments or Transfers of Control)," 52 Radio Regulation 2d 1081 (1982).

61. "Postcard Renewal," 87 F.C.C. 2d 1127 (1981), affirmed in *Black Citizens for a Fair Media v. Federal Communications Commission,* 719 F. 2d 407 (D.C. Cir. 1983).

major force in the mainstream policy arena for the first time since the early 1920s. The "market-based" or "marketplace" approach to spectrum regulation was a subspecies of that 1980s political movement, deregulation. Deregulation itself cannot be reduced to a single set of intellectual principles or political theories. It was shaped by matters ranging from voter disenchantment with politics to corporate resistance to environmental and safety regulation to the rise in popularity of transaction cost analysis and other products of the Chicago school of economics. Yet this larger movement helped bring to temporary prominence certain ways of thinking that, looked at in context, can suggest much about both the persistence and limitations of the liberal idea of property in the contemporary world.[62]

On the surface, economic competition seemed more central to the marketplace approach than private property did; the efficiencies of an unfettered marketplace were more often heralded than natural rights. Yet a faith in the marketplace alone hardly explains what was unique about the marketplace approach. Antitrust law is a profound expression of a faith in economic competition, yet it was eviscerated during the 1980s in the name of the same theories that underwrote the marketplace approach. And a policy generally favoring private enterprise in broadcasting has dominated since the 1920s, whereas the marketplace approach was typically described by its proponents as a radical departure from the last half century of broadcast regulation.[63]

The idea of property helps explain what distinguished the eighties' marketplace approach from more conventional promarket policies. Property was a central element of the logic of the marketplace fervor of the 1980s, a key to its deep structure. At moments this was explicit. As Chicago school hero Richard Epstein put it, "the grand idea of property and its principled necessity limitations provide the best guide for dealing with the complex modern issues that dominate our collective agenda today."[64] Yet it was, we shall see, more often implicit. The marketplace approach is usefully characterized as a neoformalist attempt to recreate a bright-line boundary between government and private property in the airwaves. Like the bare-bones theory, it sought to resolve ideological unease surrounding licensing by purifying the distinction between gov-

62. The best discussion of deregulation in telecommunications is found in Robert Britt Horwitz, *The Irony of Regulatory Reform: The Deregulation of American Telecommunications* (New York: Oxford University Press, 1989), especially 221–63.

63. Mark S. Fowler and Daniel L. Brenner, "A Marketplace Approach to Broadcast Regulation," *Texas Law Review* 60 (1982): 207.

64. Richard A. Epstein, "Property and Necessity," *Harvard Journal of Law and Public Policy* 13, no. 1 (1990): 9.

ernment intervention and private prerogatives. It sought to draw that line, however, between the license and government instead of between business and the license.

Underlying many of the law and economics-based proposals for a "property system of spectrum management," then, is a hope that the legal realists were wrong: property is not simply a shifting bundle of rights but something more like the nineteenth-century common law understanding of property as an absolute, natural right with a fixed content. As one spectrum-property theorist put it, "there is no middle ground" between a government-regulated system and a "pure market system" based in "freely transferable rights."[65] The contradiction *can* be transcended, the government's hand in the bundle of rights can be not only reduced but eliminated, if only we implement a full-fledged or "pure" property rights in the spectrum.

The 1980s version of this argument had its roots in discussions that began in the early 1950s. One of these discussions began in the Chicago school of economics; another appears to have taken place within the cultish right-wing intellectual movement led by Ayn Rand, who wrote an essay calling for property rights in radio frequencies in the 1950s.[66] This at first marginal trend called for the establishment of a property system that would create common law rights in the spectrum as an alternative to the current public-trustee concept.[67] This system would be superior, it was said, because the resulting market in spectrum access would allocate resources more efficiently and would in any case be more just.

One of the more interesting by-products of this movement has been a reinterpretation of the history of broadcast regulation. The existing system, it is argued, is not necessary but political. The decisions that culminated in the 1934 act were the product of a broad social and political vision. The choice to regulate the spectrum according to the criteria of

65. Milton Mueller's conclusion to Edwin Diamond, Norman Sandler, and Milton Mueller, *Telecommunications in Crisis: The First Amendment, Technology, and Deregulation* (Washington, DC: Cato Institute, 1983), 93.

66. Ayn Rand, "The Property Status of the Airwaves," in *Capitalism: The Unknown Ideal* (New York: Signet, 1967), 122.

67. Leo Herzel, "'Public Interest' and the Market in Color Television Regulation" (student note), *University of Chicago Law Review* 18 (1952): 96; Ronald H. Coase, "The Federal Communications Commission," *Journal of Law and Economics* 11 (October 1959): 1; Arthur S. DeVany, Ross D. Eckert, Charles J. Meyers, Donald J. O'Hara, and Richard C. Scott, "A Property System for Market Allocation in the Electromagnetic Spectrum: A Legal-Economic-Engineering Study," *Stanford Law Review* 21 (June 1969): 1499; Douglas Webbink, "Radio Licenses and Frequency Spectrum Use Property Rights," *Communications and the Law* 3 (June 1987): 3-29; Milton Mueller, "Technical Standards: The Market and Radio Frequency Allocation," *Telecommunications Policy* 12 (March 1988): 42-56.

the public interest was dictated less by spectrum scarcity than by the enactment of a set of political beliefs that involved nonobjective values being used to justify granting power to some groups at the expense of others.

But this does not lead to the notion that property itself is political. Instead, the period of interference in the mid-1920s, the property-rights advocates suggest, resulted not from a lack of government regulation, but from a lack of law, which they take to be something entirely different from the legislatively backed administrative rules that currently control the spectrum. In a seminal and highly respected essay, economist Ronald Coase argued that "the real cause of the [pre-1927] trouble was that no property rights were created in these scarce frequencies," and that the interference problems could have been better resolved by the introduction of property rights "without the need for government regulation." The untried alternative to the system of government intervention we have now, it is said, would have been the establishment of a full-fledged legally protected system of property rights.[68] What was needed in the 1920s was a kind of Homestead Act of the spectrum that would have given broadcaster-settlers legal protection from government, not subservience to it.[69]

The law/politics distinction generally requires some version of a state of nature, and the marketplace-approach theorists have found it in a peculiar version of broadcast history. They paint a picture of plucky commercial entrepreneurs restrained in the 1920s by the ham-handed actions of marauding government bureaucrats, as if the entrepreneurs flourished in a natural realm outside of government influence.[70] One author has argued, for example, that prior to 1926 "the radio industry operated very efficiently under a regime of saleable property rights in the spectrum. . . . No one believed that licensing in the 'public interest' was needed to allow broadcasting to function."[71]

As we have seen, this is hardly accurate. Practically speaking, the pre-1926 regime was virtually identical to the post-1926 regime, and while full-fledged property-rights advocates did exist prior to 1926, they seem to have been a minority even in the business community. It is clearly not the case that "no one" believed in the public-interest licensing. The consensus of the industry-dominated Radio Conferences was

68. Coase, "Federal Communications Commission," 14.

69. Rand, "Property Status of the Airwaves," 123.

70. For a similar characterization of the period, see Thomas Hazlett, "The Rationality of U.S. Regulation of the Broadcast Spectrum," *Journal of Law and Economics* 33 (April 1990): 133.

71. Spitzer, "Constitutionality of Licensing Broadcasters," 1045.

that some form of public-interest regulation was necessary. Similarly, describing the pre-1926 radio system as operating "efficiently" greatly exaggerates the importance of station sales in the pre-1926 period. (The data are sketchy, but it seems likely that between 1920 and 1926 the majority of broadcasters did not obtain their licenses by sale.) It is not clear, furthermore, that an infant, experimental industry frequently faced by chaos and uncertainty, where the norm was operating at a loss, is best described as "efficient." For this picture to make sense the arbitrary legal restraints that enabled the entrepreneurs' activities—the 1912 elimination of the possibility of a nonlegal means of regulation, the subsequent marginalization of amateurs and nonprofits, and the administrative creation of marketable broadcast stations—must be ignored. Similarly, it must be overlooked that the existing regulatory system was created explicitly to create and uphold a competitive, free enterprise system in broadcasting that created the very active market in broadcast stations that now exists.

Market-based theorists must be selective in their interpretation of the development of broadcast regulation because they seek to maintain the belief that the system of property provides a form of social life that can and should exist apart from the arbitrary winds of politics. Property is not merely a bundle of rights, they wish to assert, it is not just another form of privilege; it is, if not a natural right, a nonetheless neutral and legitimate bulwark against arbitrary political action. They seek to deny, in other words, the legal realist argument that property is neither natural nor logically distinct from a privilege.

There are those among the spectrum-market tradition who might be relatively uninterested in such matters of political theory. Some, most notably Levin, have approached the proposals of the marketplace-approach school as if it were purely a matter of practical economic effects: spectrum auctions conducted in such-and-such a way would have such-and-such an effect on AM radio station prices, and so forth. These questions are important, and continue to be explored in interesting ways, particularly now that the Clinton administration is showing interest in adopting some of the marketplace-approach policies for different practical and ideological reasons (such as using spectrum auctions to raise government revenue). But the principles that helped bring these theories into the policy arena need to be addressed in their own right. The marketplace approach can be reinterpreted as just another battery of regulatory techniques to add to the already existing supply that has been accumulating since the 1920s, but it did not originate that way. It began with a belief in the separability of rights from privileges, of law from politics, and the blind spots in the revised histories of the period belie the persistence of that belief. Ultimately, the legitimacy of the marketplace approach rests on the belief that "stations" can and should be neatly separated from the government actions that create and maintain them.

So it is important to look at the development of the proposals that began, at least, as promises to construct a system of formal, government-free property rights in the spectrum. The earliest pieces were vague but made it seem relatively straightforward. The natural course to follow, it was said, is the creation of freely transferable spectrum rights, created in the courts under tort law, much like the law governing ownership of land. Spectrum users would be able to stake a claim to a part of the spectrum whose "boundaries" would be defined in terms of bandwidth, time, geographical area, and transmission power.[72] Spectrum owners would be able to sue anyone who caused interference in their territory in court and collect damages (or, in extreme cases, perhaps they would be able to prosecute for trespass), thus eliminating the need for FCC regulation. Owners would be able to freely sell or rent all or any portion of their spectrum in any way they please.

Significantly, the marketplace-approach theorists have been more likely to proffer economic theorems than Latin quotations; they have tended to couch their arguments more in utilitarian than in natural-rights models. So the focus is usually less on abstract justice than on "efficiency." A favorite example concerns unused UHF frequencies. Since the UHF band was first opened up for television in the 1950s, large chunks of it have gone unused. At the same time, shortages exist in other nonbroadcast areas, which indeed suggests that the spectrum is not being used optimally. Those inefficiencies would be corrected, it is said, by the creation of a market in spectrum. Those not in need of their spectrum, such as holders of unused UHF licenses who are having trouble making a profit in television, would sell it to the highest bidder, presumably those who had the greatest need for it. The market would do a better job of regulating than the government.

While it seems fairly obvious that the UHF spectrum could be better utilized, the question here concerns whether it would be better to reallocate it in the traditional manner or try to turn it into a marketplace. Implementing a spectrum marketplace is not as simple as it might appear at first glance. It has been pointed out, for example, that the multiple negotiations and legal activity required to create and maintain such a system would be extraordinarily complex at best.[73] Interference from a broadcast signal travels much farther than the signal itself, and two non-

72. DeVany et al., "Property System," 17–25.

73. The following discussion relies heavily on Levin, *Invisible Resource*, 91–104.

interfering signals can interact so as to interfere with a third. If interference were to be prevented purely by tort law, then it seems likely that within a short time most broadcasters in a given region (e.g., the Northeast) would be involved in either negotiations with or legal action against most other broadcasters in that region. If spectrum owners chose, as would be their right, to subdivide their channels, perhaps selling narrow slices for particular uses, the result would be a proliferation of transmitters that would only compound the problem. Furthermore, at the outset any spectrum user would be faced with great uncertainty about the likely behavior of other users. Will they sue for any interference generated? Will they generate interference themselves? Such uncertainty would be likely to retard exploitation of the spectrum, thus reducing the "efficiency" of its use. In the words of one economist, common law property schemes in the spectrum are limited by the fact that "[e]nforcement and transfer costs will be too high [because] the number of transacting parties is very large and the withdrawal of any single participant can prevent a satisfactory agreement."[74]

An even more telling problem concerns the role of broadcast receivers in such a system. Logically, if property is a matter of common law principles and not of political fiat, then owners of television and radio sets have a stake in the spectrum, too. In fact, a striking characteristic of broadcasting is that the audience provides the overwhelming bulk of the capital necessary for the broadcast system through their investment in radio and television sets. If one includes the audience in a property system, however, it would become an extraordinary source of inertia against the efficient reallocations that are supposed to flow from a market system. What would motivate set owners to cooperate (i.e., agree to buy new sets) simply because a broadcaster decided to adopt a new more efficient transmission method in order to sell off part of her frequency to others?[75] Faced with this dilemma, even the more extreme advocates of a property-rights system in the spectrum agree that including the literally millions of people that make up the potential audience of a typical broadcast signal as joint "owners" in a court-based property system is a practical absurdity.[76]

As a result, if one looks at their proposals carefully, one finds that the marketplace-approach theorists who set out in search of pure property begin to introduce some impurities. For example, the scholar who boldly began from the proposition that "there is no middle ground"

74. Ibid., 96.
75. Ibid., 103–4.
76. DeVany et al., "Property System."

between a government-regulated system and a "pure market system" came upon the dilemmas of receiver rights and concluded that, although there would be reason to include receiver owners in a system of spectrum property rights, the dilemmas associated with such a system would not be resolvable within any kind of neutral, objective method. "The regulation of radio interference," he concludes, "boils down to a matter of *whose* subjective preferences will prevail. The standards of science or technical and economic efficiency cannot provide us with an answer to this question. We can answer it only by discussing whose preferences *ought* to prevail."[77] Unless the politics of this author are unapologetically feudal, one must assume that the discussion that resolves these matters of "subjective preference" would take place in the democratic political arena. After setting out to purify the spectrum of politics, the analysis leads to a quiet acknowledgment of the inevitably arbitrary, political foundations that will determine the structure of the property system it advocates.

Perhaps to avoid being swept into such murky politicized waters, most property-rights proposals suggest something more likely to be of comfort to the broadcast industry: that audience members be excluded from the property system by legislation, on the grounds of practicality and by way of other corporate liberal-era precedents, such as the limitations of the rights of shareholders vis-à-vis management.[78] That this tactic of using government power to draw protective boundaries around big capital in order to exclude the majority of private, rights-possessing individuals in society is quintessentially corporate liberal, and more in keeping with the values of Herbert Hoover than of John Locke, is not generally discussed.

The point here is this: Once one has made practical compromises with the initial ideal of a purely rights-based system, what is to distinguish qualitatively the proposed system from the compromised one we have now? How is the use of legislation to exclude audience members, the vast majority of participants in the broadcast system who collectively have made the largest investment, qualitatively distinct from the 1920s tactic of using the "public-interest" clause in the process of creating the market in broadcast stations? Once such compromises are introduced, we are back to tinkering with the mixed system created in the 1920s, not creating an alternative to it.

In sum, the conceptual problem with contemporary property-rights proposals is that the more practical they become, the less distinct they

77. Mueller, *Telecommunications in Crisis,* 113.
78. DeVany et al., "Property System," 55.

are from the existing system. In the current system, after all, the FCC's role is determinate only with new or marginal channels, whose economic value is slight. Most of the economic value in the spectrum is located in large, established, major market stations, which are already available for market exchange and have been since the 1920s; even AVCO was able to make its purchase. Thus, although the new proposals may differ greatly from the current system in their specifics, they would not be likely to really remove the politics, that is, the inevitability of value choices that favor some over others, from the enabling conditions of private commercial broadcasting. The alternative systems would, in the end, still intermingle private and public interests; the dilemma of using government to limit government interference would remain.

Conclusion: The Inevitably Political Character of Spectrum Property

The system of legal restraint extended over the radio spectrum in this century has been indisputably creative, at least in the economic sense. The marketable broadcast station created by this legal regime is a linchpin of the American broadcasting industry, which is one of the great economic success stories of the twentieth century. The results have been just as indisputably restrictive. Along the way, the force of law was used to arbitrarily eliminate a universe of possible alternatives to the corporate-centered, commercial system we have today: nonlegal means of spectrum regulation, amateur radio operators, and nonprofit broadcasters were all brushed aside or marginalized.

The case of broadcast licenses would suggest that the legal realists are correct: private property is political. It is a shifting bundle of rights with no absolute content, and thus the search for a hard and fast line between private property and government privilege is fruitless. All of the various regulatory proposals from the AVCO rule to the spectrum auctions are simply attempts to alter the content of the bundle of rights that comes with a license. Some proposals grant more sticks in the bundle to broadcasters, others less, but they are all just variations on a theme. They all involve the use of government to create a system of private control over broadcast frequencies; they may ameliorate, but cannot eliminate, the contradiction of using government intervention to limit government intervention.

Morris Cohen was one of the first to assert that property has no specific content but is rather a shifting, flexible bundle of rights, a set of contingent decisions about who gets what in what circumstances, decisions that are inevitably political.[79] Cohen and his fellow legal realists would have been unsurprised by the fact that American spectrum regulation has been political, that it has involved elaborate government action and the arbitrary exclusion of some groups at the expense of others. Nor would they have been surprised that this political activity had the effect of generating exchangeable commodities subject to marketplace forces and to interesting economic analyses. They would have seen the political character of spectrum regulation as consistent with the nature of property in general, not as the product of some exceptional condition such as spectrum scarcity or the requirements of public safety.

It is not surprising that neoliberal economists have resisted the implications of the legal realists' arguments. What is striking about the story of the property status of the broadcast license is that practically *all* participants in the debate have ignored the possibility that property is inherently political. The legal regime associated with broadcast licenses, this chapter has shown, is not the product of a simple "disintegration" of the concept of property in twentieth-century law. On the contrary, the concept of private property and the values associated with it have played a clearly visible, if contradictory, role in the regulation of broadcast channels. There is something of property in the decision to regulate in the first place, that is, to use legal force to divide the airwaves into bands and channels with access limited to certain individuals for certain purposes, as if radio frequencies were so many tracts of land. But the American system of regulation is also heavily inflected at many points, not only by the idea of the spectrum as a kind of territory, but also by the more specific vision of private property rights understood as proper limits to government action.

On the one hand, the broad faith in the value and justice of a system of "private enterprise" that has shaped regulation of commercial broadcasting throughout the century is propped upon the concept of private property. Private, for-profit entities have been favored by regulation at least in part because of the belief that creating and upholding boundaries between private ownership and government action nourishes a just and economically viable society. The practice of buying and selling government-licensed stations that originated in the 1920s seemed just and practical to regulators because of this belief. The same can be said of the decisions to refuse the navy's request for a monopoly of the spectrum in 1912 and to allocate broadcast channels in a way that favored large commercial operations in the 1920s. Subsequent efforts to limit the

79. Morris R. Cohen, "Property and Sovereignty," *Cornell Law Quarterly* 13 (1928): 8–30.

government's ability to interfere with the exchange of broadcast stations such as the 1952 amendments to the Communications Act or the 1982 elimination of the three-year rule follow the same general logic. And it seems likely that the FCC's extreme reluctance to invoke its powers of license revocation over the years also reflects a general respect for the principle of the autonomy of private property.

On the other hand, the concern for property, for protecting the boundary between government action and private ownership, has also played a role in generating many of the efforts to *limit* the propertylike character of the broadcast spectrum. The practice of regulating the airwaves in "the public interest" was itself less a decision to limit private ownership in broadcasting overall than it was a way to justify and make sense of the use of government powers to aid private ownership. By framing the licensing system as an exception to the rule of private property, the public-interest clause and its justificatory structure of technical necessity and the national interest helped maintain the meaningfulness of the rule itself. The clause upheld the belief in the coherence and value of the property/government boundary by couching the government's helpful reach across that boundary on the grounds that radio was a special case.

The same logic underlies the numerous efforts to limit the "private" character of the spectrum. The nonownership clause of the Communications Act was not introduced because of a decline in the faith in the coherence or value of private property. On the contrary, it was introduced because its authors *did* believe private property was coherent. They believed the boundary between government action and private property rights should remain uncompromised in order to prevent the granting of unfair privileges. If property is merely a shifting bundle of rights, however, then the privilege to use a channel, even in its most qualified forms, *is* a form of ownership; a blanket statement prohibiting ownership of broadcast licenses while granting use of them has little meaning. Similarly, the AVCO rule was not introduced on the grounds that the bundle of rights associated with a license should be limited in a particular way for particular policy purposes. Rather, it was introduced on the grounds that the public-interest and nonownership clauses of the Communications Act prohibited actions associated with traditional ownership, such as the direct sale of licenses. In other words, the AVCO rule assumed the existence and coherence of a traditional regime of private property, and sought to uphold the presumed boundary between that regime and government-issued broadcast licenses.

The irony of the situation is this. A belief in the principle of private property has not been abandoned in broadcast regulation; the belief informs many of the diverse regulatory innovations introduced in the system over the last seventy years, as well as the system itself. But the vigorous pursuit of the principle of property has led to a series of dilemmas. The fact that our law simultaneously has forbid ownership of the airwaves and invited their purchase and sale for more than sixty years is only the most glaring of these. The quandaries also manifest themselves in the fact that the existing system and right- and left-wing objections to it all share the belief that it is somehow unfair to allow licensees to profit from a government bequest originally obtained for free. All sides in the debate presuppose some belief in a coherent division between government activities and those of private profit-making institutions, yet reach dramatically different conclusions about the proper direction of regulation.

Jennifer Nedelsky has argued that, in the construction and interpretation of the Constitution, the protection of private property against democratic infringement became the paradigmatic instance for defining rights as limits to state action. The case of private property thus came to permeate our thinking about government and democracy in general; the habits of thought that resulted have outlasted the centrality of property itself in our legal and political systems.[80] The case of broadcast licenses bears out Nedelsky's thesis. On a broad level of justificatory discourse, the metaphors of property have played and continue to play a central role in broadcast policy. On the level of day-to-day practice, however, the simple existence of broadcast licenses fundamentally blurs the boundary between government and private interests. Government rules about the behavior of licensees may have important and beneficial effects, but none of them, not even the AVCO rule, could eliminate the fact that private interests make money off a government-created legal entity. Similarly, no efforts to limit the ability of government to interfere with the behavior of licensees can eliminate the fundamentally enabling role of government in the process. Even in the most extreme schemes, at some point, arbitrary political choices such as eliminating the audience from the property system will have to be made. At the same time that the American system for licensing broadcasting is the product of the belief in private property, in sum, it strains that belief to the breaking point.

It is fruitless to argue about whether licenses should be treated more like a right or more like a privilege. Even if licenses do confer property

80. Jennifer Nedelsky, *Private Property and the Limits of American Constitutionalism: The Madisonian Framework and Its Legacy* (Chicago: University of Chicago Press, 1990).

rights, even if those rights are indistinguishable from the rights that exist for traditional commodities, and even if the resulting economic effects can be usefully analyzed in terms of transaction costs, the rights so conferred nonetheless rely on, and thus are inevitably and properly subject to, political intervention.

SEVEN

Broadcast Copyright and the Vicissitudes of Authorship in Electronic Culture

[W]e should reexamine the empty space left by the author's disappearance; we should attentively observe, along its gaps and fault lines, its new demarcations, and the reapportionment of this void.

MICHEL FOUCAULT

Introduction

Broadcasting as we know it, particularly broadcast television, is in several senses "authorless." Many of television's most conspicuous formal textual features are determined by the impersonal bureaucratic demands of the industrial system of which television is part. Stories are dramatically structured to be conducive to the insertion of commercials, for example, and rigidly restricted to half- or one-hour-long blocks; one can accurately predict whether or not the hero will get the bad guy at the end of a scene by looking at one's watch. Television scriptwriters typically work collectively in teams, according to strict formulas and production schedules, not in isolation in moonlit towers and freezing garrets. And they feel a sharp contrast between what they do and the traditional model of the creative process associated with the literary ideal of the author. As one experienced television writer put it, "you don't have to have talent to write for television. I thought it was writing, but it's not. It's a craft. It's like a tailor. You want cuffs? You've got cuffs."[1] When television entertainment does produce moments of insight and originality—and it has many such moments—they are often the product of its anonymous assembly-line nature, such as the juxtaposing of images that results from inserting strings of commercials into the middle of programs.

In spite of its relatively "authorless" character, however, commercial television could not be what it is without copyright law, a legal institu-

Portions of this chapter originally appeared in *Cardozo Arts and Entertainment Law Journal* 10 (1992): 567.

1. Quoted in Todd Gitlin, *Inside Prime Time* (New York: Pantheon, 1985), 71. The use of the word "writer" in this quote illustrates the continued presence of the romantic construct of authorship—writing is not putting words on paper, but is an act of highly individual unique expression—even in conditions that contrast sharply with that construct.

tion that rests solidly on the image of authorship as individual creation of unique works. Copyright is in certain ways a classical liberal legal construct. As Martha Woodmansee and Peter Jaszi have pointed out, the conceptual system of copyright originated in and relies heavily on the modern construct of the author-genius, the individual who is imagined to be solely responsible for originating unique works.[2] The figure of the romantic author was to an extent an offshoot of the figure of the free, property-holding, individual capitalist entrepreneur (even if the romantic author, with its emphasis on unique expression, also contained a critique of the entrepreneur's calculating rationality).

This chapter's argument is, in brief, that the effort to make broadcasting commercial has involved a pronounced "bureaucratization" of intellectual property. More precisely, throughout the institutional and legal history of broadcasting and copyright law from the 1920s to the present, the legal, business, and political communities have repeatedly turned to corporate liberal bureaucratic terms, institutions, and procedures as a means to enact the nineteenth-century liberal values associated with traditional private property rights and free markets.[3] Bureaucratization of broadcast copyright has taken three principal forms. The legal fiction of the corporate individual has turned industrial bureaucracies into legal stand-ins for the individual author, property has been "simulated" in the statistical formulas of blanket licensing organizations, and ownership boundaries have been attenuated by the construction of an elaborate labyrinth of industry-inspired federal regulations that shape and channel the production and distribution of television programs.

Authorship, Bureaucracies, and Liberalism

Reconciling the collective, industrial nature of corporate-era culture with the individualist romanticism of copyright has been a central prob-

2. Martha Woodmansee, "The Genius and the Copyright: Economic and Legal Conditions of the Emergence of the 'Author,'" *Eighteenth-Century Studies* 17 (summer 1984): 425–48; Martha Woodmansee, "On the Author Effect: Recovering Collectivity," *Cardozo Arts and Entertainment Law Journal* 10 (1992): 293; Peter Jaszi, "Towards a Theory of Copyright: The Metamorphoses of 'Authorship,'" *Duke Law Journal* (1991): 455.

3. An important issue here is that of technology. John Frow, Bernard Edelman, and others have explored how the rise and institutionalization of new technologies of reproduction such as photography, video tape, digital sampling, and computer software have brought to the surface contradictions in copyright that were present but institutionally hidden in the medium of print on which copyright was based (John Frow, "Repetition and Limitation: Computer Software and Copyright Law," *Screen* 29 [winter 1988]: 4–20; Bernard Edelman, *Ownership of the Image: Elements for a Marxist Theory of Law,* translated by Elizabeth Kingdom [London: Routledge & Kegan Paul, 1979]).

lem for twentieth-century law in general.[4] In his famous essay "What Is an Author?" Foucault concluded with the question, "What matter who's speaking?"[5] Foucault's question can be interpreted two ways. Most obviously, the question nicely summarizes Foucault's challenge to the traditional literary and legal obsession with the author-creator; in response to all the worrying about who is the real author of a work, who deserves credit for it, what are the sources of his or her genius, and so on, Foucault cavalierly replies, "Who cares?" But the question is also a substantive one. If the answer to "What matter who's speaking?" is no longer "Because the author-genius is the source of originality," then the issue is reopened: Foucault is asking, "Why, in what circumstances, and how does it matter to us who's speaking?"

Radio and television rather vividly illustrate a peculiarity of modern cultural institutions: on one level, it seems to no longer matter who's speaking, yet, on the level of legal discourse, it most certainly does. It is a peculiarity of broadcasting (and perhaps of our day and age) that it simultaneously does and does not matter to us "who's speaking," that electronic capitalist culture is characterized by a tense mixture of indifference to and obsession with "authorship."

Copyright law is often approached in terms of debates over competing interpretations of the law: should copyright be used to protect the authors' freedom, or to encourage the public distribution of culture and information, or to turn intellectual products into marketplace commodities, or to serve the interests of corporate publishers and distributors? While it is significant that as a body of law copyright seems to oscillate between these different interpretations, a focus on the debates that emerge out of copyright's unsettled foundation can obscure some of the continuities within those foundations. As a long-term, Western historical discourse, copyright is the enactment of the dream that the disparate goals and values of individual creative freedom, commerce, and informational dissemination can be reconciled in law. Copyright, it might be said, expresses the hope that freedoms of individual authors can be protected in a way that simultaneously ensures the open distribution of ideas and the healthy functioning of a marketplace in reproduced texts.

4. Jane Gaines, *Contested Culture: The Image, the Voice, and the Law* (Chapel Hill: University of North Carolina Press, 1991); Vincent Porter, *Beyond the Berne Convention: Copyright, Broadcasting, and the Single European Market* (London: John Libbey, 1991); Jennifer Daryl Slack, *Communication Technologies and Society: Conceptions of Causality and the Politics of Technological Intervention* (Norwood, NJ: Ablex, 1984).

5. Michel Foucault, "What Is an Author?" in *Language, Counter-Memory, Practice: Selected Essays and Interviews*, ed. Donald F. Bouchard (Ithaca: Cornell University Press, 1977), 138.

Considered in this broad way, it should be acknowledged that, even in its origins, copyright law has always been something of an awkward appendage to liberal legalism, a special-case effort to extend liberal principles of property beyond their traditional realm of physical things into the intangible realm of reproducible "works." Perhaps because of this, copyright in American law has always had functionalist overtones. It developed less out of common law principles of property than out of the constitutional provision that grants Congress the power to "promote the Progress of Science and useful Arts, by securing for limited Times to Authors and inventors the exclusive Right to their respective Writings and Discoveries." From the beginning, therefore, copyright was understood more in functional than in formal or moral terms; the emphasis was more on copyright's role in encouraging the distribution of culture and information than on its inherent justice.

It is perhaps not coincidental that the legal regime of copyright, with its functionalist overtones, developed in response to what was arguably the first technology of mass production, the printing press. Copyright prefigured the functionalist legal logic whose spread throughout legal culture has paralleled the twentieth-century extension of methods of mass production into every corner of life.

The classical liberal components of copyright, however, are also crucial to its character. The connection between the romantic idea of genius and copyright in Anglo-American law is less overt than it is in the Continental tradition of the author's "moral right," but it is nonetheless clearly present. American law depends on conceptual distinctions, particularly originality and the distinction between an idea and its expression, that are derived from the romantic image of authorship as an act of original creation whose uniqueness springs from and is defined in terms of the irreducible individuality of the writer.[6]

From this broad perspective, then, what is to be made of the fact that the relatively "authorless" medium of television is constituted in part by a set of legal practices that nominally rest on a romantic notion of literary authorship? It need not suggest, in the manner of the Frankfurt school, that the genuine individual autonomy and creativity of authors has been perversely supplanted by a nightmarish, depersonalized and undifferentiated culture. The belief that television has eliminated individuality and creativity is no more true than the belief that the creations of nineteenth-century authors had nothing to do with the social and economic conditions under which they were produced. There were institutional and structural constraints then, and there are individuality and

6. Frow, "Repetition and Limitation."

creativity now.[7] What has happened is that the relations between individuality and creativity and their institutional context have undergone some transformations since the time when the modern institution of authorship first appeared.

Certainly, part of the explanation for those transformations must come from the labyrinthine history of copyright itself. The experience of a tension between the romantic image of creativity in copyright and the unromantic results of its application is by no means unique to television and radio. It has been pointed out, for example, that the author-associated concept of the "work" in nineteenth-century copyright law served paradoxically to transfer power away from authors.[8] This trend was enhanced by the extension of copyright to nonartistic works and reached an extreme in the doctrine of works for hire, wherein copyright goes to those who pay for the creation of a work, not to those who create it.[9] Because of these and related trends, it can be argued that copyright as a whole tends to serve the interests of publishers and distributors more closely than it serves the interests of either authors or users of copyrighted works. And yet these apparently antiauthorial effects are born of a legal regime that nominally exists to reward individual authors. The dependence of "authorless" television on "authored" legal constructs, therefore, may be simply an acute example of tendencies that are as old as copyright itself.

Much of this can be attributed to the legal fiction of the corporate individual. Giving corporations the status of persons under law grants them the ability to stand in for "authors" in the framework of copyright, thus transferring the bulk of control over media "works" from individual creators to large bureaucratic institutions. Programs are thus created, produced, owned, and exchanged by corporate bureaucracies. It is this fact, above all, that leads to the industry's notorious penchant for crassly

7. Some viewers such as a certain kind of media-literate fan or industry insiders take genuine pleasure in discovering the undeniable personal stamp of particular living, unique human beings—usually executive producer-writers—in programs. Some television producers such as Norman Lear are fond of pointing out the personal visions and experiences they bring to their television creations. And recent scholarship has begun to capitalize on these possibilities by advancing an "auteur theory of television." (See, e.g., Robert J. Thompson, *Adventures on Prime Time: The Television Programs of Stephen J. Cannell* [New York: Praeger, 1990].) These approaches are limited, not because the personal stamp of individual "authors" is an illusion—different individuals do make programs differently—but because it cannot begin to explain either the character of television texts or the full range of cultural experiences associated with those texts, both of which are only minimally shaped by the peculiarities of individual producer-writers.

8. Jaszi, "Towards a Theory of Copyright," 471-80.

9. Ibid., 485-91.

formulaic thinking in the creation of programs: the use of "track record" talent, "tried and true" program formulas, character stereotypes, copies, spin-offs, and what Gitlin calls "recombinants" are all administrative means of formalizing decision making characteristic of bureaucratic hierarchy.[10] The essential difference, perhaps, between what Gitlin calls today's "recombinant culture" and, say, the sonata formula of Bach's time is thus perhaps that recombinant culture is conditioned on the internal dynamics specific to a bureaucratic industrial system.

In a very real sense the classical discourse of unique individual creativity is not abandoned in this framework, but rearticulated within a new economic form. The "authorless" character of broadcast programs, a by-product of the bureaucratic social organization of broadcast corporations, is thus maintained by and (at least symbolically) reconciled with copyright and the notion of authorship that underlies it. To a large degree the bureaucratization of intellectual property in broadcasting is a product of the simple fact that large industrial bureaucracies have taken the place of individuals both in law and in the process of cultural production.

The fact that the legal construct of the "corporation" has taken the place of the construct of the "individual" in Anglo-American law, however, need not be interpreted to mean that faceless, impersonal structures have taken the place of living human beings in controlling cultural production, that the romantic "true creativity" of individual authorship has been replaced by mindless imitation. The institution of authorship and the corporate form are both ways of organizing complex human activities. They differ only in the particular configurations of human activity, the habits of thought and practice, that give them their distinctiveness. The corporate form, in other words, is less a replacement of the individual than it is a different way for individuals to think and act in relation to one another.

The Statistical Simulation of Property and the Blanket License

The remainder of this chapter, therefore, will focus not on reified institutions but on the patterns of thought and action, the imaginative workings, that help constitute those institutions in the corporate era. Corporate liberal legal logic is particularly evident in two related practices associated with broadcast copyright: the blanket copyright license and the use of federal administrative agencies to manage relations within

10. Gitlin, *Inside Prime Time,* 63.

industrial systems and to mediate industrial disputes. The technique of the blanket license erects a bureaucratic system such as the American Society of Composers, Authors, and Publishers (ASCAP) that statistically simulates a system of market exchanges of copyrighted "goods" in situations where such exchanges are unworkable. The use of federal agencies to mediate relations within industries, on the other hand, displaces classical notions of ownership with standards such as industry profitability and the public interest. Adjudicating the distribution of control between, say, cable operators and over-the-air broadcasters, therefore, is not a matter of determining, once and for all, who owns what. Rather, the problem becomes a short-term "practical" matter of negotiating a workable arrangement between competing parties, as a series of trade-offs that keep the various businesses involved happy and profitable without fully resolving the question of who, in the last instance, has proprietorship of the programs that reach home television sets.

The property status of broadcast material was first raised by the question of radio and recorded music. In 1914 sheet-music publishers founded ASCAP largely to find a way to make good on the right granted composers by the 1909 copyright act to demand payment for public, for-profit performances of their compositions. In 1917 the Supreme Court decided that payments could be demanded, not just in cases of actual commercial concerts, but in the case of any performance that was part of a profit-making operation, such as musicians hired by a restaurant to entertain its diners.[11] This decision created a potentially vast field for ASCAP to comb for royalties. ASCAP thus was faced with the simultaneously tantalizing and daunting task of trying to collect from huge numbers of often small and casual "performances" of copyrighted works in nightclubs, restaurants, and other commercial establishments across the nation.

Actually collecting payments for each individual performance of copyrighted works from, for instance, every piano player in every bar in the United States was thoroughly impractical. Instead, ASCAP turned to the device of the blanket license, which has since become a central feature of contemporary cultural production. Each establishment would pay a fee more or less like an annual subscription, determined by things like the size of the establishment but not by the specific content of the performances. The money thus collected would then be distributed to copyright holders according to a statistical formula designed to approximate the actual but unknown number of performances of each work.

The blanket license is set up to maintain a system of market relations

11. *Herbert v. Stanley,* 242 U.S. 591 (1917).

in copyrighted works, to ensure that composers and other artists get paid for the use of their "property." At first glance, this appears to be what it does. Money flows from users of copyrighted works to copyright holders, and works—in the form of sheet music or recordings—flow the other way. At second glance, however, certain constituent features of market relations are missing. Goods, even of an ephemeral kind, are not actually exchanged. Copyright holders do not get paid, and users of those works do not pay, for individual performances of works. Both parties deal first of all with a bureaucracy. On a day-to-day basis both copyright holders and users experience the process as being more like paying taxes or procuring welfare: amounts are determined by formulas and bureaucratic procedure. The technique of the blanket license, therefore, does not so much enable a full-fledged market exchange of goods as it creates a statistically grounded, bureaucratically implemented simulation of that exchange.

The blanket license thus illustrates the tendency of corporate liberal institutions to turn to bureaucracy, statistics, and expertise as a means to abstractly uphold liberal standards of property and individualism in conditions that would seem to conflict with those standards. When faced with dilemmas, in other words, our corporate liberal imagination often turns to bureaucratic institutions and statistical simulations as a means to uphold the general principles—if not the full concrete reality—of a system of property rights and market exchange.

Broadcast Music Incorporated and Tensions between Corporate Expansion and Control

In a general way, techniques such as the fiction of the corporate individual and revisionist legal logic together have quite successfully served to enact the liberal ideal of autonomous, individual creator-entrepreneurs while safely adapting that ideal to apparently conflicting twentieth-century economic and technological circumstances. Corporations have taken the place of individual authors and thus hold the lion's share of control over program production and distribution and can mold programming to internal bureaucratic requirements. And true to corporate liberal form, conflicts or contradictions that might emerge from this system can be dealt with on a case-by-case basis by various private and public bodies of experts, such as ASCAP, the NAB, or the FCC.

As with most ideological structures, however, the fit has been neither seamless nor frictionless, and maintaining it has required considerable institutional and ideological effort. Much of the activity surrounding broadcast copyright in this century has involved a kind of negotiation of the tensions and contradictions inherent to corporate liberal thought and legal institutions.

Corporations are typically caught between conflicting pressures: the drive toward expansion, the drive toward stability, and the drive toward political legitimacy. The corporate drive for stability stems from the fact that corporations are structures, not isolated elements. They are administrative systems that coordinate and rationalize the activities of numerous units of production and distribution. The concerns that dominate corporate decision making, therefore, typically involve ensuring the smooth coordination of the different parts of complex vertically integrated industrial systems. The desire for system maintenance, for stability and the smooth coordination of different parts of the processes of production and distribution permeates corporate decision making.

At the same time, however, corporations are capitalist: they seek autonomy from other institutions (such as other corporations or the state), and the growth and profits that such autonomy can enable. They are thus constantly negotiating tensions between a drive toward stability and coordination, and a drive toward growth and autonomy. Furthermore, as creatures of capitalism in a politically liberal society, corporations are limited in their drive toward stability by concerns about political legitimacy, principally expressed in antitrust.

ASCAP was, in a sense, the sheet-music industry's response to these tensions. The use of a bureaucratic organization and statistical approximation of market exchange was a means to encourage wide dissemination and sales of sheet music while maintaining control, that is, maintaining the ability to recoup profits, and avoiding the appearance of monopoly.

When broadcasting first appeared as an industrially backed fad in the early 1920s, it presented both an opportunity for outward expansion and serious problems of maintaining control. From a corporate point of view, broadcasting's ability to "cast broadly," to instantly disseminate messages to vast but unseen audiences, is a two-edged sword. On the one hand, broadcasting seems a corporate manager's dream: it can help proliferate both consumer products (e.g., receiving sets) and advertising-laden messages to potentially enormous audiences, thus expanding markets, sales, and the general penetration of consumer habits into the everyday life of the population. On the other hand, that same tendency toward indiscriminate proliferation of products and messages poses a very real threat to general corporate stability and coordination. If messages and products are too freely disseminated, the corporate system can succumb to problems of low profits, competition from small entrepreneurs, or a simple loss of control (as corporations complained was the case during the "spectrum chaos" of the 1920s). New legal and institutional arrangements are necessary in order both to tame and to exploit broadcasting, to negotiate the tension between an expansionary, centrifugal push outward and a centripetal pull toward limitation and control.

ASCAP helped inaugurate the search for a stable solution. During the first Washington Radio Conference in 1922, ASCAP sent the conference a message urging the creation of a blanket license system for radio performances of music.[12] When a response from the broadcast industry was not forthcoming, ASCAP began demanding royalties from several broadcast stations for the live and recorded musical performances that were beginning to be heard over the airwaves.[13] While at least one prominent station privately worked out a blanket license deal with ASCAP,[14] most broadcasters balked at the prospect of yet another expense in a field that was as yet largely unprofitable. ASCAP then secured its position with a court decision declaring that an over-the-air performance of *Mother Macree* during a program sponsored by L. Bamberger and Company ("one of America's greatest stores") was not eleemosynary.[15]

The music copyright problem led to the formation of the NAB in April 1923.[16] The NAB quickly became the commercial broadcast industry's principal tool for exerting the centripetal pull toward coordination, both within the industry and in relations with other industries. An immediate solution to the copyright problem, however, was not forthcoming from the April meeting. Instead, as broadcasting quickly evolved from an experimental fad into a central component of the consumer economy and profits rose, broadcasters found it easy enough to acquiesce to ASCAP's annual blanket licensing fees. ASCAP in turn became increasingly dependent on the broadcast industry for its revenues: in 1930,

12. Erik Barnouw, *A Tower in Babel: A History of Broadcasting in the United States to 1933* (New York: Oxford, 1966), 119.

13. Christopher H. Sterling and John M. Kittross, *Stay Tuned: A Concise History of American Broadcasting,* 2d ed. (Belmont, CA: Wadsworth, 1990), 88.

14. AT&T's pioneer station WEAF quickly negotiated an arrangement to pay ASCAP $500 annually, largely for public relations reasons. AT&T was at the time struggling to enforce radio patents over the objections of much of the rest of the radio industry, and could hardly afford to appear insensitive to intellectual property rights (Barnouw, *Tower in Babel,* 120).

15. *Witmark v. Bamberger,* 291 F. 776 (D.N.J. 1923). See also Barnouw, *Tower in Babel,* 120.

16. Sterling and Kittross, *Stay Tuned,* 89.

40 percent of ASCAP's income was from radio music performance fees, in 1937, 60 percent, and in 1939, roughly two-thirds.[17]

For most of the decade of the 1930s, this relatively happy arrangement between ASCAP and the broadcasters satisfactorily negotiated the tension between expansion and control felt by both industries; profits and the proliferation of electronically reproduced music expanded annually, but control was maintained. The one fly in the ointment was a by-product of the fact that the relation between copyright holders and broadcasters was a bureaucratic stand-in for market exchange, not an actual market. In a classical market, when people raise their prices too high, competition either forces them to lower prices or causes them to go out of business. In the case of ASCAP's blanket license, conventional market regulation was absent.[18] What was to prevent ASCAP from raising its prices? The problem was not one of monopoly: there were plenty of buyers and sellers—in a sense, too many of them. The problem was that prices were being set according to a theoretical model of a market where no real market existed against which to test the theoretical model's accuracy.

It thus could have been predicted that ASCAP would begin to raise its fees. Early in 1932 ASCAP asked for a royalty increase of "an estimated 300%."[19] In spite of a struggle from the NAB, the increase stuck, perhaps because the initial rates had been quite low and broadcast profits were at the time growing dramatically. However, when ASCAP in 1937 announced a further increase of 70 percent to be implemented in 1939, the NAB responded by organizing a competing licensing organization, Broadcast Music Incorporated (BMI). A large fund was established to attract copyright holders to sign on with the new organization. BMI also neatly exploited the indeterminacy of the process of distributing fees by statistical approximation. ASCAP's formula for fee distribution favored older, established composers; BMI sought to attract disaffected newer songwriters by adopting a formula favoring new entrants into the business.[20]

In spite of these efforts, BMI's library of licensed music remained slim at first. Between January and October 1941, BMI's first year, broad-

17. Ibid., 193.

18. For an economist's view of this problem, see Stanley M. Besen, Sheila N. Kirby, and Steven C. Salop, "An Economic Analysis of Copyright Collectives," *Virginia Law Review* 78 (1992): 383–411.

19. Sterling and Kittross, *Stay Tuned,* 132.

20. Erik Barnouw, *The Golden Web: A History of Broadcasting in the United States,* vol. 2, *1933–1953* (New York: Oxford University Press, 1968), 110.

cast listeners were treated to the unending repetition of the few available BMI and public-domain songs, during which the BMI-licensed "Jeanie with the Light Brown Hair" was forever engraved on American popular memory.[21] Control and limitation were being exerted at the expense of expansion. Instability ensued as listeners grew weary and various broadcast organizations considered defecting to ASCAP. To further complicate matters, the Department of Justice filed suit against both ASCAP and the broadcast networks for antitrust violations. Stability did not return until a compromise with ASCAP was reached in conjunction with an antitrust consent decree that caused license fees to return to old rates and allowed BMI to remain in existence and share the business of blanket licensing with ASCAP.[22] With small modifications, the arrangement reached in 1941 has survived to this day.

What happened to the categories of property and copyright in this process? At first glance, it seems that copyright was enforced,property rights in broadcast music created, and capitalist market relations were successfully extended into the sphere of broadcast culture. At second glance, however, the situation appears more complicated. The struggles between ASCAP and the NAB were not, technically speaking, manifestations of straightforward marketplace competition. Rather, they were more like the political struggles that often occur within or between rival bureaucratic institutions: the rivalry was expressed in terms of statistical formulas, membership lists, and general legitimacy. Profits were certainly at stake in the struggle, but profit was determined by the relative political strength of the institutions in question, not by buying more cheaply or selling more dearly. The blanket licensing organizations are bureaucratic, political entities, and they behave as such.

Copyright, in this light, can be said to have taken on a new role in relation to the process of cultural production. In classical liberal legal practice, copyright's role was formal. It was used to draw boundaries in a marketplace: one person's property rights with regards to a book or an article began and ended at a certain specific point, determined by the criteria of originality, expression, and so forth. In the corporate technique of blanket licensing, copyright's role is less formal and more like a functional standard: copyright acts as a general bureaucratic guideline, signifying the general goals of the system (capitalist profitability and expansion) to those inside it. The specific implementation of those goals depends less on boundary setting than on bureaucratic arrangements

21. Sterling and Kittross, *Stay Tuned,* 193.

22. Ibid.

that keep the system running, even if boundaries are allowed to grow quite blurry in the process. The question of who in the final instance "authored" a broadcast song, or more important, who owes whom what for it, is often left vague, but this is unproblematic as long as the general goals of the system are being served, as long as the industries involved are profitable, expanding, and relatively stable.

FCC Regulation of Program Ownership and Control

A second transformation in the role of intellectual property is evident in the elaborate role of federal broadcast regulation in shaping the control and ownership of broadcast programs. Against the commonsense view of private property as a bright-line boundary, blurry boundaries are the norm in program ownership and control. Besides the fact that network executives regularly force changes in plots and lines of dialogue in programs owned by nominally independent program producers, television corporations face an elaborate labyrinth of regulations that shape and channel the production and distribution of program "property." Television stations, for example, cannot broadcast network entertainment programs between 7:30 and 8:00 P.M.; networks cannot contractually obligate their affiliates to broadcast network programs; program producers and distributors cannot grant networks distribution rights for program reruns; cable operators must blank out certain cablecast programs that duplicate local over-the-air offerings. All of these rules have generated vociferous debate over the rules' fairness and efficiency, yet few object to them on the grounds that they violate property rights.

In the day-to-day workings of the television industry, the concepts of ownership, property, and copyright have become increasingly residual categories, supplanted by considerations of efficiency, fairness, and the overall functionality of the system. Although broadcast executives and lobbyists are fond of publicly bemoaning their "second-class status" under the First Amendment implicit in this system of regulation, at regular intervals throughout the system's history they also have embraced it with quiet enthusiasm. Most of the existing regulations, in fact, originated in suggestions or complaints from industry members.

Both this maze of regulations and the industry's deeply ambivalent attitude toward it can be understood in terms of the highly bureaucratic nature of broadcasting, particularly broadcast television, as a system of production and distribution. Before the networks can compete with each other, relations between networks and affiliates, advertisers and broadcasters, and independent producers and broadcasters all need to become formalized and regularized, and the values of stability and predictability prevail. It is extremely rare for a broadcast station to change its network affiliation, and the concerns that dominate decision making in the industry tend to involve, not just short-term profits, but ensuring the smooth coordination of the different parts of a complex, vertically integrated industrial system. This internal focus on system maintenance helps account for what Bernard Miege has characterized as television's "flow culture," the character of a system where "programming must be uninterrupted, constantly renewed and therefore produced on an unbroken conveyer belt."[23]

Government regulation of broadcast program production and distribution has proven useful as a means for system maintenance and coordination, and dates back to the beginning of broadcasting. When hobbyists, entrepreneurs, and corporations first began using radio to send news, music, and entertainment to mass audiences in 1920, Herbert Hoover's Department of Commerce proved an invaluable tool in sorting out fundamental relations in the industry. The Department of Commerce helped resolve the institutional questions in the process of solving the interference problem. Hoover, as we have seen, firmly established broadcasting as a commercial, corporate activity by directly and indirectly shaping what kinds of materials could be sent over different channels, initially by forbidding radio amateurs from "broadcast[ing] weather reports, market reports, music, concerts, speeches, news or similar information."[24]

Since the 1920s, government involvement in programming generally has concerned regulating two linked flows through the system: the flow of programs and the flow of profits between program producers, networks, and network-affiliated stations. Predictable and relatively stable relations among these three different industry elements are necessary to the profitable operation of the system, yet, given the numerous players and interests involved, difficult to maintain. Over the years, as minor disputes among participants in the system have erupted, the industry has gotten in the habit of turning to the FCC to serve as a moderator.

This process began to take on its contemporary form during the

23. Bernard Miege, "The Logics at Work in the New Cultural Industries," *Media, Culture, and Society* 9 (July 1987): 276.

24. Marvin R. Bensman, "Regulation of Broadcasting by the Department of Commerce, 1921-1927," in *American Broadcasting: A Sourcebook for the History of Radio and Television,* ed. Lawrence W. Lichty and Malachi C. Topping (New York: Hastings House, 1975), 548.

chain broadcasting investigations that began in 1939. Although the resulting rules that had the most impact focused on station and network ownership, some were directed at the flow of programming and profits as well. Networks were prohibited from controlling an affiliate's advertising rates and from contractually obligating affiliates to broadcast network programs. Affiliates, in turn, were required to allow other stations to broadcast network programs that the affiliates refused to air. The justification for the rules was openness and competition. Yet the rules also led to a situation in which a government agency was requiring private agencies to make copyrighted programs available to others so as to maintain a steady flow of programming through the system.

Since the controversy over the chain broadcasting investigation, FCC regulation of program and profit flow among industry participants settled into a relatively quiet, ongoing, and routine pattern. Industry leaders and lobbyists for broadcast industry factions became accustomed to using the FCC as a terrain for settling factional disputes and pie-sharing struggles. As we have seen, the *Barrow Report* and subsequent investigations and rulemaking in the 1960s led to the syndication, financial-interest, and prime-time access rules in 1970. As result, the FCC forbade networks from syndicating independently produced programs, and from obtaining any financial or proprietary rights in independently produced programming beyond the right for first-run network broadcast, and from broadcasting more than three hours of entertainment programming during prime time. The 1970 rules, most would agree, did not dramatically change the character of the system. Networks continued to dominate the system long after the rules were in place, and to this day network executives exert detailed, line-by-line control over the scripting and production of television programs. The rules did shift some of the profits to the Hollywood producers, however, and create space for low-budget independent programs such as *Wheel of Fortune* and *PM Magazine* in the prime-time access slot. The rules, in sum, usefully helped maintain equilibrium in the system.

The rise in this century of the use of federal administrative bodies as interindustry dispute-resolution mechanisms is a long and elaborate story. What is significant here is the transformed role of copyright and the principle of intellectual property. In blanket licensing, intellectual property is simulated by bureaucracies. In FCC regulation of program flows among industry subdivisions, on the other hand, the question of property is very nearly abandoned and replaced by the goal of a smoothly functioning and profitable system for program production and distribution, justified by standards such as efficiency and the "public interest."

Combining Federal Regulatory Management with the Blanket License: Cable Television and the Copyright Royalty Tribunal

The television industry's biggest equilibrium-upsetting event in the last twenty years has been the appearance of cable television as a fourth element of the system.[25] When cable first began to grow in small markets in the 1960s, the FCC, under prompting from over-the-air broadcasters, put a halt to cable's expansion: at the time the new technology seemed too threatening to the existing system's stability. As the FCC began to reverse itself in the early 1970s and cable began to expand again, numerous questions of program and profit flow arose: if a cable system carries a local television station, for example, is some form of payment called for, and if so, should the cable system pay the television station for the signal or does the local station owe the cable system for the privilege of being retransmitted? What if a cable system imports a distant signal to compete with local television channels, perhaps with some of the same programs that local stations contracted for an exclusive right to broadcast? To a classically liberal mind, these questions might be construed as questions of property boundaries: Who owns what? Where do the boundaries fall? Although the history and nature of the regulations dealing with these issues is too labyrinthine to detail here, the important point is that the underlying regulatory patterns (if not always the official rhetoric) followed the general corporate liberal principle of systems consciousness, of negotiating differences and maintaining overall profitable functioning within the television industry as a whole.

In cable the question of copyright has resurfaced as a central issue, alongside the more general questions of system maintenance characteristic of the electronic media as a whole. The programs that appear on a local cable system typically have had multiple "owners" and have passed through many hands along the way. Syndicators, sports and music interests, local stations, and others all can be thought to have property rights associated with the material distributed on cable systems. Given the huge volume of programming that fills a typical cable system, however, most seem to feel that regular direct payments to the thousands of individual copyright owners would be a practical impossibility.

As a result, when Congress rewrote the Copyright Act in 1976, they created the Copyright Royalty Tribunal (CRT), one of whose functions

25. Thomas Streeter, "The Cable Fable Revisited: Discourse, Policy, and the Making of Cable Television," *Critical Studies in Mass Communication* 4 (June 1987): 174–200.

was to operate a blanket licensing system for cable television.[26] The CRT collected money from cable operators, put that money in a "pot," and then redistributed it to program copyright holders. Cable operators paid an amount set by a formula based on their subscriber rate, and program copyright holders received payments according to a similar formula. The CRT, in sum, combined the strategy of the blanket license with the use of federal regulation as a form of system maintenance.

As with ASCAP and BMI, the CRT was set up to maintain a system of property rights and market exchange relations for the ephemeral, electronically reproducible good of television programming. And as with the music licensing agencies, in a general way this appears to be what it did. Money flowed from cable operators to program producers, and programs flowed the other way. Again, however, many of the constituent features of classical market relations were absent. Costs and payments were set, not by supply and demand, but by politically established bureaucratic formulas. Entry into this particular "market" involved having the appropriate qualifications and filling out the appropriate forms, not offering to buy or sell a product. Increasing one's profit was a matter of lobbying the CRT and Congress, not of buying more cheaply or selling more dearly. The CRT, furthermore, had numerous "side effects" uncharacteristic of markets but characteristic of bureaucracies. First, nonmarket rationale often played a crucial role in the process. In a frank if modest redistributive effort, for example, Congress allocated an extra-large percentage of the CRT "pot" to PBS, and none to the three networks. Even when "market criteria" were used to make decisions, the formulas used to calculate payments inevitably favored some at the expense of others and thus became matters for intraindustry political disputes. The CRT was a bureaucratic, political entity and behaved like one; it just happens that one of its directives, one of its administrative functions, was to simulate a system of private property and market exchange.

The CRT was heavily criticized from a number of angles. Copyright-holding organizations predictably lobbied and litigated for higher rates and higher shares of the distributed royalties. These efforts, in turn, made the workings of the CRT convoluted and lumbering. Between 1979 and 1982, for example, squabbles between industry groups and the ensuing CRT tinkering with distribution formulas postponed finalized

26. The cable television compulsory licensing scheme has been described as "undoubtedly the most complicated provision of the new Copyright Act" (Edward W. Ploman and L. Clark Hamilton, *Copyright: Intellectual Property in the Information Age* [Boston: Routledge & Kegan Paul, 1980], 104).

distributions for this period until 1986, in spite of the fact that the amounts involved were often relatively small.[27] The CRT did not complete its distribution proceedings for 1987 until 1989.[28] Noting the perhaps predictable tendency of the CRT's bureaucratic machinery to generate an ever burgeoning and, on the surface, inefficient series of proceedings and hearings, some critics have called for the creation of a more "pristine" form of property relations, based on simple direct contractual relations between copyright owners and cable operators.[29]

In the fall of 1993 the CRT was abolished, and its functions were shifted to the Librarian of Congress, who convenes ad hoc arbitration panels for determining payment formulas.[30] While the current system might be more efficient, according to one or another definition of the term, significantly the change did not involve turning matters over to the marketplace. The habit of turning to bureaucratic practices as a means to obtain goals such as fairness, efficiency, and "free" markets remains.

Conclusion

Liberal ideas of property and selfhood remain central to corporate liberal broadcasting. In particular the idea that culture emanates from the uniquely creative soul of gifted individuals, and the related belief that the product of that emanation can and should be treated as a kind of property, continues to shape both popular consciousness and legal decision making. Program production happens largely because producers hope to be rewarded for their efforts through the sale of programs, and our legal system continues to enforce the process in terms of originality of expression rooted in the assumption of an isolated creative subject.

Yet the pattern in this century has been a gradual eclipse of formal, bright-line approaches to property by various mechanisms shaped by systems consciousness: the emphasis has increasingly shifted toward the bureaucratic coordination of flows of programs and profits with an eye toward maintaining the system overall. Patterns of coordination characteristic of inter- and intracorporate organization in general, and copyright collectives and government regulation in particular, render the

27. Fred H. Cate, "Cable Television and the Compulsory Copyright License," *Federal Communications Law Journal* 42, no. 2 (1990): 211–12.

28. Ibid., 214.

29. Ibid., 222 and passim.

30. The Copyright Royalty Tribunal Reform Act of 1993 (Public Law 103–98 [December 17, 1993], 107 Stat. 2304); see "Copyright Royalty Tribunal Bill Becomes Law," *Computer Lawyer* 11 (January 1994): 29.

question of who owns what at any particular moment increasingly less important. The creation and distribution of broadcast culture is not a matter of individual authors transmitting unique works, not simply because production is collective, but because it is collectivized in a particular way: in terms of the imagined needs of coordinated industrial bureaucracies.

As a result the clearest boundaries in the program production industry arguably are no longer the legal boundaries between autonomous creators or the boundaries between various owners of copyrighted materials. The key boundaries are those that delineate the "system," the cultural and social barriers separating industry insiders from outsiders. Today, the job seeker "breaks into" the industry as a whole, not into this or that company. Becoming a successful writer is a matter, not of establishing oneself as a unique literary figure, but of proving one's ability to conform to the needs of the system.

Significantly, however, this general collectivization of cultural production is conducted, not in the name of the social over the individual, but in the name of individualism and free markets. Depersonalizing bureaucratic relations are created in the name of "personalizing" legal institutions.

Viewing as Property: Broadcasting's Audience Commodity

[Since World War II] historical *traditional and* linguistic *unity were recast as a broader and deeper determinant. . . . this memory or symbolic common denominator concerns the response that human groupings, united in space and time, have given not to the problems of the* production *of material goods (i.e., the domain of the economy and of the human relationships it implies, politics, etc.) but, rather, to those of* reproduction, *survival of the species, life and death, the body, sex, and symbol.*

JULIA KRISTEVA, "Women's Time"

Introduction

Industry apologists sometimes speak as if the broadcast audience were simply a democratic polity, "the public," and they its faithful servants. Neo-Marxist and other critics have countered with the charge that audience members are more like unwitting factory workers on the assembly line of the culture industries. This chapter argues, in turn, that the broadcast audience is less like either of these than it is like the legal category of wife. Specifically, the audience bears a relation to commercial broadcasting similar to that of "women's work" to the political economy: a precondition to the economic system, but invisible and unthinkable from within its constitutive ideology (except as a kind of natural process happening outside of human will). This chapter explores the social conditions that enable the creation of a marketable audience that can be sold to advertisers, that is, the commodification of viewership. After locating some of the contradictions in conventional views of consumer sovereignty and the industry's use of the broadcast ratings, the chapter goes on to advance an alternative approach to understanding the audience that draws on varieties of feminist scholarship.

Much of the material for this chapter, including many of the ideas and some of the prose, was developed in collaboration with Wendy Wahl, and presented as "Feminist Theories of Power and the Construction of the TV Audience," to the conference "Console-ing Passions: Television, Video, and Feminism," Los Angeles, April 1-4, 1993.

To begin, some caveats. First, the argument here is certainly not that the audience is inherently feminine or feminized (though I do discuss why some people may think of it that way). The argument is that there are illuminating parallels in the legal and ideological mechanisms by which both domestic labor and the work of interpreting advertising have been defined as nonwork. Second, the discussions of dichotomies such as production/consumption, work/home, public/private, and masculine/feminine are not intended as descriptions of empirical social structures (which are more complicated) but as descriptions of ways that people in our society, particularly elites in decision-making capacities, *imagine* social relations as they make decisions within and about institutions. Third, the goal of this chapter is not to provide a single answer to the riddle of the audience, not to construct an exhaustive theoretical model of the economics of viewing, a direct competitor to utilitarian-based liberal theories of consumer sovereignty or class-based Marxist theories. The goal is to open up the question to its multiple determinants, to point to the complex, sociohistorical conditions upon which the industry-audience relation rests.

"We Give the Audience What It Wants" versus "We Sell Audiences to Advertisers": The Contradictions of Consumer Sovereignty

One of the central incongruities of American broadcasting is that the audience is construed simultaneously as both subject and object of the system, both the buyer and the thing sold. It is sometimes suggested that the problem with commercial broadcasters is that they imagine their audience only as a commercial market, and that this is all they mean when they assert that they "give the audience what it wants."[1] Broadcast executives also like to assert, however, "we don't sell programs to audiences, we sell audiences to advertisers." Commercial broadcasting in the United States requires its operators to imagine their viewers and listeners as property, as a commodity, as if the viewing and listening public were produced, harvested, and marketed like some kind of mushroom. At the same time that liberal political ideology encourages broadcasters to construct their activity as giving the audience what it wants, then, commercial broadcasting also requires that they imagine the audience less as a market of buyers than as a form of property to be bought, as the audience commodity.

1. For example, Ien Ang, *Desperately Seeking the Audience* (New York: Routledge, 1991), 26.

The Audience as a Determinate Fiction

Central to the peculiarity of audience commodification is the distant, abstract way in which broadcasters relate to their audience. Broadcasters do not view the audience with the indifference sometimes imagined by conspiracy theorists. On the contrary, they constantly hunger, with desire and anxiety, for knowledge about the invisible multitudes that they hope will tune in. But broadcasters see their audiences only at a great distance, as if across a great chasm, and through a haze of statistical approximation and corporate imperatives. Their blurred, remote view flattens differences, hides social contexts, and obscures subtleties in such a way as to allow blunt abstractions to stand in for more subtle understandings of people: families, in all their variety, are reduced to "households using television" (HUT to industry insiders), viewing is reduced to set tuning, social difference to demographics, culture to "lifestyles," and ultimately, the public to that vaguest abstraction of all, the quantified audience rating. The ratings numbers that are the lifeblood of commercial broadcasting are a classic example of the common twentieth-century habit of thinking of others in terms of "masses."

That the idea of the "masses" is limited has been a refrain in media criticism from the Frankfurt school to contemporary cultural studies. The ratings, and the image of the "masses" that they embody, encourage an image of people as something outside of human control to be measured, probed, and analyzed like a distant constellation, and then only to be better managed by impersonal institutions. The complex realities of real people's lives are thereby glossed over, and the audience is abstracted into something "out there," at once natural and unknown, a mysterious "other." As Raymond Williams put it, "[t]here are in fact no masses; there are only ways of seeing people as masses."[2]

There is much truth to the argument that the "audience" is an oversimplified fiction, something more imagined than real. Fictions, however, can be determinate, can have effects. They are part of why things are the way they are. For the broadcast industry in particular, the audience is not merely an abstraction, but an abstraction that is crucial to doing business; the audience *as fiction* is part of the production of broadcasting, of why broadcasting is the way it is, of how and why it plays a role in contemporary culture and power relations. It is important that what Ien Ang calls "institutional approaches" to the audience obscure the complexities of the "everyday life" of actual television watchers, yet it remains the case that the institutional view bears a pro-

2. Raymond Williams, *Culture and Society, 1780-1950* (New York: Harper & Row, 1958), 300.

foundly important material relation to that everyday life. Ang's romantic valorization of everyday life over the institutional, as if the two were separable in contemporary society, cannot in itself get to the heart of the matter.[3] Much can be learned, therefore, by focusing on how different groups imagine the audience and how those ways of imagining in turn shape the organization of the institution.

The problem should not be framed too linearly, however. The effects of imaginary constructs of the audience are inseparable from and reciprocally related to their causes. Hence, it is also important to ask why, if it is "inaccurate" to imagine the broadcast audience members as a passive and undifferentiated mass, do people then experience them as masses in the first place? In other words, what are the conditions of possibility of the chasm that separates broadcasters from their audiences?

The answers to this second question are more complex than some might think. The chasm between broadcasters and their audience is not a matter of simple physical distance. A network executive in New York is likely to have rather detailed knowledge of the tastes and concerns of the community of industry insiders whose names fill her Rolodex, even though they are widely dispersed over both coasts. Her knowledge of the tastes and concerns of the millions of viewers that live within a few miles of her Manhattan office, on the other hand, is likely to be little more precise than her knowledge of viewers in rural Arkansas. And although the chasm is technologically enabled, it is not technologically caused: if, as was the case in broadcasting before 1921 and is the case with the Internet today, broadcasting were more interactive and audience members could transmit as well as receive, the problem of audience feedback would be greatly reduced.

The separation between broadcasters and their audience, therefore, is not so much technological or spatial as social and historical. To a large degree, the mysterious otherness of the audience is a mystery of our own making, a by-product of the peculiar organization of social life and communications in this century. Particularly, the sociopolitical choices to make broadcasting strictly one-way and to fund the system indirectly through advertisers (instead of through, say, license fees) do as much to separate broadcasters from their audience as does the physical fact of geographical separation.[4]

The chasm across which broadcasters try to make sense of their

3. Ang, *Desperately Seeking the Audience.*

4. Before broadcasting was legally segregated from two-way amateur activities in 1921, amateurs who transmitted news and entertainment to large numbers of listeners could get feedback directly over the airwaves from the other amateurs who were their primary audience. See chapter 3.

audience is a segment of the more general divide between production and consumption, a social division built out of relations of class and gender. The character of that chasm and the audience commodity that it enables can be understood less as a mechanical by-product of broadcast privatization than as a sociopolitical *precondition* of privatization that rests on noneconomic, cultural, and social foundations.

The Sociology of Ratings (In)accuracy

The legitimacy of the audience-commodity system is not frequently contested: in American mainstream political culture, selling audience attention is generally thought of as a strictly marketplace, if crass, means of funding broadcasting. And it is frequently construed as democratic by assuming that the ratings are unmediated representations of the self-evident "wants" of the audience, which are then transparently transmitted by advertisers to the broadcasters. As Fowler and Brenner put it, broadcasters are able to "determine the wants of their audiences," by way of broadcasting's "indirect market mechanism [wherein] the advertiser acts as the representative for consumers."[5]

This utilitarian vision of consumer sovereignty in broadcasting rests on a logic of unmediated transmission that understands communication in terms of a metaphor of electrical technology. It is a classic case of what James Carey calls "the transmission view of communication."[6] It is based on the assumption that desires are communicated through the labyrinth of "the television-industrial complex"[7] as precisely as digital data over electronic networks. The advertising system, of course, is hardly a neutral and transparent mediator; it has its own agendas, pressures, and limitations that filter, warp, and cloud whatever information about audience desires might be obtained.[8] After all, if information about what people wanted was really the only issue, one could simply ask them, perhaps through some kind of vote (as has been done in the Dutch broadcast system).[9]

5. Mark S. Fowler and Daniel L. Brenner, "A Marketplace Approach to Broadcast Regulation," *Texas Law Review* 60 (1982): 210, 232.

6. James W. Carey, "A Cultural Approach to Communication," in *Communication as Culture: Essays on Media and Society* (Boston: Unwin Hyman, 1989), 13–36.

7. Todd Gitlin, *Inside Prime Time* (New York: Pantheon, 1985), 115.

8. For a description of some of the effects of institutional and economic pressures on ratings methodologies, see Peter V. Miller, "Made-to-Order and Standardized Audiences: Forms of Reality in Audience Measurement," in *Audiencemaking: How the Media Create the Audience,* ed. James S. Ettema and D. Charles Whitney (Thousand Oaks, CA: Sage, 1994), 57–74.

9. Anthony Smith, *The Shadow in the Cave: The Broadcaster, His Audience, and the*

One powerful legitimating device for the American system, a large part of its claim to being neutrally in touch with audience desires, is its reliance on quantitative audience ratings. Numbers and statistics in our culture have powerful connotations of science and irrefutable precision, against which references to values often seem shallow and sentimental. It is worth pointing out at the outset, therefore, that, even within its own terms, the audience-for-dollars system rests upon audience ratings that are empirically tenuous. Scientifically speaking, the audience ratings are not worthless. They do tell us that many more people watch *Roseanne* than *MacNeil-Lehrer NewsHour,* and most would agree that, particularly on a national level, they are capable of considerably finer discriminations. Yet it is also the case that the industry overall is oddly cavalier about the accuracy of the ratings, and regularly makes buying and programming decisions based on distinctions that often skirt and at times fall below what many would argue are minimally acceptable levels of accuracy.

The limited accuracy of the ratings has been well documented. The Nielsen national sample has attracted the most attention. Its most important weakness, most would agree, is that its sample is not representative; the elderly, minorities, and other groups are systematically undercounted. Serious questions have also been raised regarding the precision of the "people meter" system currently in use; there is reason to believe that some types of people are more likely to enter information into the meters than others, which would lead to systematic undercounting of some groups. The more striking problems with ratings accuracy, however, are found at the less-discussed local level, in the ratings of local television and radio stations, particularly in smaller markets. In general, because the cost of using sufficiently large and frequent samples in local markets would be prohibitive, the ratings in these cases are typically statistically weak. The system of estimating audience sizes based on ratings during selected "sweeps weeks," furthermore, has led to efforts to inflate ratings by offering blockbuster programming during those periods. On occasion the industry will mount attacks on the more egregious cases of such ratings-inflating "hypoing" during the sweeps (such as a station that ran a series on Nielsen families during its evening newscast).[10] The fact remains, however, that in a strictly scientific sense

---

State (Urbana: University of Illinois Press, 1973), 264–78; for an update on the current structures of Dutch broadcasting, see A. J. Nieuwenhuis, "Media Policy in the Netherlands: Beyond the Market?" *European Journal of Communication* 7 (1992): 204–14.

10. "Lines Blur between Hype and Distortion in Local Sweeps," *Broadcasting,* June 29, 1987, 40.

the entire industry practices a form of "hypoing" in its sweeps weeks programming as a matter of course.[11] Throughout the history of the American broadcast industry, executives have bought and sold advertising time based on numbers that are tenuous, in a few cases even fraudulent.

The most striking fact about the statistically mediocre character of the ratings is how little industry members care about it. When there have been pressures for ratings reform, they have been just as likely to come from government regulators and academic survey methodologists as from the industry itself. The sociology of this phenomenon has only begun to be explored.[12] In general, the industry's explanation for the problem is cost: more accuracy in the ratings would be nice, but too expensive to be worth it. While there is a certain fundamental truth to this answer, it does not explain everything. For example, if the ratings are vague, one might expect that profit-seeking executives would acknowledge the uncertainty and then turn to other means for decision making, particularly when it is to their advantage in bargaining. Yet buyers and sellers in the market for audience attention tend to accept the ratings numbers at face value even when this is statistically not justified. This is true of both broadcasters and advertisers. As one ABC executive said, "[Y]ou can't really take into consideration everyday, 'boy do I have confidence in that number or not?' . . . It wouldn't really be practical to think about whether these numbers are right or not." Similarly, J. Walter Thompson executive Alice Sylvester noted, "The Nielsen ratings right now represent an agreement between the buyers and sellers to use them as the currency of negotiation. . . . for what we're doing with the ratings and the way we plan, buy, and execute, it works just fine."[13] As Gitlin describes the situation, "Once managers agree to accept a measure, they act as if it is precise. They 'know' there are standard errors—but what a nuisance it would be to act on that knowledge. And so the number system has an impetus of its own."[14]

The key point here is that to be "practical," the ratings must conform to the perceived needs and interests of the industry *as a whole.*

11. Sydney W. Head and Christopher H. Sterling, *Broadcasting in America: A Survey of Electronic Media,* 6th ed. (Boston: Houghton Mifflin, 1990), 380.

12. For a compelling though not fully theorized analysis, see Gitlin, *Inside Prime Time,* 19–55. See also Eileen R. Meehan, "Why We Don't Count: The Commodity Audience," in *Logics of Television: Essays in Cultural Criticism,* ed. Patricia Mellencamp (Bloomington: Indiana University Press, 1990), 117–37.

13. Alice Sylvester and ABC executive Ted Harbert, interviewed on *Nova,* "Can You Believe the Ratings?" first broadcast in the United States on PBS, February 19, 1992.

14. Gitlin, *Inside Prime Time,* 53.

They are not so much a product of marketplace competition as they are an "agreement" or social convention without which competition could not take place.[15] The ratings, in other words, are as much a matter of collective sensibility as of science. When industry members say the ratings are "practical" or that they work "just fine," they are referring to the pressures and constraints of the culture and institutions within which they work, not just to a certain level of accuracy determined on scientific grounds. As survey researcher Percy Tannenbaum put it, "[i]n its own way, it's all very American. . . . In the ratings business, all you need is to be *known* as more accurate, not actually to be more accurate. . . . We are dealing here with a business where the network client doesn't care about the things you and I care about. They are not buying accuracy. They're buying credibility."[16]

The role of the ratings as collective agreement was suggested in 1986 and 1987, when the industry was faced with the appearance of two, and for a brief period three, different and inconsistent ratings reports that accompanied the introduction of a competitor to Nielsen's Audimeters, Audits of Great Britain's (AGB's) people meters. Differences between Nielsen and AGB numbers for the same programs were large, sometimes as much as 15 percent, and when Nielsen introduced its own people meter system to compete with AGB's, discrepancies surfaced between Nielsen's two systems.[17] As one trade magazine of the time put it, "[t]his confusion and the realization that the television business may have been relying on inaccurate ratings for years have introduced an almost intolerable degree of uncertainty into the $20 billion television advertising business. . . . Up to now, the Nielsen's ratings monopoly was the glue that held the television advertising business together."[18] At the time the main problem for industry members seemed to be, not the chronic inaccuracy made evident by the disparities, but the difficulty of doing business when more than one number is available because of the uncertainty this introduces into the buying and selling of advertising time. The issue was less confusion about the audience, than it was confusion in the day-to-day affairs of the corporate system, in the relations

15. Miller, "Made-to-Order and Standardized Audiences," 57.

16. Quoted in Michael Couzens, "Invasion of the People Meters," *Channels of Communications* 6 (June 1986): 42.

17. Marianne Paskowski, "The Good, Bad, and Ugly of the New People Meters," *Electronic Media,* January 4, 1988, 48. For example, in one week in the fall of 1987, *Spencer for Hire* according to the Nielsen Audimeters received a 24 share, according to Nielsen's people meters, a 21 share, and according to AGB's people meters, a 19 share (Adam Buckman, "People Meter Era Arrives in Confusion," *Electronic Media,* September 7, 1987, 1).

18. Couzens, "Invasion of the People Meters," 41–42.

between industry segments. The ratings, then, are conditioned not by science or simple economic self-interest alone, but by the social and political contours that join and integrate the institutions of electronic media: the ratings and their interpretation, use, and abuse point to the shared structures of corporate liberal broadcasting as a whole.

The relationship of industry consciousness to statistical knowledge was also illustrated by the reversal in fortunes of country and western music following *Billboard Magazine*'s adoption of the Soundscan company's figures for music sales. When in May 1991 *Billboard* replaced its traditional method for compiling its charts (based on phone data from record store managers and sales people) with Soundscan's figures based on actual record sales, to the industry's surprise Garth Brooks suddenly became the number-one pop artist in the country—not because more people were suddenly buying his records, but because it appeared that the old data was biased by store owners' assumptions that country western is always secondary to rock and pop. Yet the industry responded, not by learning to be more skeptical of the way it uses data like *Billboard*'s charts, but by adjusting its conventional wisdom: country western now suddenly was becoming a mainstream craze that was "sweeping the nation" (which in turn generated massive promotional efforts that probably helped make the belief a self-fulfilling prophecy). The fact that country and western had been popular for a long time and was now merely sweeping industry awareness was hardly acknowledged.[19]

The Political Roots of the Audience Commodity

As we have seen, the principal decision makers of the early 1920s did not have the direct product advertising system in mind. The corporate liberal elite of the 1920s were wary of importing the direct advertising model into broadcasting, and as a group leaned toward Herbert Hoover's more genteel vision of patrician sponsorship.[20] The decision to fund broadcasting through direct advertising thus seems to have been arrived at indirectly. The stated intentions of industry leaders cannot of themselves adequately explain the origins of the system. Nor can simple functionalist explanations that attribute the ratings system to the logic of capital, a conspiracy on the part of the ruling class, or the market. To be

19. Chuck Philips, "Rock'n'Roll Revolutionaries: Soundscan's Mike Shallett and Mike Fine Have Shaken Up the Record Industry with a Radical Concept—Accurate Sales Figures," *Los Angeles Times,* December 8, 1991, "Calendar" sec., 6.

20. Robert McChesney, *Telecommunications, Mass Media, and Democracy: The Battle for the Control of U.S. Broadcasting, 1928–1935* (New York: Oxford University Press, 1993), 14–18; Roland Marchand, *Advertising the American Dream: Making Way for Modernity, 1920–1940* (Berkeley: University of California Press, 1985), 89–93.

sure, the language of money is numbers, and so capitalism will perhaps always provide someone the motivation to try to quantify activities so as to be able to extract profit from them. And in the 1920s the advertising and manufacturing complex already represented a formidable force: advertising and marketing were well on their way to becoming central components of corporate strategy, and the general principle of funding media by selling advertising time priced according to circulation had been pioneered in newspapers and magazines. Furthermore, as Meehan has pointed out, the ratings themselves are salable commodities: Nielsen and similar companies profit by selling ratings to advertisers and broadcasters.[21] But these pressures toward quantification and advertising support need not be characterized as evidence for ineluctable economic laws of motion that in and of themselves explain the creation of the audience commodity. For there were numerous countervailing forces working against the direct advertising approach.

On the one hand, the indirectness of the advertising approach to funding presented obstacles that were arguably just as formidable as those presented by alternative funding mechanisms. In the 1920s methods for measuring audience attention to broadcast advertisements were almost nonexistent. For an industry that was prospering under government-sanctioned monopolies like those of RCA and AT&T, therefore, license-fee systems and receiver taxes had their attractions, and did not involve the perplexing organizational and methodological problems associated with selling advertising time.

On the other hand, the obvious capitalist approach might have been to simply work out a way to directly charge audiences for programs. Though the omnidirectionality of radio waves makes erecting a ticket booth in the airwaves more difficult than in the case of, say, cinema, solutions to the problem were experimented with as early as 1931.[22] License-fee systems and receiver taxes, furthermore, in spite of their associations with public-service broadcasting, in a sense do involve audience payment for programming (to roughly the same degree as, for example, copyright collectives). Even in later years, the American broadcast industry would be faced with numerous opportunities to implement the comfortably capitalist idea that people pay for what they get, such as "subscription television" proposals in the 1950s and 1960s, yet

21. Meehan, "Why We Don't Count."

22. Zenith Radio Corporation, for example, began working on a system of pay television called phonevision in 1931, involving scrambled broadcast signals that could be unscrambled by way of signals delivered over the phone lines (Richard A. Gershon, "Pay Cable Television: A Regulatory History," *Communications and the Law* 12 [June 1990]: 3–4).

the industry repeatedly resisted what would seem to be one of the most obvious means of applying marketplace principles to radio and television.[23] And in today's cable television, where a ticket booth does come naturally with the technology, most cable offerings are nonetheless advertising-supported.

One might argue, finally, that although audience measurement was vague in the early years, things have improved a great deal since then; whether or not direct advertising support was the obvious economic solution in the 1920s, it has long since proven itself as such. Indeed, audience measurement techniques have gradually improved in breadth and sophistication throughout the century. Yet the exchange of audiences for advertising has accelerated ahead of audience measurement, not in its wake. As a rule the industry has made decisions to buy and sell an audience segment *before* (and sometimes long before) there was an established means of measuring that segment. For example, in the late 1920s William Paley's fledgling CBS radio network played a key role in encouraging the trend toward advertising support. Paley's secret, it has been said, is that "[h]e could envision the audience at a time when there was in fact no audience."[24] The most dramatic major improvements in accuracy, moreover, were arguably prompted not by economic pressures from industry members, but by government investigations of ratings accuracy in the early 1960s.[25] The industry seems driven by a desire to buy and sell audiences *whether or not* it has very clear information about what it is that is being bought and sold.

So it seems likely that there were and are other sources of pressure toward the selling-audiences-to-advertisers formula. Again, the historical evidence on these matters is as yet unclear. Yet it is nonetheless reasonable to suspect that one force that drives the industry toward the audience-commodity solution is political pressure, or more specifically, the liberal fear of politics. As we have seen, one way to avoid the pain of

23. In the 1950s and 1960s the broadcast industry waged a steady and successful campaign against "subscription television," using FCC regulations to hobble a series of efforts to sell programming directly to audiences by way of scrambled signals and metered sets. See ibid., 3–26.

24. David Halberstam, *The Powers That Be* (New York: Dell, 1979), 40, quoted in Sut Jhally, *The Codes of Advertising: Fetishism and the Political Economy of Meaning in a Consumer Society* (New York: Routledge, 1990), 70.

25. Following congressional hearings into the ratings in 1963 that revealed extensive statistical distortions in Nielsen and other ratings services, a mixture of FTC consent decrees and industry self-regulation prompted a major industrywide improvement and clarification of the way the ratings were collected, interpreted, and used by advertisers. See Karen S. Buzzard, *Chains of Gold: Marketing the Ratings and Rating the Markets* (Metuchen, NJ: Scarecrow Press, 1990), 102–23.

democratic political choice is with practices of bureaucratic expertise: in the 1920s, problems of spectrum access and copyright were resolved this way. License fees or receiver taxes overseen by private or public professional structures might have functioned similarly.

But problems of policing the spectrum and copyright were relatively easy to construct as strictly technical matters, as policy, and thus of relevance only to experts and industry insiders. The question of extracting money from the audience, in contrast, directly involved the broad public that was already familiar with this newly popular medium. Progressive legislators and others suspicious of what was then called the "radio trust," furthermore, were on the lookout for ways to influence developments in radio. In the 1920s most of the practical alternatives to selling advertising time thus raised the specter of political, particularly congressional, involvement in broadcast structure and the potential political strains such involvement might produce. License fees, consumer taxes, and funding from general tax revenues, as attractive as these might have seemed to corporate leadership, all had the disadvantage of requiring some form of direct government intervention in funding, at least to the point of specifying mechanisms by which rates would be set. Direct advertising support, on the other hand, had the advantage of being outside government purview because it involved buying and selling, that is, a traditional market (even if it was difficult to determine precisely what it was that was going to be bought and sold). From a corporate executive's point of view, license-fee systems and the like could well have opened up a Pandora's box of political difficulties. Selling airtime to advertisers, on the other hand, had the advantage of at least appearing to be a straightforward market exchange and thus to be exempt from political responsibility; even if it raised severe problems of method and organization and invited accusations of bad taste, at least it circumvented the messy difficulties of political intervention.

The audience commodity, then, may have emerged, not because of market pressures, but because of political pressures to create a market. Political involvement, when shielded from democratic politics by the practices of bureaucratic expertise, was acceptable in the form of spectrum licensing. When it came to the question of funding, however, the corporate liberal polity sought to construct a market that would be more insulated from political intervention. Precisely because the advertising system renders the commodity form the primary relation of broadcasting to its audience, advertising support and the ratings that make it possible have largely escaped the labyrinthine administrative, legal, and political struggles that have surrounded the interpretation and implementation of the public-interest clause, the licensing system, and copyright collectives. With the exception of the stormy 1963 congressional hearings on the ratings, the mechanics of the system have gone largely unquestioned in the political arena, and criticisms have been limited to concerns about "lowest-common-denominator" programming largely rooted in middle-class anxieties about mass culture. Hence, from the point of view of law and policy, the audience commodity appears as relatively straightforward market exchange and has generated little of the legal and regulatory confusion characteristic of spectrum licensing and copyright.

Theorizing the Production/Consumption Distinction

Because American law and policy generally accepts the basic economic logic of the audience commodity as a fait accompli and limits its concerns to minor after-the-fact regulation, locating the audience commodity's contradictory heart requires looking at things from a broader view: from the point of view of the general productive relations in twentieth-century capitalism. The fact that it seemed commonsensical to authorities to legally segregate transmission from reception at the moment radio became a popular medium of communication was reflective of a larger trend: the broad separation of production from consumption associated with the industrial revolution and the creation of a privatized domestic space—the home as haven in a heartless world of industry and market competition. In theory this social innovation separated the public world of work, markets, and politics from the world of the nuclear family. Concretely, it enabled new arrangements of space, time, gender, and labor, with complex ramifications. Both the practice of funding broadcasting by "selling audiences to advertisers" and the radical separation of transmission and reception that constitutes broadcasting itself are premised on the production/consumption divide. As Jody Berland puts it, "[t]he process that produces [broadcast] audiences is in fact indissoluble from the process that produces the spaces which they inhabit."[26]

Mobile Privatization

As industrial production became ever more pervasive, the private, domestic space evolved toward and within that social complex that Raymond Williams dubbed "mobile privatization": a pattern of social life

26. Jody Berland, "Angels Dancing: Cultural Technologies and the Production of Space," in *Cultural Studies,* ed. Lawrence Grossberg, Paula Treichler, and Cary Nelson (New York: Routledge, 1992), 39.

characterized by high mobility, a consequent uprootedness, and the construction and valorization of privatized family homes whose contents were both easily transportable and conducive to isolation from the surrounding community. "Broadcasting in its applied form," Williams states, "was a social product of this distinctive tendency." In a characteristic effort to avoid functionalist explanations, Williams emphasizes that mobile privatization was not simply a mechanical effect of the need of industrial capitalism for a mobile, self-reproducing workforce. It was "at once an effective achievement and a defensive response," achieved through "intense social struggle" that led to "improvement of immediate conditions."[27]

The creation of the domestic mobile privatized space, in any case, was part of a process that erected a divide between certain aspects of life and others—between child rearing and manufacturing goods, between personal and public life, and between consumption and production. Though not a matter of simple physical distance, the separation involves arrangements of space: the separation of home from workplace, suburb from city, and so forth. It also involves separations in time: for waged or salaried men, for example, the separation of leisured evenings from working days. It is above all a *social* divide, as much a matter of hearts and minds as of physical, economic, and technological constraints. But it is a massive divide nonetheless, both on the level of experience and in terms of the distribution of power and resources within society.

Several aspects of the production/consumption divide are particularly important to the constitution of broadcasting. It is in part the separation of the domestic space from the public world that produces a need to bring information about that world into the home, that is, a need for mass media such as radio and television. Just as significant is the fact that, once the production of goods has been removed from the home and the farm to the factory, then the principal relationship of the home to those goods becomes one of "consumption": the use of goods becomes radically separated from their production, and consumption becomes an activity in its own right, helping to define the social character of the home both in terms of domestic life—shopping and consumption become defining categories for organizing everyday life of the home—and for producers of goods, who envision their markets in terms of domestic households. And the creation of the domestic space occasioned the widespread development of "personal life" and the idea that

27. Raymond Williams, *Television: Technology and Cultural Form* (New York: Schocken, 1977), 26–27. For a recent elaboration of the concept of mobile privatization, see Shaun Moores, "Television, Geography, and 'Mobile Privatization,'" *European Journal of Communication* 8 (1993): 365–79.

the search for satisfaction beyond the fulfillment of basic needs is a private, personal matter (as opposed to, say, a religious or workplace one). Hence broadcasting itself, advertising support, and the notion that broadcasting is a leisure activity all make sense to middle-class Americans largely because so many of us live in a world characterized by the dissociation of consumption from production and by the mobile privatized domestic space.

On the one hand, this broad social divide is one of social expectation; one feels different, and relates differently to oneself and to others, when one crosses from the world of work to the world of the home. It is also a divide of knowledge; if a society separates consumption from production in space and time, if using things is isolated from the production of things, it is little wonder that those who produce goods lack knowledge about how, when, and by whom those goods are consumed, and develop elaborate forms of marketing research intended to send information back across the sociohistorical divide. Commercial broadcasting and more generally advertising and marketing work to maintain lines of communication across the social distance of the production/consumption division.

Significantly for understanding the audience commodity, the divide also involves complex forms of power, of control and access to resources. This is to a large degree a matter of basic socioeconomic status: on both sides of the divide, an individual's power to influence events correlates rather closely with that individual's wealth; it is what early advertisers once called a "democracy of the dollar" (as opposed to a democracy of individuals).[28] It is also a matter of social class: people who labor in factories have much less power to influence affairs than those who manage them, not only because of their lower incomes but also because their power to exercise choice is almost entirely limited to the activity of consumption, whereas managers wield considerable influence on the side of production as well.

The Blind-Spot Theory

One important attempt to theorize the relation between broadcasters and their audience, "the blind-spot debate," operates by extending this class analogy. The debate began with an essay titled "Communications: Blindspot of Western Marxism," wherein Dallas Smythe argued that critical theorists need to analyze the media, not just as a propaganda organ, but in terms of its functioning for the economy. And the principal function of the media for the economy, Smythe suggested, could be found by

28. Marchand, *Advertising the American Dream*, 64.

extending Marx's labor theory of value to the media audience: the industry's principal product was the audience commodity. The audience's relation to the industry, then, was much like the relation of factory workers to factory managers: the audience was "working" for advertisers when it viewed advertisements. Watching television is less simple leisure than it is a kind of productive labor conducted in the interest of the media and the consumer-products industries.[29]

In one intriguing effort to expand on Smythe's thesis, Jhally and Livant have suggested that "[w]atching is a form of labour. . . . when the audience watches commercial television it is working for the media, producing both value and surplus value. This is not meant as an analogy. Indeed, watching is an extension of factory labour, not a metaphor."[30] This effort to apply Marx's labor theory of value to commercial broadcasting is compelling for many reasons: it captures much of the "industrial" quality of broadcast content and organization, for example, and offers an interesting way to characterize the power imbalance between broadcasters and their audience that goes beyond a simple instrumental, intentional model of power.

Yet if the parallel between the television watcher and the factory laborer is more than metaphorical, Jhally and Livant must come up with some equivalent to the factory wage. They suggest that the wage offered audiences is the program, and the work for which it is exchanged is watching advertisements. There's certainly some truth to this: some of us may very well sit through the ads because we want to watch the programs. Yet the parallel is still far from exact: the boundaries between programs and ads are in many ways blurry (ads are often entertaining, for example, and most programs celebrate consumerism if not specific products), and the exchange is hardly like the quantified, strictly moneyed exchange characteristic of wages. As Maxwell has pointed out, there is a difference between what Marx called the "commodity form" and commodities; just because the industry imagines the audience through a fetishized commodity form, does not mean that the audience is a simple commodity identical in economic character to straightforward manufactured objects.[31]

29. Dallas W. Smythe, "Communications: Blindspot of Western Marxism," *Canadian Journal of Political and Social Theory* 1 (fall 1977): 1-27.

30. Jhally, *Codes of Advertising,* 83. On page 64 Jhally acknowledges Bill Livant as coauthor of this chapter. See also Sut Jhally and Bill Livant, "Watching as Working: The Valorization of Audience Consciousness," *Journal of Communication* 3 (summer 1986): 124-43.

31. Richard Maxwell, "The Image Is Gold: Value, the Audience Commodity, and Fetishism," *Journal of Film and Video* 43 (spring-summer 1991): 29-45.

The Other Side of the Paycheck: Lessons from Feminism

Maxwell hints that the solution might be in thinking of viewing and listening as "unwaged" labor.[32] In U.S. history, along with slavery, the most important forms of unwaged labor have been what might loosely be called "women's work." In the early nineteenth century, domestic activities involved production of most goods and other necessities of life: the manufacture of clothes and other essentials occurred in the home, most often by women, whereas men were increasingly working outside the home for wages. As industrial production gradually expanded to include more and more goods associated with everyday life, however, the unpaid labor of women in the home, particularly middle-class women, increasingly became a matter of child rearing and nurturing and—crucially for understanding commercial broadcasting—shopping for goods instead of manufacturing them.

But this line of thought in turn opens the door to questions about the meaning and coherence of terms like labor, production, consumption, and reproduction. Like most traditional economists, and because he was analyzing production in a capitalist context, Marx drew a line between productive and nonproductive activities in a traditional place: the deliberate and useful transformation of physical objects was productive, everything else was not. So, in a famous formulation in the *Grundrisse,* building a piano was productive, playing it was not. Raymond Williams departed from Marx at exactly this point, on the grounds that this boundary is not nearly as clear as Marx assumed: labor in this traditional sense is often made possible by kinds of effort—for example, cultural, social—that would be unproductive in Marx's sense.[33]

The most powerful critiques of the coherence of traditional understandings of "labor" and "production" have come from certain strains within feminist scholarship. In a series of discussions extending from the domestic labor debates of the 1970s to ongoing discussions about the economic and social character of child bearing, child raising, family nurturing, and similar activities, a variety of feminist scholars, particularly Marxist and socialist feminists, have criticized both conventional and Marxist economic theories for ignoring or undervaluing the fundamental work of creating and maintaining human life traditionally done by women in a family setting. Such activity, after all, is both necessary to human existence and a form of work; it is not a "natural" process that happens of its own accord. For although capitalism's world of market

32. Ibid., 44 n. 9.

33. Raymond Williams, "Base and Superstructure in Marxist Cultural Theory," in *Problems in Materialism and Culture* (London: New Left Books, 1980), 34.

competition and private property presents itself as the entirety of economic life, to a large degree, there is much productive activity that is marginalized or excluded from the world of markets. The argument is not just that there's more to life than economics, but that historically, at least in the American and British context,[34] capitalist social relations developed in large part because they were afloat on extramarket patriarchal institutions, particularly the unpaid labor of homemakers.[35] To some degree this may be still the case. Although assigning dollar value to nonmarket work is methodologically problematic, it has been estimated that, as recently as 1947, nonmarket work was worth approximately $169 billion compared to the total national income that year of $228 billion.[36]

What this suggests then, is that there is an invisible social and historical exterior to traditional economics, of which the socially constructed divide between production and consumption is a primary example. The social condition of the production/consumption divide, in other words, is necessary to the world of markets and private property, but obscured by the constructs of the natural role of the family and women. The fact that it is typically unwaged does not mean that it is unworthy of consideration, and any materialist analysis must take it into account. As one feminist classic put it, economic theory needs to give full attention to "the other side of the paycheck," to all the humanly necessary productive activities associated with domestic labor such as shopping and generally making a house into a home.[37]

34. I do not wish to suggest that capitalism's reliance on unpaid domestic labor and markets is inevitable. As Charles Sabel has pointed out, French industrial development did not rely all that much on domestic-market creation (*Work and Politics : The Division of Labor in Industry* [New York: Cambridge University Press, 1982], chap. 2). The point is rather that capitalism has existed in complex relations with noncapitalist ways of organizing work, authority, and resource distribution.

35. The standard entry point into this literature is the set of essays collected by Lydia Sargent, ed., *Women and Revolution: A Discussion of the Unhappy Marriage of Marxism and Feminism* (Boston: South End Press, 1981). For a more recent review of the literature, see Julie Matthei, "Marxist-Feminism and Marxist Economic Theory: Beyond the Unhappy Marriage," paper presented at the Harvard Seminar in Non-Neoclassical Economics, November 21, 1988. A recent major contribution to the discussion is Nancy Folbre, *Who Pays for the Kids? Gender and the Structures of Constraint* (New York: Routledge, 1994).

36. Nordhaus and Tobin, "Is Growth Obsolete?" in *The Measurement of Economic and Social Performance,* ed. Milton Moss (New York: National Bureau of Economic Research, 1973), 508–64, cited in Fred Block, *Postindustrial Possibilities: A Critique of Economic Discourse* (Berkeley: University of California Press, 1990), 60.

37. Batya Weinbaum and Amy Bridges, "The Other Side of the Paycheck: Monopoly Capital and the Structure of Consumption," *Monthly Review* 28 (July–August 1976): 88–

Sometimes the argument tends to take the form of efforts to enlarge traditional economic categories: reproduction should be taken into account as well as production; perhaps women should demand wages for domestic labor. Yet there is also a strain within the tradition that, with varying degrees of explicitness, questions traditional economic categories themselves. Matthei, for example, points out that

> [f]amily relations involve a level of love and sharing and indeed of fusion of personality absent from production relations. . . . Women's work of mothering is the perfect example of an activity which is neither material nor ideal. . . . women's actual physical reproduction of children, through pregnancy and childbirth and breast-feeding, is at the same time a social and ideal process, in which children develop their consciousness and identities.[38]

Mies summarizes the implications of this strain of thought. "Whereas the old movement and the orthodox left had accepted the capitalist division between private housework or—in Marxist terminology—reproductive work, and public and productive work—or wage-work," she writes, "the feminists not only challenged this division of labour but also the very definitions of 'work' and 'non-work.' "[39] The feminist discussions of labor, in other words, have not only questioned the valuation of production over reproduction, but also the coherence of the boundary between the two.

The feminist scholarship has, then, two implications for understanding the broadcast audience. First, so-called consumption, the category of activities into which broadcast viewing falls, might be in a broad sense a kind of productive work; to this extent the feminist scholarship parallels the arguments of scholars such as Smythe, Jhally, and Livant. Second, whereas the "blind-spot" scholars have tried to fit broadcast viewing into fairly conventional (in their case, Marxist) economic categories based on an analogy with wage labor, the feminist scholarship suggests that one

103. For a discussion of some of these issues in relation to newer technologies, see Andrew Calabrese, "Home-Based Telework and the Politics of Private Woman and Public Man: A Critical Appraisal," in *Studies in Technological Innovation and Human Resources,* vol. 3, *Women and Technology,* ed. U. E. Gattiker (Berlin and New York: De Gruyter, 1995).

38. Matthei, "Marxist-Feminism and Marxist Economic Theory," 26. Ann Ferguson has proposed the category of "sex/affective production" to refer to this complex set of socially and psychologically ordered and ordering productive activities (*Sexual Democracy: Women, Oppression, and Revolution* [Boulder: Westview Press, 1991], 69).

39. Maria Mies, *Patriarchy and Accumulation on a World Scale: Women in the International Division of Labor* (London: Zed Books, 1986), 31.

question the categories themselves, and look to the broad social conditions that form the conditions of possibility of the conventional economic activities in the first place.

Gender, Values, and "Nature"

The power relations constituted by the production/consumption divide, then, are not just matters of differential wealth or of the unequal power of management and labor. They also involve relations of gender. The chasm across which broadcasters try to make sense of their audience is created, in other words, not simply by separating production and consumption, but by concurrently *valuing* some activities as productive parts of the market economy and thus worthy of pay, while *devaluing* other activities—notably those traditionally engaged in by women—as either not part of the economy or as consumption and thus not worthy of pay. In other words, the divide is created in part by the process of excluding "women's work" from society's definition of economic activity.

This exclusion of "women's work" has been enabled and legitimated by the notion that the family and activities necessary to it such as child rearing and shopping are part of nature, not willed actions. And because the family was part of nature, both women and a set of essential human activities were defined as naturally outside of economic relations, that is, as not willful production but as a basically involuntary working of nature.[40] And this then became coupled to the complex Victorian constructions of the feminine, sentimentality, and "the cult of domesticity," which together tended to valorize this construction of gender while upholding its limiting boundaries: women were to be admired, seduced, put on a pedestal, even worshipped, but not allowed an equal role in decision making. They might have obtained various kinds of "influence," particularly in the private sphere, but were not allowed power of the most direct sort.

In sum, the chasm across which broadcasters attempt to make sense of their audience is embedded in sociohistorical unequal relations of class and gender. On one side of the chasm human work is valued as a willful productive activity; on the other it is relegated to the category of a natural process. The character of this divide is crucial because those relations not only shape broadcasters' ways of seeing the audience—what they notice, what they ignore, and how they construct their interpretations—but are also part of, and thus contributory to, the dynamic that creates the chasm in the first place. The interpretive prac-

40. Thanks to Wendy Wahl for making this point clear to me.

tices, the ways of seeing implicit in the practice of selling audiences to advertisers, in other words, help to create that which they purport to describe.

Culture as "Women's Work": Defining Interpretation as Private Consumption

Watching as Working Revisited: Semiotic Labor

Like school, church, political rallies, and psychotherapy, watching television and listening to the radio are interpretive activities. They do not principally involve food, shelter, or physical objects of any sort. They are acts of collective meaning creation and interpretation that involve using learned knowledge of symbolic systems to make sense of texts. And they involve work, applied ability, and effort: whether a text is a lecture, a dream, a mass, or a situation comedy, for both the text's creators and its audience, making meaning out of it involves both background knowledge built up through a lifetime of cultural experience—for example, mastery of English syntax, or the knowledge that sitcom fathers wear sweaters[41]—and an active effort of interpretation, the work of building meaningful associations and connections out of a stream of symbols.

Different interpretive activities, however, are characterized by different structures of constraint and inducement, which in turn help shape their cultural meaning. Some activities are publicly coerced: children are forced by law and their parents to attend school, and in theocratic societies skipping mass or the mosque could be, if not illegal, at least deleterious to one's social and material standing. And those activities that are couched within the realm of private market exchange are quite varied in terms of who gets paid for what: psychiatrists get paid for interpreting patients' life stories, whereas moviegoers pay for the privilege of interpreting the life stories told to them in films.

The Construction of Semiotic Work as Consumption

Television is often compared (usually negatively) with religion, education, traditional politics, and therapy, yet the comparisons often fail to take into account the structures of social constraint and inducement within which broadcasting is couched. Among interpretive activities, entertainment generally and television in particular are socially classified

41. The cultural code that sitcom fathers wear sweaters is noted in Ellen Seiter, "Semiotics and Television," in *Channels of Discourse,* ed. Robert C. Allen (Chapel Hill: University of North Carolina Press, 1987), 17–41.

as passive activities. The interpretive work involved, in other words, is categorized as nonwork. This is sometimes attributed to the fact that television and radio communicate meaning iconically; the signifiers of television and radio visually resemble that which they signify, and thus do not require the same knowledge of a specialized symbolic code as does, say, reading a book or making sense of a Catholic mass. Watching television is easy, we think, reading a book is hard. Yet the iconic aspect of broadcast entertainment is easily exaggerated. A television image of a person physically resembles a real-life person, but whether that person is wealthy or poor, attractive or unattractive, good or bad, and so forth—that is, the aspects of the person that matter to the narrative—can only be understood with the help of subtle forms of informally learned cultural codes.[42]

It is worth considering, therefore, an alternative explanation for why our society categorizes watching television and listening to the radio as passive "nonwork": because of the time and place in which they take place, in the home during unpaid hours, on the "nonproductive" side of the production/consumption divide. Broadcasting, once it was incorporated into the corporate liberal system in the early 1920s, was legally, politically, and culturally structured as an act of privatized consumption, as a "leisure" activity that belonged outside of the workplace, and in the home. It was placed in the same space, both physically and ideologically, as "women's work," physically in the mobile privatized domestic space, and ideologically as a part of nature. Interpreting broadcast messages, the work of making sense out of them, was defined as something that happens, not something that is done.

This was not unique to broadcasting, of course. The history of the categorization of a host of cultural activities in general as feminine, passive, and trivial is complex. It began to take on its basic contours by the late eighteenth century: a classic case would be the scorn heaped upon novels when they first appeared; intellectuals of the day decried novel reading as a passive, unproductive activity engaged in largely by women and involving "mere" imagination instead of rational understanding. Since then, discourses surrounding culture have in a variety of ways entered into dialogue with this tendency toward the dismissal of culture: the cult of sentimentality, for example, revalorized "intuition" and attention to the subtleties of life as one of women's special virtues; this, in turn, provided many of the key elements of the emerging consumer cul-

42. John Fiske and John Hartley, *Reading Television* (New York: Methuen, 1978), 39–40.

ture. The Arnoldian tradition, alternatively, valorized certain forms of elite culture as "the best that has been thought and said," and thereby relegated much of "women's" culture to the status of the vulgar.

Whether denounced as simply irrational and trivial or devalued as lowbrow, by the mid-nineteenth century large parts of cultural activity were being classified as both feminine and passive, a natural act of reception unmediated by human will, and thus as distinct from creative labor. The work of interpreting culture was thus placed ideologically beyond or outside the market, and physically in nonpublic domestic spaces and/or leisure time.

The Work of Making Meaning with and through Goods

For commercial broadcasting in particular, both the work of interpretation associated with viewing and listening and its categorization as nonwork take place within the larger sociocultural complex of the consumer society: viewing and listening is encouraged to be, and in several ways is, a part of the activity of shopping. Like novel reading or listening to music, shopping in a consumer society is a cultural activity: we shop not just to satisfy material needs, but to participate and contribute to shared systems of cultural meanings. When Thorstein Veblen coined the term "conspicuous consumption," he was not only expressing a Protestant disdain for upper-class ostentation and pretentiousness and for the nonfunctional in general.[43] He was also articulating a critique of the utilitarian logic that dominates popular economic explanations of personal behavior in a market society: the assumption that goods simply satisfy the self-interest of buyers. By showing how the value of many goods could only be understood by way of their symbolic relation to other goods, Veblen demonstrated the fundamentally relational and symbolic character of marketplace value, that is, that the value of things to individuals is determined only by way of the meaning of those things to *other* individuals: we buy things in the hope that they will communicate solidarity with, admiration for, or—Veblen's main focus—superiority to others. So the problem is not that we don't buy things to satisfy our self-interest, but that self-interest is not a complete explanation: what we are interested in ultimately hinges on what we have learned that others are interested in.

Anthropologists tell us that all societies assign cultural meanings to goods, and that studying and learning about the social meanings and

43. Thorstein Veblen, *The Theory of the Leisure Class: An Economic Study of Institutions* (1899; reprint, New York: Mentor, 1953), 60-79.

values associated with goods—developing the skills to recognize the significance of fashions, styles, and so forth—are part of the process of consuming them.[44] What is unique about a consumer economy is not that people acquire things for their symbolic value, but that the creation and transformation of the meanings of goods becomes complexly linked to the prerogatives of large-scale industrial production. According to some economic theorists, marketing and advertising patterns have become one of the central corporate strategies for overcoming the problem of overcapacity in industrial production: the cycles of fashion, as monitored and to some extent guided by the advertising industry, help coordinate shifting cultural tastes with the rhythms of industrial production.[45] And it is in this context that shopping has been constructed as gendered.

Shopping and the cultural skills it requires are not limited to women, of course, but they were first integrated into the modern economy with the case of middle-class women shoppers, and as an activity our culture tends to label them as feminine even when performed by men. Full participation in a consumer culture calls for attention to subtle details such as slight variations in colors, home decoration, and fabric, and, just as importantly, to the cultural values associated with these details in various combinations; it requires a mastery of cultural codes as well as their creative application. With a few stereotypical exceptions (cars, sports equipment) women are generally considered to be the experts at making these kinds of discrimination in our culture. Of course, women's "mastery" of shopping in our culture has nothing to do with anything inherent to female biology. Rachel Bowlby and others have pointed out that the cultural skills associated with shopping have their origins in women's experience of adapting to the strictures of the industrial-era nuclear family and "femininity." The interpretive competencies that bolster the advertising/consumer complex are a product of and response to the situation of women whose social role is constrained to, in Bowlby's

44. Mary Douglas and Baron Isherwood, *The World of Goods* (New York: Basic Books, 1978); William Leiss, *The Limits to Satisfaction* (London: Marion Boyars, 1976); Marshall Sahlins, *Culture and Practical Reason* (Chicago: University of Chicago Press, 1976).

45. Alfred D. Chandler emphasizes the role of advertising and marketing as a rationalizing mechanism central to twentieth-century corporate enterprises, focusing more on what might be called "demand control" or "demand management" than "demand creation" (*The Visible Hand: The Managerial Revolution in American Business* [Cambridge: Harvard University Press, 1977], 290–92). Stuart Ewen's is the classic statement of the view that advertising serves to harness people's psyches, particularly their dissatisfactions, to the needs of industrial capitalism (*Captains of Consciousness: Advertising and the Social Roots of the Consumer Culture* [New York: McGraw-Hill, 1976]).

phrase, "just looking," that is, to women whose principal role outside the home is one of shopping.[46]

Advertising and Market Research as a Way of Seeing: The Consumer and the Feminine

So why does our culture treat the considerable effort and skill required to interpret mass media messages and to shop for the goods they advertise as mere leisure, as unworthy of recompense? The short answer is that consumption in our culture is placed in roughly the same category as homemaking: it is imagined to be more a natural process than a human accomplishment. It is easy, however, to oversimplify matters into a rigid set of dichotomies: consumption, culture, women, unpaid labor, and domestic private space on one side, production, science, men, paid labor, and public spaces on the other. Of course, real men and real women live lives that are much more complex than this, and the boundaries between these categories have tended to be shifting, unstable, and often blurry. Young teenagers, for example, often devote more intellectual and interpretive effort to learning the subtleties of contemporary fashion than they do to the subtleties of Shakespeare or algebra because they rightly recognize that the former is where our society puts more of its economic and cultural energy. The problem is not that fashion is unimportant in our culture, but that one doesn't get formally rewarded for learning to interpret it; teenagers are correct about cultural values but have yet to be persuaded of the rationality of the production/consumption divide.

For the purposes of this argument, however, the significant thing is that people who work in modern industries of communication often *think* about what they are doing in terms that are almost this simple, and organize their activities accordingly. So even though ordinary people do not live entirely according to the dictates of simple production/consumption, public/private dichotomies, the conditions in which they live, particularly those shaping the cultural environment, are heavily determined by those categories because the people with the most power to influence matters use those categories to determine what they do.

The assumption that the production/consumption divide is a natural one, then, has been a formative operating principle in the construction of many fundamental structures of American commercial broadcasting, which in turn has served to reinforce and spectacularize

46. Rachel Bowlby, *Just Looking: Consumer Culture in Dreiser, Gissing, and Zola* (New York: Methuen, 1985), especially 1-34.

the divide. The assumption was present in the exclusion of amateurs from broadcasting in 1921 and the resulting rigid divide between transmission and reception that largely defines the medium. The assumption also helps underwrite the broad acquiescence to the idea that the public's principal role in broadcasting is a matter of listening and viewing. Even for political liberals who assert that the public-interest clause should mean that, as the *Red Lion* decision put it, the "rights of listeners and viewers are paramount," these are rights of *listening* and *viewing* (not of transmitting or production).[47] Mainstream legal and political discourse takes for granted that listeners and viewers are naturally of a different category than broadcasters. The former are consumers, the latter, producers; as far as mainstream discourse goes, that is all there is to say about the matter.

A more subtle but nonetheless central example is the assumption that viewing takes place in the domestic space. This notion has become much more than a simple cultural bias: it has become an industrial organizing principle, a necessary element to the workings of the system of broadcasting, of the television-industrial complex. In particular, it is one of the principal ways that the industry solves what Gitlin calls "the problem of knowing," that is, the difficulty of organizing centralized program production in the face of the invisibility and diversity of the broadcast audience.[48] Chronic uncertainty about the invisible audience, Gitlin argues, presents networks, program producers, and advertisers with a dilemma: on what basis are they to arrive at agreements about programs and their value? Without some sort of collective consensus, business could not proceed. Gitlin shows that the television production community has a variety of ways of generating and asserting the validity of a kind of common wisdom among industry insiders, a sense of what everyone "knows" to be true. This common wisdom need not be all that accurate; Gitlin demonstrates that at least in some cases it has been dead wrong.[49]

47. *Red Lion Broadcasting Co., Inc., et al. v. Federal Communications Commission et al.*, 395 U.S. 367 (1969). Although the now-defunct fairness doctrine has some elements of access to it, it involves news coverage of issues, not direct citizen access, and has a decidedly consumerist orientation. See Steven Douglas Classen, "Standing on Unstable Grounds: A Reexamination of the WLBT-TV Case," *Critical Studies in Mass Communication* 11 (March 1994): 73–91. Similarly, section 315 of the Communications Act grants a limited form of access to politicians, but not to citizens. Cable access and community radio are the only notable exceptions to the general consumerist assumption that dominates the regulation of the electronic media, and they are really political concessions, not legal creations.

48. Gitlin, *Inside Prime Time,* 19–30.

49. Gitlin discusses in detail, for example, the industrywide belief in 1981 that the American audience that had just elected Reagan would be interested in more law-and-

But for industry survival over the short term, it is only necessary that everyone in a leadership position agree that something is true. The assumption that viewing occurs in a privatized domestic space by enthusiastic consumers, whether or not it is an accurate generalization, serves as a tool of industrial organization of production, as a way of understanding and reaching consensus about the industry's imagined audience.

The social fact of everyday life in privatized domestic settings is the cornerstone of a host of other industry practices as well: it is how the industry finds an audience that will "sit down and be counted."[50] Life in living rooms is not all that predictable, but it is a safe sociological generalization that it is much *more* predictable than most other parts of everyday life outside the workplace. Relatively speaking, the audience "holds still" because within the domestic space it can be counted on to return to the set in fairly predictable ways (which is partly why the ratings industry prefers to ignore nondomestic settings such as college dorms and the workplace, where behavior is even less predictable). The schedule is organized into segments designed around domestic habits (the weekly programming cycle, daytime, prime time, etc.) which in turn allows for the cultivation of "reliable" audiences with series programming. (It is telling that the electronic media industries continue to search for target audiences that are physically and socially "trapped" in relatively passive positions around which they can organize programming and advertising: the drive-time radio audience of traffic-bound commuters, for example, or current efforts to introduce narrowly targeted broadcast systems into classrooms, supermarkets, and doctors' waiting rooms.)[51] The ratings system is even today based on the unit of "households," not on individual viewers; people meters aside, audience measurement is still generally predicated on the use of television in living rooms. (The contemporary spread of television sets into other rooms of the house is gen-

---

order programming. Pilots were made, program deals struck, and a fall lineup of programs was developed around this assumption. Subsequent events proved the assumption wrong, but this did little to shake the industry's faith in its ability to make such judgments (ibid., 221–46).

50. The phrase is from the title of an interesting book on a rather different topic: Lelia Doolan, Jack Dowling, and Bob Quinn, *Sit Down and Be Counted: The Cultural Evolution of a Television Station* (Dublin: Wellington, 1969).

51. For the effort to exploit captive audiences in schools, see "Back to School: Whittle Explains 'Channel One' Show," *Advertising Age*, January 23, 1989, 49. Other venues for the current "captive audience" craze include tour buses (see "Tour Buses: Another New Venue in the Drive to Place-Based Media," *Advertising Age*, January 18, 1993, 12), truck stops ("'No Necktie' Network for Truckers Ready to Roll," *Electronic Media*, September 14, 1992, 8), and airport lounges ("Turner's New Gamble: How Marketers View 'Airport Channel,'" *Advertising Age*, April 22, 1991, 3).

erally imagined in the industry in terms of a radiation of viewing outward from the living room, more an extension of the habit of living-room viewing than its abandonment.) In sum, the maintenance, the institutional reproduction, of television requires the reproduction of living rooms, of the domestic scene.

Making Sense of the Audience Commodity: Consumer Sovereignty and the Contours of Private Choice

The Industry: The Consumer on a Pedestal

If one takes into account the full connotations of the word "consumer," the phrase "consumer sovereignty" captures much of the nature of the advertising system's relation to the audience. It must be said that the "consumer" is an open category. It is in the nature of capitalism, with its built-in desire for the new and the different, that consumers are not only expected but encouraged to change, and to explore new pleasures, activities, and identities. But just because the category of the consumer is infinitely variable does not mean that it can be any and all things. Addressing people as consumers is not the same as addressing them as, say, citizens or souls. The variability of the meaning of "consumer" takes place within certain boundaries and is shaped by certain pressures. Sometimes those constraints and pressures involve matters that are highly culturally specific, such as local cultural traditions concerning sexual explicitness, which make certain American ads unacceptable in India and certain European ads unacceptable in the United States. And there are general patterns that emerge from the mechanics of persuading people to buy products, such as the tendency to idealize the consumer and to imagine everyday life, as Michael Schudson puts it, "not as it is, but as it should be" in a capitalist society.[52]

Generally, though, the key defining characteristic of "consumers" is that they are not invited directly into decision-making processes concerning production. Advertisers are desperate to know whether consumers want this or that kind of toothpaste, and quite curious about consumers' views on things like sex, respect, the ideal family, and even the ecology. But the advertising system is quite incapable of inquiring seriously into consumers' views on, say, whether or not the factory that manufactures the toothpaste should be moved overseas. The same is true of decisions about public versus advertising-supported television,

52. Michael Schudson, *Advertising, the Uneasy Persuasion: Its Dubious Impact on American Society* (New York: Basic Books, 1984), 215.

or about the involvement of citizen representatives on corporate boards. This kind of decision—about which many ordinary people, were they asked, would most probably have something to say—cannot be raised within the advertising system because to do so would cross the production/consumption boundary.

As American industry evolved increasingly toward a consumer economy in the early decades of this century, the intertwined professions of advertising and marketing emerged to establish lines of communication and control across the growing gap between factory and home. Because of that gap, modern advertising operates from a structural position of incomplete knowledge of its audience. Advertisers thus have had to fill the lacunae in their understanding with a variety of suppositions and guesses that are more or less educated, more or less reflective of the actual people they seek to address. The speculative character of advertisers' knowledge of their audience ensures not only some imprecision, but also that their ideas involve a certain amount of projection: audience research can tell us almost as much about advertisers as it does about audiences.

It seems likely that advertisers have drawn their ideas about the audience as much from conversations with acquaintances, stereotypes, personal tastes, and other media as from surveys and other modern research techniques.[53] Roland Marchand's account of the history of the development of advertisers' collective guesses about their audiences between 1920 and 1940 shows how the industry in its formative years had to engage in considerable struggle to create a coherent image of their audience, vacillating between images that overestimated the audience's wealth and cultural similarity to advertisers and cynical, derogatory stereotypes—all the while engaged in a difficult struggle to maintain professional self-image and respectability. Yet over the years the industry has achieved what Leiss, Kline, and Jhally call a kind of "practical knowledge" of what works and what doesn't, gleaned sometimes from research but just as often from practical experience such as the rare wildly successful ad campaign or surprising media hits such as the success of *True Stories* magazine in the early 1920s.[54] But this practical knowledge can be quite rough; as Marchand has pointed out, even advertisers who chastised their colleagues for projecting false images onto

53. Marchand, *Advertising the American Dream,* 74.

54. "The advertising agency is the repository for an unmatched collection of skills and knowledge accumulated and honed over almost a century" (William Leiss, Stephen Kline, and Sut Jhally, *Social Communication in Advertising: Persons, Products, and Images of Well-Being,* 2d ed. [Scarborough, Ontario: Nelson Canada, 1985], 181). See also Marchand, *Advertising the American Dream,* 53-63.

their audiences would go on to make their own projections, such as systematically overestimating the incomes and class status of their targets.[55]

Advertisers' tendency to project abstract notions onto consumers has not disappeared as their practical knowledge and market research techniques have grown increasingly sophisticated since the 1920s and 1930s. It remains the case that the role of advertisers is hardly neutral: their first purpose is to serve the interests of those on the production side of things, to sell products for clients in a way conducive to corporate system maintenance, not to share ideas with or even to please the audience toward whom they direct their communications. This nonneutral goal of advertisers shows its face in a number of ways. One of the more obvious is the quite predictable pressures on them to manipulate and distort in their effort to sell products, which has necessitated truth-in-advertising regulation. But the limits to advertisers' vision are more importantly constrained by the structures from within which they operate. As Michael Schudson puts it, "the consumers the marketers are listening to are not persons, not citizens, but thin voices choosing from among a set of predetermined options. The 'people' the marketers are concerned with are only those people or those parts of people that fit into the image of the consumer the marketer has created."[56] The "consumer" is not simply a person, but a very particular way of understanding a person. Advertisers' address their audience strictly as *consumers,* and only consumers.

It is not coincidental, then, that the general character of the constraints and pressures that shape the category of the consumer can be illuminated by parallels with the way our culture has imagined femininity. From the advertiser's point of view, consumers are, in a sense, on a pedestal. Advertisers tend to view the consumer as something to be desired, pleased, catered to, and seduced—but not given direct power. Advertisers tend to imagine themselves as rational, calculating, and active individuals in the position of communicating to the emotional, relatively passive masses—in much the same way our culture has imagined the difference between men and women.[57] On the one hand, the parallel between the consumer and the feminine stems from the simple fact that most consumers *are* women. The gendered division of labor coupled to the domestic space structured as the primary site of consumption ensures that much of the time the people making decisions

55. Marchand, *Advertising the American Dream,* 77–80.
56. Schudson, *Advertising,* 235–36.
57. Marchand, *Advertising the American Dream,* 66–69.

about buying consumer products will be women. On the other hand, the parallels between the consumer and the feminine are also structural. Both "women's work" and the activities of learning about, selecting, and buying products have been excluded from the category of production and naturalized, imagined as something passive or at least driven more by impulse than by rational deliberation. It is perhaps because of these structural parallels that the consumer culture in general generates many of the same cultural anxieties generated around women: within the mass-culture tradition, the audience of mass culture is imagined as passive, manipulable, impulsive, and irrational.[58]

The Audience: Private Life inside and outside the Market

Viewers and listeners, on the other hand, are not all that likely to see themselves as the passive seductees imagined by advertisers, and are more likely to construct themselves as active rational individuals. Polls find that most respondents think advertisements are an insult to their intelligence (even if entertaining), and believe that advertising does not influence consumer choice.[59] Yet the range of activity available to viewers and listeners is not infinite; it, too, has its contours.

In general, television viewing falls under the umbrella of those activities we label "private." But the "private" in this context has two sets of connotations attached to it. In the first, viewers see themselves through the lens of classical liberalism: as freely acting individuals making self-interested choices in a marketplace. This perspective is most evident in the way we look at the television set itself: an object of ownership, a consumer product. Its possession and use, furthermore, is a matter of individual choice. One is "free" to choose channels or to not view at all.

In the second connotation of "private," viewing television is couched within the social practices associated with the experience of the domestic or leisure sphere. Television is part of feeling "at home," that is, it is a component in the creation and maintenance of a private world that, in deeply contradictory ways, serves experientially and ideologically not as

58. Patrice Petro, "Mass Culture and the Feminine: The 'Place' of Television in Film Studies," *Cinema Journal* 25, no. 3 (1986): 5–35; Andreas Huyssen, "Mass Culture as Woman: Modernism's Other," in *Studies in Entertainment: Critical Approaches to Mass Culture,* ed. Tania Modleski (Bloomington: Indiana University Press, 1986), 188–208; William Boddy, "Archaeologies of Electronic Vision and the Gendered Spectator," *Screen* 35 (summer 1994): 105–22.

59. Canadian Radio-Television and Telecommunications Commission, *Attitudes of Canadians toward Advertising on Television* (Ottawa: Supply and Services Canada, 1978), cited in Leiss, Kline, and Jhally, *Social Communication in Advertising,* 2.

the marketplace, but as its counterpoint. The domestic space, after all, originated in the nineteenth-century understanding of the home as a haven *from* the heartless world of the marketplace. It on this level that television intersects with the habits and values of the family, leisure, and so forth. Here the connotations of the word "freedom" are not so much "free to choose" as free *from* the constraints and pressures of the outside world.

From within both of these constructions of the "private," in any case, the broadcast industry, particularly its structure and organization, tends to appear to the viewer as unalterable, as a given. It is not just that viewers feel they don't have a choice on this level—practically speaking, they don't—but that the culture encourages them to believe that *no one* does. The organization of television is like a fact of nature, a product of inexorable forces like technology, economics, and the market. For this reason it is possible from the point of view of audience members to experience the organization of the system as wholly legitimate: one may or may not like it, but it has to be the way it is.

The theory of consumer sovereignty, in sum, is in several ways analogous to the argument that confining women to the role of husband-dependent homemakers is in their own best interests. In both cases the assumption that the domestic arrangement of family life is not a political structure but a natural state of affairs allows restriction to be redefined as agency, dependency as freedom. The audience is imagined by industry executives as active in its passivity in much the same way that homemaking wives have been construed as having special authority in the domestic sphere: the authority, not to coercively control resource distribution or decision making, but, in the traditional Victorian vision, to gently persuade, nurture, and set examples for men and children. Because in both cases this "active passivity" is understood as a natural state of affairs, the classical liberal axiom that all people are created equal is reconciled with systematic inequality—the inequality, in one case, of women with men regarding control over material resources and, in the other, of viewers with corporations regarding control over cultural production. In neither case is the inequality absolute; neither homemakers nor audience members are completely without power. Yet in both cases the power that does exist is not equally distributed.

Conversely, what viewers construct as natural and unalterable is not their own behavior but the structure of the television industry, particularly its profit-driven, centralized character. Like many homemakers, audience members "make do" within the constraints of their situation in part because they are able to construct themselves as active *within* the bounds of what they imagine to be the "natural" conditions under which they operate. As Stuart Hall says, "[P]eople have always had to make something out of what the system is trying to make them into."[60] But in the end what this means is that viewers are catered to by those who profit from their participation but have unequal power to influence the content of what is broadcast. The cultural work that viewers provide the system may have parallels to unpaid domestic work because it is made possible by cultural differences (gendered experience, etc.) that are technically outside the marketplace, but necessary to its functioning. Both activities are ideologically rendered outside the activity of market exchange in being categorized as "leisure." Just as shopping, housework, and so forth go unpaid, leisure-time television viewing, understood as a passive reception of a transmitted signal, is thought of as not involving work, as not worthy of pay. Both the work of women in the home and the "work" of interpretation accomplished by the audience as viewers and shoppers, then, fall under the extramarket categories of the private (the domestic sphere and the sphere of leisure) and thus in economic terms are not counted as work at all.

Conclusion: Sit Down and Be Counted

What the preceding analysis suggests is that viewership is turned into a marketable commodity, not by simply extending commodity relations into the sphere of culture, but by systematically excluding or marginalizing several kinds of activity, activities integral to the contemporary industrial system, from the category of productive work. Commodification in this case is less a matter of expanding than of circumscribing the market. Or, rather, the extension of market relations into the sphere of culture requires an elaborate bifurcation of human life into categories of consumption and production. Including some things within the market requires the simultaneous exclusion of other, matched activities from the market.

The legitimatory effects of this bifurcation are powerful. In categorizing broadcast viewing and listening as a subcategory of consumption called leisure, broadcasters can reconcile for themselves the contradiction between "giving the audience what it wants" and "selling audiences to advertisers." The former is true within the sphere of consumption, the latter, of production. If one accepts the inevitability of the way we have constructed the production/consumption division, then

60. Stuart Hall, "The Culture Gap," *Marxism Today,* January 1984, 19.

the system is likely to appear relatively legitimate, and the commodification of the audience is consistent with both economic logic and democracy. But it only makes sense if one takes for granted the notion that the cultural work of viewing, listening, shopping, and so forth is nonwork, leisure, a passive activity that just happens without any prerequisite knowledge, intellectual activity, or social conditions.

Toward a New Politics of Electronic Media

Plus Ça Change . . .

> [T]he stage is being set for a communications revolution . . . there can come into homes and business places audio, video, and facsimile transmissions that will provide newspapers, mail service, banking and shopping facilities, data from libraries and other storage centers, school curricula and other forms of information too numerous to specify. In short, every home and office will contain a communications center of a breadth and flexibility to influence every aspect of private and community life. . . . [we should make] a commitment for an electronic highway system to facilitate the exchange of information and ideas.[1]

The preceding passage was published, not in the last few years, but a quarter century ago, in 1970. The new technology that was supposed to bring about this communications revolution was not the information superhighway, but cable television. In this century, breathless descriptions of revolutionary new communications technologies have become a time-honored tradition. Talk of the new has become old.

This needs to be kept in mind when considering media policy today, where everywhere the talk is of change. We are told that computers, fiber optics, the breakdown of industry barriers between broadcast, cable, and telephone companies, and a new preference for the marketplace over government regulation are sweeping away the old technologies like broadcast television, old institutional arrangements like the networks, and old regulations like the public-interest principle. Certainly, real changes are taking place. Besides the availability of new devices for delivering media messages, the production of those devices has become increasingly internationalized, and the larger political climate in the United States has changed considerably, not the least because of the end of the cold war. But encouraging and coping with a certain

1. Ralph Lee Smith, "The Wired Nation," *Nation,* May 18, 1970, 582–602.

amount of change has always been part of corporate liberal expectation and planning. While technologies change, the way in which the corporate liberal polity imagines and responds to those changes remains remarkably constant.

Current policy discussions and arrangements, for example, continue to be characterized by an artful negotiation of the tensions between cooperation and competition, between the perceived need for collective, particularly governmental, efforts on behalf of corporate industrial coordination and the principles of business autonomy—tensions between the "corporate" and the "liberal." The similarities between today's patterns of behavior and those of Herbert Hoover's "associational" activities of the 1920s are still more striking than the differences. Republicans insist that they are more for competition and market forces than Democrats, yet it was a Republican administration that, in 1987, created an Advisory Committee on Advanced Television Service for coordinating industry efforts to develop HDTV standards, and for a period suggested a suspension of antitrust laws to facilitate industry cooperation on the matter. As George Bush's secretary of commerce described the policy, "we in the Government need to clear the underbrush out of the way so that the private sector can get into this. . . . the Government can . . . make it easier for private companies to get together in groups, consortiums."[2] And early in 1995, Republican members of Congress organized closed-door meetings with leading corporate heads to cooperatively hammer out the terms of boundary-setting telecommunications legislation satisfactory to industry members.[3] Democrats, meanwhile, though a little more frank about their willingness to use the legal powers of government, carefully circumscribe their proposals with references to competition and minimalist government and—perhaps less frankly than Republicans—circumscribe their actions to make sure those legal powers are largely used on behalf of the industry as a whole.

Again, this is not to suggest simply that corporations are all-powerful. Like radio broadcasting in the teens and twenties, the most exciting of today's new media technologies for the most part originated outside corporate walls. In particular, as corporations in the 1980s tried to bring computer communications into the everyday lives of Americans by investing in elaborate experiments with videotex and commercial online services, the Internet developed within nonprofit research uni-

2. Robert Mosbacher on *MacNeil-Lehrer NewsHour* broadcast of Monday, June 26, 1989.

3. Brock N. Meeks, "Closed Hearings Dent Bipartisan Alliances," *Inter@ctive Week*, January 30, 1995. Available at http://www.ziff.com/intweek/web/issues/950130/pol1.htm.

versities, and surprised the corporate world by dramatically surpassing commercial efforts in size, effectiveness, and popularity in the late 1980s and the 1990s.

But, also like radio broadcasting in the early days, the pressures to incorporate the possibilities suggested by the Internet into a corporate framework are strong, and most people in decision-making positions seem to think and talk about the Internet in corporate liberal-inflected ways. The Internet's explosive growth in size and sophistication was enabled in part by National Science Foundation funding, but just as importantly was also fostered by an academic culture of free intellectual exploration, collaboration, and sharing of work, which enabled the wide low-cost or free distribution of software and protocols; TCP/IP, ftp, gopher, Mosaic, and similar systems rapidly became international standards in part because their distribution was not restricted by the proprietary tendencies of for-profit research and development. Inside the beltway, however, the implications of the nonprofit origins of the Internet are largely ignored, and its eventual privatization is taken for granted.

Although the question of commercial control is not on the agenda in policy discussion, there are, as always, areas of debate. Today's Young Turks in media policy are the celebrants of the iconoclastic, exploratory, anarchic, high-tech computer culture fostered by the Internet's "unregulated" openness. Like the radio amateurs early in this century, the Internet subculture itself is a highly diverse and intriguing phenomenon. But when its values and ideas are filtered into the policy-making community, most of the argument seems to work by emphasizing liberal terms over corporate ones, not by stepping outside the bounds set by corporate liberal dichotomies.

If the mainstream media are to be believed, for example, most of the major controversies today center on matters of censorship and privacy. The Young Turks, we are told, are against regulation of computerized pornography, and against giving government agencies the power to "wiretap" digitally encoded messages (the "Clipper Chip" debate). These are important issues, but they also are easily understood in terms of the quintessential liberal problem of whether to use government power to constrain the actions of individuals. That same problematic, meanwhile, renders matters of private censorship and structural influence on content invisible. The dominant interpretation of the Internet's success in policy circles, in other words, reflects the liberal habit of interpreting matters in terms of the familiar binary of abstract individual freedoms juxtaposed against legal and governmental restraints. Implicitly, the Internet succeeded because of "individual initiative"—not the initiative of particular *kinds* of individuals (e.g., educated males) in particular social contexts (e.g., research universities) but individuals in the abstract, individuals who were simply "free."

When objections to a corporate takeover of the Internet are expressed, they are similarly often framed by abstract individualism. Sometimes this is simply a matter of assuming, in a gesture that Marx identified as characteristic of bourgeois consciousness, that one's own particular experience and vision are universal: it's surprising how often university professors say they wish that the Internet would remain "free" of cost, as if the current system were not heavily funded by university computing budgets, and how often they talk as if everyone in the world would, if given a chance, share their personal enthusiasms for E-mail discussion groups and downloaded files. But there have also been more sophisticated criticisms. Elements of the Internet subculture have become increasingly aware of policy issues and formed inside-the-beltway policy organizations such as the Electronic Frontier Foundation (EFF) and created magazines like *Wired.* From this group, forceful objections to some aspects of corporate plans have been raised, particularly the corporate tendency to try to adapt computer networking technologies to an advertising-supported, one-way, broadcast model: so-called 500-channel television. Against the corporate tendency toward bureaucratic centralization and consumerism, Internet enthusiasts advocate a fully two-way, open, relatively anarchic system more like the current Internet.

These objections, however, generally get reduced to calls for business entrepreneurialism, for a safely liberal vision of private, profit-oriented individuals in an open, competitive marketplace—a capitalism of the many against the capitalism of the few. As an influential essay by EFF founder Mitch Kapor puts it, the alternative to conventional, centralized corporate models of new media is a "Jeffersonian vision of diversity, openness, and decentralization of control." Yet Kapor asserts, "[T]he private sector, not the government, will build and operate" this "open" system. That there might be other alternatives besides the government and the private sector, such as nonprofit private organizations (like the universities that created the Internet in the first place), is not mentioned, lost behind the liberal government/business dichotomy.[4]

The point is not that a more entrepreneurial, open, private Internet necessarily would be a bad thing, but that arguments against centralization on behalf of the open, the local, and the competitive can themselves become part of conventional corporate liberal discourse, not alterna-

4. Mitchell Kapor, "Where Is the Digital Highway Really Heading? The Case for a Jeffersonian Information Policy," *Wired,* July–August 1993, 53–59, 94.

tives to it. The breakup of NBC's network duopoly in the forties, the UHF television channel allocations of the 1952 *Sixth Report and Order*, the prime-time access rule, cable reregulation, and similar policy actions were all motivated and justified by concerns for the autonomy of local broadcast stations, local communities, and the values of openness and competition. Yet, as we have seen, those efforts generally functioned, not to introduce dramatic new levels of openness, but to readjust center-periphery relations and allow for the negotiation of tensions between the twin drives toward expansion and stability endemic to the corporate form of organization.

Open Internet enthusiasts like Kapor, then, seem likely (though not inevitably) to become successors to the Young Turks of earlier eras of policy making: trust-busting New Dealers of the 1940s, the "blue sky" cable policy activists of the late 1960s, and the neoclassical economists of the 1980s. If history is any guide, the current generation's efforts may shift a little power from the corporate center to the more competitive peripheries of the media industries, but will not fundamentally disrupt the center-periphery economic pattern itself. In the end, such efforts may only function to remind the corporate liberal system to remain dynamic—as it must for its own survival—without challenging its underlying principles.

Theory and Politics

Readers who are generally sympathetic to my argument so far might nonetheless express a number of reservations. Those interested in concrete policy activism, for example, might wonder if my skepticism about the possibilities for substantial reform within the existing terms of debate is perhaps too pessimistic. Are legal and political efforts on behalf of the public interest, open access, fairness, and the like really that limited by corporate liberal interpretations? Isn't there something still alive and vital in these terms, something important that is being masked by an easy armchair cynicism in my analysis? A certain kind of Marxist political economist, conversely, might be entirely comfortable with my skepticism about conventional broadcast policy, but feel that the analysis is too idealist, overemphasizing the power of ideas and ideologies at the expense of the underlying material forces. Some critical theorists, in turn, might appreciate my emphasis on discourse, complexity, and contradiction instead of imagined underlying economic laws or conspiracies, but wonder whether I have left matters undertheorized, leaving notions like bureaucratic market simulation too much in the background and staying a little too close to the conventional discourse I am criticizing, perhaps reflecting a residual liberalism of my own. And finally, adherents of cultural studies may grant the validity of my critiques of law and policy, but may wonder about the value of expending so much effort criticizing "dominant" values and structures, and perhaps wish that my brief references to gender, to the domestic/public boundary, and to other marginalized categories had been used as a starting point instead of appearing near the end of the book.

Among media scholars in the United States who share an interest in expanding democracy in the media, reservations like these appear with some regularity. While the reservations as I have presented them here are no doubt simplifications, I have heard real, more subtle versions of them many times. The concerns they express deserve to be taken seriously; they are too important either to be indiscriminately mixed (thus ignoring the issues raised by the differences between them) or to be reduced to competitive debating positions (where one simply adopts one of the positions and treats the others as foils). So this book has brought to bear this range of concerns on the historical and institutional details of commercial broadcasting. And while this book certainly cannot resolve all these concerns, it has suggested some avenues for thinking through their differences and connections.

Marxism

Any book that emphasizes property relations and their arbitrary character is of course indebted to the Marxist tradition. This book brings to that tradition an illustration of the importance of law and policy as a creator—not just a legitimator or enforcer—of markets and property relations. The history of property creation and regulation in commercial broadcasting nicely illustrates legal realist Robert Hale's argument that "[t]he market value of a property or of a service is merely a measure of the strength of the bargaining power of the person who owns the one or renders the other, under the particular legal rights with which the law endows him, and the legal restrictions which it places on others."[5] Much of what broadcast regulation is about in the United States is crafting the mix of rights, privileges, and restrictions that form the conditions of operation, the bargaining power, and thus the market value of stations, copyrights, and audiences. It is inherent in the institution that a change in license regulation, copyright distribution formula, or ratings proce-

5. Robert Hale, "Bargaining, Duress, and Economic Liberty," *Columbia Law Review* 603 (1943): 625, quoted in Duncan Kennedy, "The Stakes of Law, or Hale and Foucault!" in *Sexy Dressing Etc.* (Cambridge: Harvard University Press, 1993), 84.

dure can change the market value of a station, a program library, or an audience demographic; this is not because "government" or other "externalities" interfere with markets, but because government actions create markets in the first place. If this book emphasizes ideology and discourse, then, it is not because of a simple romantic fondness for ideas over economic realities, but because of an emphasis on the materially constitutive role of laws and policies, which are themselves principally, materially, discursive or intellectual practices, practices of language and ritual.

Law is not only constitutive, but also one of the arenas in which internal contradictions, as some Marxists call them, are evident and played out. The incongruity between the social character of production and private ownership of the means of production certainly resonates throughout many aspects of the legal regulation of commercial broadcasting. It is evident, for example, in the tensions between the nonownership and transfer clauses of the Communications Act, which in turn reflect the difficulty of legitimating political intervention on behalf of private interests. It is also expressed in the efforts to specify creative author-owners within the collective, assembly-line cultural production machinery of corporate broadcasting by way of such legal contortions as works for hire, corporate individuals, and copyright collectives.

Contradiction, Discursive Practices, and Bureaucratic Market Simulation

But the fact of commercial broadcasting suggests that such tensions are not best understood, in the fashion of some classic Marxist narratives, as surface manifestations of fundamental contradictions festering away in the soul of the system, harbingers of its impending collapse. In the same way that property is not disintegrating in the face of its legal incoherence, commercial broadcasting does not exactly "suffer" from these contradictions. It was built on them and has flourished in their presence; to the extent that they profoundly shape its character, it is constituted by them. The situation of the nonownership clause coexisting with the legally sanctioned selling of stations, for example, may generate intellectual discomfort, but it has also survived as an institutional practice for three-quarters of a century. The use of copyright collectives to simulate market exchange in conditions that prohibit the real thing may be awkward, but it also has successfully served to integrate a broad range of creative practices with corporate industries as a whole. So I have taken Foucault's advice, and looked at contradictions as "neither appearances to be overcome, nor secret principles to be uncovered [but] objects to be described for themselves." The purpose of analyzing contradictions, he suggests, "is to map, in a particular discursive practice, the point at which they are constituted, to define the form that they assume, the relations that they have with each other, and the domain that they govern."[6]

It is in this context that I have approached the role of bureaucratic practices in the constitution of commercial broadcasting: not simply as resolutions or expressions of capitalism's contradictions, but with attention to their specific "constitution, forms, interrelations, and domains." Bureaucratic practices appeared in the first instance because they were intellectual habits characteristic of the professional and managerial classes that did most of the decision making. Managers, jurists, and politicians do turn to administrative agencies, copyright collectives, audience ratings, and the like to resolve problems of organization, legitimacy, and expansion, but they do so as much out of vision and hope as out of need. And if, in many cases, their efforts seem to resolve contradictions and tensions, they also create new ones. The practice of administering the spectrum in the public interest, for example, helped legitimate the government's helpful reach across the public-private boundary, but also set the stage for an ongoing set of quandaries and struggles over the appropriate extent and character of regulation that continue to this day.

The idea that property and markets are bureaucratically simulated in the electronic media provides a characterization that neither reduces institutions to falsehoods nor accepts their bureaucratic logics at face value. While the term "simulation" is Baudrillard's, my use of it is less sweeping than his. Bureaucratic market simulation is not something that happens throughout the electronic media in a uniform way. It is not just a fancy way to refer to the social character of property and markets. Rather, it is a specific discursive practice available in our era to deal with institutional difficulties, particularly of industrial coordination. It is the accomplishment of professional managers whose mode of operation is bureaucratic and whose horizons are shaped by the liberal vision of productive activity as comprising autonomous firms competing in a marketplace.

Cultural Studies

This book is about a portion of elite culture; it does not investigate the complexities of everyday life and culture for the majority. But in suggesting a way to interpret the complexity of power relations between audi-

6. Michel Foucault, *The Archaeology of Knowledge*, translated by Alan Sheridan (New York: Pantheon, 1972), 151–56.

ence members and the media, it does point to some connections between the dominant structures and everyday life, and offers a corrective to tendencies to treat the two as radically separable. Cultural studies' explorations of the marginal, alternative, and transgressive currents in contemporary culture are important but, in the contemporary political climate, also pose a danger. In the mainstreams of American society there is a strong if mistaken tendency to read such activities through a liberal lens: subcultures are envisioned as acting in their own self-interest as they struggle against static social constraints and norms much like the selfish abstract individuals of more conventional liberal narratives. Cultural studies thereby risks being reduced in the broader imagination to yet another version of the theory of consumer sovereignty.

The industrial strategy of consumerism forms a key context and connection between the institutions of broadcasting and the lives of its viewers and listeners. But the realities of consumerism, and the complex relations of power it exploits and encourages, are not best explained functionally, in terms of the power of the integrated advertising-broadcasting-manufacturing complex, as if power radiated outward from the "system." That system, I hope to have shown, is not so much the reality of the institution but a managerial vision of that reality, a product of the habits of thought and practice characteristic of managerial culture. Audience ratings, for example, do not mechanically link broadcasters and advertisers to the desires of the audience, but the assumption that they do smooths the day-to-day lives of executives. The power relations associated with the institutions of broadcasting and consumerism are best approached, then, not in terms of an idealized functional system, but in terms of the gaps and contradictions associated with the *vision* of a system. Hence the value of the analogy with the feminist critique of domestic labor: the effect of the overconfident managerial assumption of the system's seamlessness is seen not so much in what it includes, but in what it excludes, in the way it defines important aspects of life out of the picture. The economic life of broadcasting depends on the naturalization, and thus exclusion from the category of production, of the labor, imagination, and energy that goes into life on "the other side of the paycheck." That condition, in turn, forms much of the context of the interpretation of broadcasting, the structures of constraint and inducement within which cultural meanings are produced.

Activism

But what is the value of all this for policy activism, for concrete efforts to effect change? Certainly, efforts directed at progressive reform over the near term are important, and it is perhaps inevitable that some of these will adopt inside-the-beltway terminology for strategic reasons. Those interested in increasing democracy in media and social life, however, also need to think about the longer term. At this point in history, the principal question for media policy in the United States should be, How do we, as a matter of democratic choice, want to organize our popular communications, our means of producing and distributing culture and information? This question has yet to be fully raised, I hope to have shown, because political discussion of broadcast organization has chronically been both impoverished and encumbered with some ideological weak links, associated with conventional forms of liberalism. For example, since the introduction of the public-interest principle, reformers repeatedly have clung to the term in hopes of enacting limits to corporate control, and have repeatedly watched their efforts be either trivialized or co-opted for corporate ends; the dismay of the legislators who voted for the 1927 Radio Act because they imagined the public-interest clause would prevent commercialism was just the first in a long line of similar experiences. This pattern is partly a result of the way the public interest is generally construed in terms of a public/private dichotomy: as opposite to and autonomous from private interests rather than a generalized outcome of structural decisions made within and on behalf of the "private" sphere. Progressive uses of the public-interest principle thus easily become either after-the-fact palliatives, like adult-content warnings before programs, or subsumed within the practice of government regulation on behalf of industry coordination, like the prime-time access rule and channel licensing in general.

The limits of liberal discourse are evident, moreover, not just in the older regulatory principles that we are so often told are outmoded. In recent years the FCC has been moving ever farther into the brave new world of spectrum auctions, wherein channels are sold to the highest bidder instead of licensed according to the public interest. This has attracted notable public attention, including an article in the *New Yorker*.[7] (As I write, only nonbroadcast spectrum has been auctioned, but auctioning of broadcast spectrum is being explored.) Debate over this new approach has tended to revolve around the government/business dichotomy: conservative market purists herald this trend as a move in the direction of market principles away from government

7. Ken Auletta, "Selling the Air," *New Yorker*, February 13, 1995, 36–41. Auletta's use of my title is undoubtedly a coincidence; I first used it several years ago, in Thomas Streeter, "Selling the Air: Property and the Politics of U.S. Commercial Broadcasting," *Media, Culture, and Society* 16 (January 1994): 91–116.

interference, while some on the left worry about the FCC's abrogation of its duties to uphold the public's interest. Yet it takes little reflection to show that both positions are somewhat beside the point; the two sides are arguing about the location of a formal boundary between private and public that does not exist.

Significantly, one of the earliest mentions of the idea of charging for licenses came from Marxist Dallas Smythe, who argued in 1960 that the FCC would be better off simply charging for broadcast licenses and giving the money to public television instead of struggling to extract public-service programming from commercial licensees.[8] It was only later in the decade that the concept was revived by conservative, neoclassical economists for entirely different reasons: annoyed by the fact that valuable economic commodities like broadcast licenses originated in the noneconomic activities of government licensing, they imagined that if the government sold or auctioned the spectrum, these impure origins would be effectively eliminated.

Spectrum auctions will hardly remove the government's hand from electronic communications altogether, however. Technical standards, the definition of channel characteristics in spectrum allocations, the timing and mechanisms of auctions, antitrust enforcement, and other market-defining actions will continue to be determined by some combination of Congress, the courts, and the FCC. By the same token, auctions will not dramatically change the FCC's principal relation to industry, which has always been one of generalized support and protection of propertied interests in the spectrum; auctions are not an abrogation of responsibility so much as a minor adjustment of the way it is enacted. The central difference is merely that, traditionally, channels were sold only by private entities; now the government is selling them too. As it is conventionally presented, in sum, the sound and fury over auctions ignores the underlying consistency between auctions and public-interest licensing: in each case, the larger goal is to use government legal powers to enable and support corporate media. It is telling that in the last few years the tactic of spectrum auctions has been most aggressively pursued, not by economic conservatives in the name of market purity, but by a centrist Democratic administration that sees auctions simply as a way to raise government revenues, as a convenient tax in disguise.

It is not just a naive faith in dialogue, therefore, that motivates me to

8. Dallas Smythe, "Apply Revenue from the Rental of Frequencies to Commercial Broadcasters to the Support of a Public Service Broadcasting Agency," in *Counterclockwise: Perspectives on Communication,* ed. Thomas Guback (Boulder: Westview Press, 1994), 85–90.

suggest that broader discussion of structural choices is an important priority. One cannot speak meaningfully of fairness, the public interest, free speech, or economic efficiency and growth without reference to the basic goals and purposes of the institutions within which those values are to be enacted. To be silent about structural choices is merely to leave them implicit, and thus to ratify them without making a democratic decision to do so.

A Politics of Structure, a Politics of Property

So one goal of efforts for change should be to put the basic tenets and patterns of corporate liberalism itself on the agenda, to make them available for debate. The term "corporate liberalism" itself is not the central issue; one could make both strategic and analytical arguments on behalf of some other terminology. What is important is that the electronic media be identified, not just as "private," "unregulated," or "commercial," but as a deliberate political choice to foster a consumerist, oligopolistic, for-profit electronic media, with fundamental structures underwritten by government and stabilized by a mixture of private and public administrative arrangements. Only then can the naturalness of that choice be questioned, and a full-fledged discussion of its value be undertaken.

Of course, if one calls commercial broadcasting "greed-driven" or a "monopoly," everyone understands, but call it a "consumerist oligopolistic administratively stabilized mixture of private capital and government power" and they might not. If the character of corporate broadcasting is to be made accessible outside the ranks of academic political economists, it should be emphasized that it is not just a system, an invisible set of relationships accessible only through sophisticated analyses. Corporate liberalism is also about familiar matters like work and the home. It is a set of hopes as well as principles; a matter of everyday imagination, of passions, as well as institutions and arcane legal arguments. Certainly, it has its complexities, and any discussion of it must address them, but those complexities need not obscure the basic value choices at work. So if corporate liberal broadcasting is to be made a matter of broad discussion, what should be foregrounded is not just abstract relations and systems but fundamental values.

There are many value-laden elements of corporate liberalism that deserve to be put up for discussion. Some, such as free speech or the belief in the morality and effectiveness of administrative technique, already are being discussed to various degrees, though largely in sterile, formalist terms: that is, should we qualify the market and free speech with administrative regulation or protect them with strict, formal barriers? Others, such as the fiction of the corporate individual, remain buried within the realm of legal expertise, outside of public view. Others, such as the feminist critiques of the conventional constructions of the boundaries between home and work, are discussed but have yet to be connected to matters of communication in the public mind.

While all of these issues deserve broad discussion, what this book has to offer the larger project of putting fundamentals on the public agenda is a reconsideration of property. Property is not just a technicality of the law. It is a feature of everyday experience as well as a key element of many aspects of the American political imagination; property in many ways remains the defining archetype of rights in general. Property remains an ideological linchpin of contemporary social relations. Consciously or not, property touches on central questions of justice and social vision.

The field of electronic technologies, moreover, seems to be one of the only areas where the troubled, ambiguous character of property is widely acknowledged. A computer magazine headline declaring that "Everything You Know about Intellectual Property Is Wrong" is typical.[9] Neoclassical economists began their march to center stage by reacting with alarm at what they took to be the muddled, nonpropertylike character of practices like broadcast licensing. And a recent conservative policy manifesto on telecommunications widely circulated inside the beltway noted that "to create the new cyberspace environment is to create *new* property—that is, new means of creating goods (including ideas) that serve people. . . . two questions that must be answered [are] first, what does 'ownership' *mean*? What is the nature of the property itself, and what does it mean to own it? Second, once we understand what ownership means, *who* is the owner?"[10]

Of course, what's missing from most of these discussions is the legal realist insight that the question, What does ownership mean? is raised not just in the spheres of new technologies. Property has been unsettled and shifting throughout its career. If property serves as a reigning metaphor for justifying the distribution of resources, power, and privileges in our society, foregrounding both the importance and the oddity of property might help open up democratic discussion of relations of power in

9. John Perry Barlow, "The Economy of Ideas: Everything You Know about Intellectual Property Is Wrong," *Wired,* March 1994, 84–90, 126–29.

10. Esther Dyson, George Gilder, George Keyworth, and Alvin Toffler, "Cyberspace and the American Dream: A Magna Carta for the Knowledge Age," release 1.2, August 22, 1994, Progress and Freedom Foundation. Available from http://www.pff.org/position.html.

general. And concerns about the regulation and character of the electronic media provide a fertile context for encouraging such a debate.

In a sense, what I am suggesting is that we revisit the concerns of the 1930s and 1940s, when regulators were worried about reconciling the nonownership clause of the Communications Act with the buying and selling of stations. In the 1930s and 1940s, however, the progressive approach to this question was formalist: the AVCO rule and its cousins, like today's "marketplace-approach" neoclassical economists, sought to draw a bright-line boundary between private property and public licenses to broadcast. The problem with formalist approaches is not only that they are the province of the Right today, but, as we have seen, that they are generally indeterminate; under close inspection, bright lines become blurry and contingent, and "practical" policy decisions sneak in between the lines of crystalline principles. What is needed, then, is a more realist approach to questions of property in the electronic media.

Foregrounding a realist notion of property in the electronic media might help provide a context in which to bring discussion to matters of structure that would simultaneously appeal to important and popular values; property carries strong connotations of individuals' power, of privacy, and of personal liberties in general. At the same time, if the approach were clearly realist, if the question were, What kind of property rights on behalf of what kind of social relations do we want to create? not, How should we protect natural or preexisting property rights? then questions of structure and organization would more easily come into play.

One place to start might be to take a cue from the important success of the noncommercial channel reservations strategy of the 1940s, which took advantage of the creative, as opposed to regulative, power of government action. It is probably well worth the effort to advocate, as a recent proposed bill in Congress put it, "a public lane on the information superhighway."[11] Like the noncommercial channel reservations, carving out a space today might enable the development of alternative media forms later on.

Yet the metaphor of a "public lane," while catchy, implies that the "public lane" would be an after-the-fact addition to some preexisting private or natural highway. This obscures the inherently public character of

11. People for the American Way and the Media Access Project, "Public Interest Groups Hail Introduction of Bill to Provide 'Public Lane' on the Information 'Superhighway,'" press release, June 15, 1994, in support of S. 2195, legislation introduced by Senator Daniel Inouye (D-HI).

the "highway" as a whole. Just as broadcasting invariably requires extra-market cooperative arrangements to create and define the character of electronic "territories" and the institutional relations that grow from them, the Internet today and electronic media of the future—whether or not they are privatized—will require a framework of collective agreements about operating principles and the distribution of control, enforced by some mixture of legal, legislative, and administrative power.

Over the long term, little will be gained by reacting to this fact with liberal alarm and trying to draw a bright line between rights-bound private property and politically beholden public privileges, by trying to protect public forums from private control with formal legal prohibitions. Government action does not just protect the public interest from the private; it creates and allocates privileges and power in the electronic media overall. A realist approach, then, might readily grant the fact that property in some sense is being created. It is not wrong or scandalous to admit that the government creates property in the electronic media, non-ownership clause or no. The question should not be, How can we protect public interests from private interests? but, What *kind* of property relations do we, as a democratic community, wish to create?

Such a question need not be disguised as a matter of social engineering, as an exception to the rule necessitated by questions of efficiency, technical necessity, or scientifically measurable social problems. Government involvement in either broadcasting or the Internet, for example, is not required by the technological characteristics of either: it is a political means to decide, create, and enforce the institutional arrangements and associated social values involved with any kind of elaborate nationwide communications system, including private ones. Similarly, children's programming could be treated primarily as a moral issue instead of a scientific one: instead of asking, What do the experts say is the best kind of television for children? the question could be, What kind of cultural values do we want our children to be raised with, and How should we build a system that encourages those values? Notice that, in the context of a discussion about property creation, this would not be a matter of creating regulations that would constrain what existing media producers can do; it wouldn't necessarily involve matters like limits on obscenity, commercials per hour, or violence. It would involve questions of *creating* basic relations within which program creators would work—shaping funding structures, relations of exchange, and so forth—which is something the government does already, via law, legislation, and administrative policy.

A realist sense of property would not and could not replace existing discourses, but its introduction might help realign them. Free speech might be understood as a structural outcome that is more encouraged by some constructions of property relations than by others. Rather than being seen as constraining free speech, therefore, the regulatory question would be, How should things be arranged on a fundamental level so as to encourage free speech and broad public dialogue? The public interest, if it were no longer seen as a qualification to preexisting private control, might be opened up to more substantive, and specific, interpretations: for example, in terms of the public sphere, of places and situations conducive to involvement of all citizens in full public dialogue on matters of interest and importance. Access would no longer be understood as a counterweight or exception to the rule of private control, but would become coterminous with it: creating property rights would be considered similar or even identical to granting access provisions.

The Coming Homestead Act for Cyberspace

As this is being written, the media landscape is undergoing some shifts, and some of the basic terms of media regulation are being reconsidered. Decisions are being made about new networks of communication, decisions about who will get to use those new networks for what kinds of services and how they will be able to interact. The structural decisions being made are not just matters of engineering, economic efficiency, and competitiveness. The structure of future electronic media will shape the quality and character of media content, of media's role in culture and politics, more profoundly than any content-oriented regulations such as the fairness doctrine or antiobscenity provisions. Over the next several years, either purposely or by default, the FCC, in concert with other regulatory bodies and corporate management, will be drawing the lines that will determine where the fences and walls of the electronic future are going to be. Regulators are going to be creating boundaries in the world of electronic communication, boundaries that encourage some technologies and some uses while discouraging others. In creating those boundaries, the regulatory system will be creating the private property of the future, the equivalent of a Homestead Act for cyberspace.

When, as they so often have in this century, corporate representatives talk of new technologies and technological revolutions, it is usually a safe bet that they are exaggerating both the newness of the technologies and the technological character of the innovations they predict, which are more a matter of institutional arrangements than of technologies per se. Coaxial cables, computers, and even fiber optics have been around for some time; the Internet itself is in a sense not even a technology but a set of protocols, a language or a set of codes and rules for interactions between people using computers. All the hype, however, is a symptom of the fact that the media industries are facing a series of major structural decisions in the coming years: the level of corporate uncertainty is higher than usual. And quite predictably, as the uncertainty has risen, corporate management's principled antipathy to government has declined considerably. The question is no longer the question asked in the 1980s, Should government regulate? but, What kind of regulation is best?

It seems relatively certain at this point that, for better or worse, the "information society" will be developed largely on the basis of marketplace exchange. The buying and selling of information need not be discussed as if it were the natural outgrowth of economic or technological imperatives: the creation of commodities in information involves an elaborate act of collective imagination. Images, symbols, and creative works are not self-evidently property. The fundamentally social world of shared signs and symbols contains no obvious natural boundaries that tell us where to draw the property lines, where to put the picket fences (or whether they should be picket fences or brick walls). There is little or nothing inherent in the concepts of property and information that dictates exactly how, when, and to what extent information should be commodified. The creation of property in symbolic "goods," therefore, involves profound moral, social, and political choices.

The creation of property through legal means is one of the principal ways in which social boundaries have been and continue to be drawn in the world of electronic media. Most obviously, the commodification of electronic ephemerals involves designating who owns what, and thus allocates power of the sort most fundamental to capitalism. But it also involves designating what ownership entails, what powers are created by the conditions of ownership, and thus helps shape relations between institutions, organizations, and individuals. Careers, corporations, and entire industries hinge on the shifting terms of property in electronic ephemerals. In a sense, property defines who under law "originates" an act of communication and thus determines who controls the process; it plays an integral role in constructing fundamental communicative relations. The overworked terrain of First Amendment and libel laws involve after-the-fact tinkering with existing communicators' ability to communicate, whereas property defines who is a communicator in the first place.

At the same time, it should be remembered, information property also involves matters far less prosaic: property plays a key role in the construction of subjectivities. In an era where acts of communication are increasingly constructed as property, property does more than determine who can communicate and who cannot. It also defines and shapes the character of communicators. In the context of electronic media, it goes a long way toward providing definitions of individuality, definitions of what it means to be a self. In the late eighteenth century, it is said, both the cult of creative genius and the modern publishing industry emerged from the efforts of writers who sought to secure and justify an income through the legal enactment of the cultural theory of romantic genius, which, among other things, provided the historical rationale for copyright law.[12] In the 1920s, when Herbert Hoover drew a legal boundary between broadcasting and amateur two-way uses of radio, he foreclosed important social possibilities and helped ensure the corporate, consumer-oriented centralization of the production of electronic culture. What new cultural and institutional constructs might emerge as the links between technologies, law, and culture continue to change into the next century?

Julia Kristeva once wrote, "I speak and you hear me, therefore we are."[13] If this is true, the choices that shape property in media, insofar as they shape what it means to be a speaker and a listener in an electronically mediated environment, and hence subjectivity, may influence the character of social existence. The law of ephemeral property is thus becoming a principal terrain for constructing the contours of contemporary cultures. Ongoing developments in "information" law and policy will draw boundaries that will undergird the development of social life well into the next century.

What kinds of activities will be encouraged and what kinds discouraged, which made central and which marginalized? What kinds of relations between work and home will be encouraged? Which communities will be dissolved, which formed, and which transformed? What might be the fate of Western individuality in commodified electronic regimes? Of collective modes of cultural production? Are there information-age analogs to the Native American cultures that were destroyed by the

12. Martha Woodmansee, "The Genius and the Copyright: Economic and Legal Conditions of the Emergence of the 'Author,'" *Eighteenth-Century Studies* 17 (summer 1984): 425–48.

13. Julia Kristeva, *Desire in Language: A Semiotic Approach to Literature and Art*, translated by Thomas Gora, Alice Jardine, and Leon S. Roudiez, ed. Leon S. Roudiez (New York: Columbia University Press, 1980), 74. She attributes the phrase to Francis Ponge, but gives no reference.

imposition of European property schemes on the North American interior? Might there be electronic parallels to the Great Plains or rain-forest ecosystems that have been devastated under the force of the property system? What influence might the choices made have on the character of life and culture? These are the questions we should be asking as the outlines of the cultural industries of the future are drawn. This book, I hope, provides, if not answers, some ways of thinking that might ensure that these questions get asked.

Index